T0122340

Communications in Computer and Information Science 1303

Editorial Board Members

Joaquim Filipe
Polytechnic Institute of Setúbal, Setúbal, Portugal
Ashish Ghosh
Indian Statistical Institute, Kolkata, India
Raquel Oliveira Prates
Federal University of Minas Gerais (UFMG), Belo Horizonte, Brazil
Lizhu Zhou
Tsinghua University, Beijing, China

More information about this series at http://www.springer.com/series/7899

Minrui Fei · Kang Li · Zhile Yang ·
Qun Niu · Xin Li (Eds.)

Recent Featured Applications of Artificial Intelligence Methods

LSMS 2020 and ICSEE 2020 Workshops

Workshops of the 6th International Conference
on Life System Modeling and Simulation, LSMS 2020
and 6th International Conference on Intelligent Computing
for Sustainable Energy and Environment, ICSEE 2020
Shanghai, China, October 25, 2020
Proceedings

Springer

Editors
Minrui Fei
Shanghai University
Shanghai, China

Kang Li
University of Leeds
Leeds, UK

Zhile Yang
Shenzhen Institute of Advanced
Technology, Chinese Academy of Sciences
Shenzhen, China

Qun Niu
Shanghai University
Shanghai, China

Xin Li
Shanghai University
Shanghai, China

ISSN 1865-0929   ISSN 1865-0937 (electronic)
Communications in Computer and Information Science
ISBN 978-981-33-6377-9   ISBN 978-981-33-6378-6 (eBook)
https://doi.org/10.1007/978-981-33-6378-6

© Springer Nature Singapore Pte Ltd. 2020
This work is subject to copyright. All rights are reserved by the Publisher, whether the whole or part of the material is concerned, specifically the rights of translation, reprinting, reuse of illustrations, recitation, broadcasting, reproduction on microfilms or in any other physical way, and transmission or information storage and retrieval, electronic adaptation, computer software, or by similar or dissimilar methodology now known or hereafter developed.
The use of general descriptive names, registered names, trademarks, service marks, etc. in this publication does not imply, even in the absence of a specific statement, that such names are exempt from the relevant protective laws and regulations and therefore free for general use.
The publisher, the authors and the editors are safe to assume that the advice and information in this book are believed to be true and accurate at the date of publication. Neither the publisher nor the authors or the editors give a warranty, expressed or implied, with respect to the material contained herein or for any errors or omissions that may have been made. The publisher remains neutral with regard to jurisdictional claims in published maps and institutional affiliations.

This Springer imprint is published by the registered company Springer Nature Singapore Pte Ltd.
The registered company address is: 152 Beach Road, #21-01/04 Gateway East, Singapore 189721, Singapore

Preface

This book constitutes the proceeding of the 2020 themed workshops on smart energy systems, intelligent manufacturing, and bioinformatics and life system modeling, as part of the 2020 International Conference on Life System Modeling and Simulation (LSMS 2020) and 2020 International Conference on Intelligent Computing for Sustainable Energy and Environment (ICSEE 2020). Due to the pandemic situation, the Organizing Committee decided to postpone the two conferences to October 2021, while themed workshops were still organized online on October 25, 2020. The themed workshops aimed to bring together international researchers and practitioners in the fields of advanced methods for smart energy systems, intelligent manufacturing, and intelligent biological systems and information systems. These events are built upon the success of previous LSMS conferences held in Shanghai, Wuxi, and Nanjing, China, in 2004, 2007, 2010, 2014, and 2017, and ICSEE conferences held in Wuxi, Shanghai, and Nanjing, China, in 2010, 2014, and 2017 respectively, and are based on large-scale UK-China collaboration projects on sustainable energy.

At the themed workshops, technical exchanges within the research community took the form of keynote speeches, and oral presentations. The themed workshops received over 165 submissions. All papers went through a rigorous peer-review procedure and each paper received at least three review reports. Based on the review reports, the Program Committee finally selected 38 high-quality papers for presentation at the themed workshops for LSMS 2020 and ICSEE 2020. These papers cover three topics, and are included in one of CCIS proceedings published by Springer.

The organizers of LSMS 2020 and ICSEE 2020 would like to acknowledge the enormous contribution of the Program Committee and the referees for their efforts in reviewing and soliciting the papers, and the Publication Committee for their editorial work. We would also like to thank the editorial team from Springer for their support and guidance. Particular thanks are of course due to all the authors, as without their high-quality submissions and presentations the conferences would not have been successful.

Finally, we would like to express our gratitude to our sponsors and organizers, listed on the following pages.

October 2020

Minrui Fei
Kang Li
Qinglong Han

Organization

Sponsors

China Simulation Federation (CSF)
China Instrument and Control Society (CIS)
Chinese Association for Artificial Intelligence (CAAI)
IEEE Systems, Man & Cybernetics Society Technical Committee on Systems Biology
IEEE CC Ireland Chapter

Organizers

Shanghai University, China
University of Leeds, UK
China Jiliang University, China
Swinburne University of Technology, Australia
Life System Modeling and Simulation Technical Committee of CSF, China
Embedded Instrument and System Technical Committee of China Instrument
and Control Society, China

Co-sponsors

Shanghai Association for System Simulation
Shanghai Instrument and Control Society
Zhejiang Association of Automation (ZJAA)
Shanghai Association of Automation

Co-organizers

Queen's University Belfast, UK
Nanjing University of Posts and Telecommunications, China
University of Essex, UK
Queensland University of Technology, Australia
Central South University, China
Tsinghua University, China
Peking University, China
University of Hull, UK
Beijing Jiaotong University, China
Nantong University, China
Shenzhen Institute of Advanced Technology, Chinese Academy of Sciences, China
Shanghai Key Laboratory of Power Station Automation Technology, China

Complex Networked System Intelligent Measurement and Control Base,
Ministry of Education, China
UK China University Consortium on Engineering Education and Research
Anhui Key Laboratory of Electric Drive and Control, China

Honorary Chairs

Wang, XiaoFan   Shanghai University, China
Umezu, Mitsuo   Waseda University, Japan

General Chairs

Fei, Minrui   Shanghai University, China
Li, Kang   University of Leeds, UK
Han, Qing-Long   Swinburne University of Technology, Australia

International Program Committee

Chairs

Ma, Shiwei   China Simulation Federation, China
Coombs, Tim   University of Cambridge, UK
Peng, Chen   Shanghai University, China
Chen, Luonan   The University of Tokyo, Japan
Zhang, Baolin   China Jiliang University, China
McLoone, Sean   Queen's University Belfast, UK
Tian, Yuchu   Queensland University of Technology, Australia
He, Jinghan   Beijing Jiaotong University

Local Chairs

Aleksandar Rakić   University of Belgrade, Serbia
Cheng, Long   Institute of Automation, Chinese Academy of Sciences,
                            China
Ding, Jingliang   China
Ding Ke   Jiangxi University of Finance and Economics, China
Duan, Lunbo   South East University, China
Fang, Qing   Yamagata University, Japan
Feng, Wei   Shenzhen Institute of Advanced Technology, Chinese
                            Academy of Sciences, China
Fridman, Emilia   Tel Aviv University, Israel
Gao, Shangce   University of Toyama, Japan
Ge, Xiao-Hua   China
Gu, Xingsheng   East China University of Science and Technology,
                            China
Gupta M. M.   University of Saskatchewan, Canada
Han, Daojun   Henan University, China

Hong, Xia	University of Reading, UK
Hunger, Axel	University of Duisburg-Essen, Germany
Jia, Xinchun	ShanXi University, China
Jiang, Zhouting	China Jiliang University, China
Lam, Hak-Keung	King's College London, UK
Li Juan	Qingdao Agricultural University, China
Li, Ning	Shanghai Jiaotong University, China
Li, Wei	Central South University, China
Li, Yong	Hunan University, China
Liu, Kang	China Jiliang University, China
Liu, Wanquan	Curtin University, Australia
Liu, Yanli	Tianjin University, China
Ma, Fumin	Nanjing University of Finance & Economics, China
Ma, Lei	Southwest University, China
Maione, Guido	Technical University of Bari, Italy
Na, Jing	Kunming University of Science and Technology, China
Naeem, Wasif	Queen's University Belfast, UK
Park, Jessie	Yeungnam University, South Korea
Qin, Yong	Beijing Jiaotong University, China
Su, Zhou	Shanghai University, China
Tang, Wenhu	South China University of Technology, China
Wang, Shuangxing	Beijing Jiaotong University, China
Xu, Peter	The University of Auckland, New Zealand
Yan, Tianhong	China Jiliang University, China
Yang, Dongsheng	Northeast University, China
Yang, Fuwen	Griffith University, Australia
Yang, Taicheng	University of Sussex, UK
Yu, Wen	National Polytechnic Institute, Mexico
Zeng, Xiaojun	The University of Manchester, UK
Zhang, Jianhua	North China Electric Power University, China
Zhang, Tengfei	Nanjing University of Posts and Telecommunications, China
Zhang, Wenjun	University of Saskatchewan, Canada
Zhao, Wenxiao	Chinese Academy of Sciences, China
Zhu, Shuqian	Shandong University, China
Hou, Weiyan	Zhengzhou University, China

Members

Aristidou, Petros	Ktisis Cyprus University of Technology, Cyprus
Azizi, Sadegh	University of Leeds, UK
Bu, Xiongzhu	Nanjing University of Science and Technology, China
Cai, Hui	Jiangsu Electric Power Research Institute, China
Cai, Zhihui	China Jiliang University, China
Cao, Jun	Keele University, UK
Chang, Xiaoming	Taiyuan University of Technology, China

Chang, Ru	Shanxi University, China
Chen Xiai	China Jiliang University, China
Chen, Qigong	Anhui Polytechnic University, China
Chen, Qiyu	China Electric Power Research Institute, China
Chen, Rongbao	Hefei University of Technology, China
Chen, Zhi	Shanghai University, China
Chi, Xiaobo	Shanxi University, China
Chong, Ben	University of Leeds, UK
Cui, Xiaohong	China Jiliang University, China
Dehghan, Shahab	University of Leeds, UK
Deng, Li	Shanghai University, China
Deng, Song	Nanjing University of Posts and Telecommunications, China
Deng, Weihua	Shanghai University of Electric Power, China
Du, Dajun	Shanghai University, China
Du, Xiangyang	Shanghai University of Engineering Science, China
Du, Xin	Shanghai University, China
Fang, Dongfeng	California Polytechnic State University, USA
Feng, Dongqing	Zhengzhou University, China
Fu, Jingqi	Shanghai University, China
Gan, Shaojun	Beijing University of Technology, China
Gao, Shouwei	Shanghai University, China
Gu, Juping	Nantong University, China
Gu, Yunjie	Imperial College London, UK
Gu, Zhou	Nanjing Forestry University, China
Guan, Yanpeng	Shanxi University, China
Guo, Kai	Southwest Jiaotong University, China
Guo, Shifeng	Shenzhen Institute of Advanced Technology, Chinese Academy of Sciences, China
Guo, Yuanjun	Shenzhen Institute of Advanced Technology, Chinese Academy of Sciences, China
Han, Xuezheng	Zaozhuang University, China
Hong Yuxiang	China Jiliang University, China
Hou, Guolian	North China Electric Power University, China
Hu, Qingxi	Shanghai University, China
Hu, Yukun	University College London, UK
Huang, Congzhi	North China Electric Power University, China
Huang, Deqing	Southwest Jiaotong University, China
Jiang, Lin	The University of Liverpool, UK
Jiang, Ming	Anhui Polytechnic University, China
Kong, Jiangxu	China Jiliang University, China
Li MingLi	China Jiliang University, China
Li, Chuanfeng	Luoyang Institute of Science and Technology, China
Li, Chuanjiang	Harbin Institute of Technology, China
Li, Donghai	Tsinghua University, China
Li, Tongtao	Henan University of Technology, China

Li, Xiang	University of Leeds, UK
Li, Xiaoou	CINVESTAV-IPN, Mexico
Li, Xin	Shanghai University, China
Li, Zukui	University of Alberta, Canada
Liu, Kailong	University of Warwick, UK
Liu, Mandan	East China University of Science and Technology, China
Liu, Tingzhang	Shanghai University, China
Liu, Xueyi	China Jiliang University, China
Liu, Yang	Harbin Institute of Technology, China
Long, Teng	University of Cambridge, UK
Luo Minxia	China Jiliang University, China
Ma, Hongjun	Northeastern University, China
Ma, Yue	Beijing Institute of Technology, China
Menhas, Muhammad Ilyas	Mirpur University of Science and Technology, Pakistan
Naeem, Wasif	Queen's University Belfast, UK
Nie, Shengdong	University of Shanghai for Science and Technology, China
Niu, Qun	Shanghai University, China
Pan, Hui	Shanghai University of Electric Power, China
Qian, Hong	Shanghai University of Electric Power, China
Ren, Xiaoqiang	Shanghai University, China
Rong, Qiguo	Peking University, China
Song, Shiji	Tsinghua University, China
Song, Yang	Shanghai University, China
Sun, Qin	Shanghai University, China
Sun, Xin	Shanghai University, China
Sun, Zhiqiang	East China University of Science and Technology, China
Teng, Fei	Imperial College London, UK
Teng, Huaqiang	Shanghai Instrument Research Institute, China
Tian, Zhongbei	University of Birmingham, UK
Tu, Xiaowei	Shanghai University, China
Wang, Binrui	China Jiliang University, China
Wang Qin	China Jiliang University, China
Wang, Liangyong	Northeast University, China
Wang, Ling	Shanghai University, China
Wang, Yan	Jiangnan University, China
Wang, Yanxia	Beijing University of Technology, China
Wang, Yikang	China Jiliang University, China
Wang, Yulong	Shanghai University, China
Wei, Dong	China Jiliang University, China
Wei, Li	China Jiliang University, China
Wei, Lisheng	Anhui Polytechnic University, China

Wu, Fei	Nanjing University of Posts and Telecommunications, China
Wu, Jianguo	Nantong University, China
Wu, Jiao	China Jiliang University, China
Xu, Peng	China Jiliang University, China
Xu Suan	China Jiliang University, China
Xu, Xiandong	Cardiff University, UK
Yan, Huaicheng	East China University of Science and Technology, China
Yang, Aolei	Shanghai University, China
Yang, Banghua	Shanghai University, China
Yang, Wenqiang	Henan Normal University, China
Yang, Zhile	Shenzhen Institute of Advanced Technology, Chinese Academy of Sciences, China
Ye, Dan	Northeastern University, China
You, Keyou	Tsinghua University, China
Yu, Ansheng	Shuguang Hospital, China
Zan, Peng	Shanghai University, China
Zeng, Xiaojun	The University of Manchester, UK
Zhang, Dawei	Shandong University, China
Zhang Xiao-Yu	Beijing Forestry University, China
Zhang, Huifeng	Nanjing University of Posts and Telecommunications, China
Zhang, Kun	Nantong University, China
Zhang, Li	University of Leeds, UK
Zhang, Lidong	Northeast Electric Power University, China
Zhang, Long	The University of Manchester, UK
Zhang, Yanhui	Shenzhen Institute of Advanced Technology, Chinese Academy of Sciences, China
Zhao, Chengye	China Jiliang University, China
Zhao, Jianwei	China Jiliang University, China
Zhao, Wanqing	Manchester Metropolitan University, UK
Zhao, Xingang	Shenyang Institute of Automation, Chinese Academy of Sciences, China
Zheng, Min	Shanghai University, China
Zhou, Bowen	Northeast University, China
Zhou, Huiyu	University of Leicester, UK
Zhou, Peng	Shanghai University, China
Zhou, Wenju	Ludong University, China
Zhou, Zhenghua	China Jiliang University, China
Zhu, Jianhong	Nantong University, China

Organization Committee

Chairs

Sun, Jian, China
Li, Ni, China
Li, Xin, China
Sadegh, Azizi, UK
Zhang, Xian-Ming, Australia

Members

Chen, Zhi, China
Du, Dajun, China
Li, Xin, China
Song, Yang, China
Sun, Xin, China
Sun, Qing, China
Wang, Yulong, China
Zheng, Min, China
Zhou, Peng, China
Zhang, Kun, China

Special Session Chairs

Wang, Ling, China
Meng, Fanlin, UK
Chen, Wanmi, China
Li, Ruijiao, China
Yang, Zhile, UK

Publication Chairs

Niu, Qun, China
Zhou, Huiyu, UK

Publicity Chair

Yang, Erfu, UK

Registration Chair

Song, Yang, China

Secretary-General

Sun, Xin, China

Contents

Intelligent Manufacturing and System

Intelligent Biology and Information System

Smart Energy Systems and Devices

Operation and Dispatch of Distribution Network with Seawater Desalination Considering Source-Grid-Load Interaction Index

Peiyu Chen[1], Lu Jin[2(✉)], Jinlu Zhang[3], Ling Cheng[2], and Jiancheng Yu[3]

[1] Electric Power Research Institute, State Grid
Tianjin Electric Power Company, Tianjin 300380, China
[2] China Electric Power Research Institute Co., Ltd., Beijing 100089, China
au_1012@sina.cn
[3] State Grid Tianjin Electric Power Company, Tianjin 300010, China

Abstract. In the smart grid environment, flexible load participation plays a very cost-effective role in improving the source-grid-load interaction capability of the distribution network. The essence of the source-grid-load interaction with seawater desalination is to maximize energy utilization in the distribution network. The desalination load is combined with distributed renewable energy or the grid to operate in a coordinated manner to maximize the renewable energy utilization rate and to reduce system operating costs. A source-grid-load interaction comprehensive evaluation index system is initially established for source-grid-load systems, and the interaction indicators are determined from the power source side and the grid side in order to evaluate the flexibility and the interaction effect. The objective is to minimize the operating cost of the distribution network. A source-grid-load interactive comprehensive scheduling model that considers the interaction costs is constructed, and an optimization algorithm is used to solve the problem. The case studies using real project data validate the rationality and practicability of the proposed model.

Keywords: Smart grid · Source-grid-load interaction · Evaluation index · Demand response · Desalination load

1 Introduction

With the continuous change of global climate, there is a serious shortage of fresh water resources worldwide, which has become a bottleneck restricting economic and social development. As a solution to the shortage of fresh water resources, seawater desalination is playing an increasingly important role, especially in coastal water-scarce areas and islands where desalination is very urgent [1, 2]. The load demand such as coastal seawater desalination increases continuously, which seriously affects the load balance of the power grid. Combining the development potential of distributed renewable energy sources such as coastal wind energy [3] and solar energy [4], the energy interconnection technology is used to realize joint coordinated control of distributed renewable energy

© Springer Nature Singapore Pte Ltd. 2020
M. Fei et al. (Eds.): LSMS 2020/ICSEE 2020 Workshops, CCIS 1303, pp. 3–17, 2020.
https://doi.org/10.1007/978-981-33-6378-6_1

sources [5] and various loads including desalination. This paper focuses on the study on the flexible interaction of "source-grid-load" including various controllable loads such as seawater desalination loads, which is of great significance for reducing the cost of seawater desalination operation and promoting the large-scale desalination technology [6].

At present, some literatures have focused on load dispatching and coordinated operation of active power distribution networks. Literature [7] proposed a distributed demand-side energy management strategy based on game theory, which improved user satisfaction. Literature [8] considered the network security constraints and modeled the electric vehicle aggregator with limited storage capacity. A large number of unit combination models with security constraints for the coordinated control of electric vehicles and wind power output were proposed, and examples showed the model could improve the operating efficiency of thermal power units and better accommodate wind power. Literature [9, 10] emphasized that power systems showed significant random characteristics on both sides of supply and demand, and the increase in random disturbances led to reduced system controllability and increased security risks. At present, analytical methods based on deterministic theory are difficult to meet these new challenges. Therefore, it is necessary to address the actual, complex, changeable, and uncertain needs of the interactive environment, and consider the interaction between source, network, and load to better utilize the temporal and spatial complementarity of renewable energy and controllable loads.

Traditionally, the power balance is achieved by regulating the controllable generators. However, this approach may lead to higher power generation operating costs and pollution emissions. In order to overcome the limitations of the conventional power generators to achieve power balance, controllable loads have drawn increasing attention. Literature [11] emphasized a multiagent-based distributed method to minimize the operation cost in AC microgrids, in order to obtain an optimal power dispatch. Literature [12] proposed the technology linked with electric vehicles, the charging infrastructure and their impact on the performance of the grid. Literature [13] delivered a comprehensive approach for evaluating the impact of different levels of plug-in electric vehicles penetration on distribution network investment and incremental energy losses.

Taking into account the seawater desalination, Literature [14] proposed an optimal operation strategy of a coastal hydro-electrical energy system with consideration of seawater desalination, to make best use of available coastal renewable energy and to meet local freshwater demand. The virtual energy storage characteristics of the seawater desalination was described and introduced in the multi-objective optimal model, which was solved by the third generation of the constrained non-dominated sorting genetic algorithm (NSGA-III) algorithm. Literature [15] reviewed desalination technologies - membrane distillation, forward osmosis, and capacitive deionization. Literature [16] proposed a real-time energy management scheduling strategy based on ultra-short-term wind speed prediction for an independent microgrid system for desalination of seawater containing wind energy, and the genetic algorithm-back propagation neural network (GA-BP) was used to predict wind speeds. Literature [17] proposed a new optimization algorithm combining harmonic search and chaotic search. According to the load requirements of the desalination system, the number of installed power generation systems was optimized to obtain the best cost-effectiveness. Literature [18] introduced an optimal capacity planning method for renewable energy sources-pumped storage-seawater desalination (RES-PS-D) system, in order to make better use of water resources, desalination and storage systems. Through the use of all cost and benefit components for cost-benefit analysis, the objective function for determining the maximum economic benefit of the RES-PS-D system was established, and finally the mixed integer linear programming algorithm was used to solve the optimal function.

First of all, the existing literatures on source-grid-load interaction is more about other kinds of flexible loads such as electric vehicles and air conditioning, while ignoring the desalination load as a controllable load, which cannot only participate in grid regulation, but also effectively solve the global water resources shortage problem. As a demand side resource, the desalination load cannot only obtain the economic benefits of "peak-shaving and valley-filling", but also increase the output of water resources. Moreover, most of the existing research on the coordinated operation of active distribution networks focuses on the interaction between the power grid and the load or the interaction between the power supply and the load. While, there is no in-depth study on the interaction between the source-grid-load interaction. More importantly, although very few literatures propose the "grid-load" interaction index or the "source-load" interaction index, they fail to establish a relationship between the interaction index and the overall scheduling model, and cannot effectively quantify the model interaction effect, which is not convincing enough.

In order to solve the above problems, this paper mainly establishes the "source-grid-load" interactive comprehensive evaluation index system, and determines the indicators for evaluating the effect of "source-grid-load" from two perspectives of the power source side and the grid side. The main contributions of this paper can be summarized as follows: 1) the connotation of the source-grid-load interaction is introduced and the indicators to describe the interaction are proposed; 2) the optimal operation model considering source-grid-load interaction indicators is established, and the indicators are integrated as part of the objective function; 3) Taking into account the seawater desalination as controllable loads, the desalination load is introduced into the "source-network-load" interactive model to achieve the effect of "peak-shaving and valley-filling". At the same time, to a certain extent, it has increased the output of freshwater resources and provided some help to solve the global water shortage.

The reminder of this paper is as follows: Sect. 2 firstly introduces the connotation of source-grid-load interaction. Next, Sect. 3 introduces two indicators for the source-grid-load interaction. Afterwards, Sect. 4 introduces the source-grid-load flexible interactive operation model. Then, Sect. 5 presents the case studies based on the data from an eastern city in China. Finally, Sect. 6 is the conclusion of this paper.

2   Connotation of Source-Grid-Load Interaction

The source-grid-load interaction refers to the goal of more economically, efficiently, and safely improving the power system's dynamic balance capability through a variety of interaction forms among the power supply, grid, and load [19]. The source-grid-load interaction is essentially a way to maximize the use of energy resources. As shown in

Fig. 1, the traditional power system operation mode is adjusted by the power supply tracking the load change. As shown in Fig. 2, the power, grid, and loads in the future power grid will have flexible characteristics, in order that a comprehensive source-grid-load interaction will be formed, presenting multiple new interaction methods such as source-source complementation, grid-load interaction, source-load interaction, and source-grid coordination.

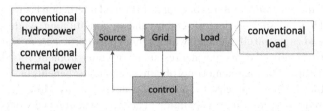

Fig. 1. The traditional power system [19].

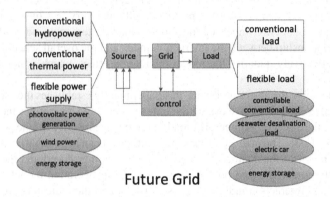

Fig. 2. The connotation of source-grid-load system interaction [19].

2.1 Source-Source Complementation

Source-source complementation is to promote the consumption of renewable energy by complementing the output temporal characteristics and frequency characteristics between different power sources. Although conventional power plants coordinated renewable energy output can better achieve renewable energy integration into the grid, conventional power plants have high reserve capacity and high costs. The primary energy of the future power grid is diverse, such as hydropower, wind power, photovoltaic solar power, biomass power, tidal power, etc. There is a certain correlation and wide-area complementation in their spatial and temporal distribution and dynamic characteristics. Complementation of sources can overcome the shortcomings of a single renewable energy source which is easily affected by geographical, environmental, meteorological and other factors, thereby effectively improving the utilization efficiency of renewable energy sources, reducing the rotating reserve of the power grid, and enhancing the system's autonomous regulation capability.

2.2 Grid-Load Interaction

Grid-load interaction is the use of flexible loads to participate in the auxiliary services of the grid under the demand response mechanism to promote the optimal operation of the grid, such as participating in reactive power optimization for voltage regulation, active power optimization for frequency regulation, active power fluctuation mitigation to the exchange power with the main network, and network loss reduction, etc. Grid-load interaction means that when a contingency occurs in the power grid or is about to occur, the load flow is actively adjusted to change the load flow distribution to ensure the safe, economic, and reliable operation of the power grid. As a part of the power system's instantaneous balance, the load characteristics and behaviour characteristics largely determine the safety and economy of the power grid. Different loads have different requirements for power supply reliability. With the gradual accommodation of the demand side, the electricity tariff policy can be used to encourage power-side resources to actively cut peaks and fill valleys and balance power, which will improve the economics and stability of power system operations. As another form of backup, the interruptible load is the emergency reserve "power generation" resource that can be dispatched by the power grid, and it can also economically and effectively cope with insufficient reserve capacity with small probability and high risk to ensure the safe and reliable operation of the power grid.

2.3 Source-Load Interaction

Source-load interaction means that both power and load can be used as dispatchable resources to participate in power supply and demand balance control. The flexible change of load becomes one of the important means to balance power supply fluctuations. The future power grid is composed of multiple power sources and loads that are widely distributed in temporal and spatial scales. Both the power source side and the load side can be used as dispatchable resources. Controllable loads such as energy storage, electric vehicles, and seawater desalination on the load side can participate in the active power regulation of the power grid. Industrial loads, commercial loads, and air-conditioning and refrigerators in the residential life load of power users can respond to grid demand in real time and participate in power supply and demand balance through effective management mechanism. Flexible loads will be able to become an important means of balancing intermittent energy power fluctuations, where the desalination load will be of significant importance in coastal areas and islands.

2.4 Source-Grid Interaction

The existing power grid operation control is based on the comprehensive application of load forecasting, unit combination, day-to-day planning, online scheduling, and real-time control, according to different time scales to achieve coordinated control of power supply and power grid. With the large-scale centralized grid connection of intermittent energy and small-capacity distributed access to the power grid, source-grid coordination is mainly reflected in two aspects [19]. On one hand, the large-scale renewable energy, hydropower, thermal power, especially pumped storage and other conventional energy will be divided into divisions of labor and cooperation, and jointly bundled for power delivery; On the other hand, according to the balance of supply and demand of the power grid, a large number and diverse forms of distributed power sources can be flexibly and efficiently combined through microgrids and intelligent active distribution networks. With the development of source-grid coordination technology, the predictable, dispatchable and controllable capabilities of intermittent renewable energy will be greatly improved, in order to overcome its "unfriendly" characteristics and to increase the utilization rate of renewable energy.

3 Indicators of Source-Grid-Load Interaction

3.1 Power-Side Indicator

The power-side indicator is mainly considered by the unit power supply reliability rate, which reflects the interaction effect of the power supply side. The duration of the effective power generation and the average power outage time of each unit in the system when the source and load are coordinated is a critical factor affecting system reliability, coordination, and system interaction. The unit's power supply reliability rate reflects the reliability of each unit on the power supply side, providing sufficient energy protection for the entire system. It can be calculated according to the equation:

$$r = \frac{t_e}{T} \times 100\% = (1 - \frac{t_s}{T}) \times 100\% \tag{1}$$

where r is the unit power supply reliability rate, t_e is the unit effective power supply time, t_s is the unit failure power time, and T is the statistics time. The closer the value of power supply reliability rate is to 1, the better the power supply reliability of the unit.

3.2 Grid-Side Indicator

The grid-side indicator is mainly measured by the system load peak-to-valley difference. In order to evaluate the interaction effect on the grid side, the peak-to-valley difference ratio of the system is introduced to reflect the balance between the peak and valley power consumption of the system. The larger the peak-to-valley difference ratio, the greater the pressure of the system to cut the peaks and to fill the valleys, and the lower the economy. It can be calculated according to the equation:

$$q = \frac{\Delta P}{P_{\max}} \times 100\% \tag{2}$$

where q is the system load peak-to-valley difference, ΔP is the peak-to-valley difference, and $P_{\max}$ is the maximum of the system load. The lower the value, the lower the pressure of the system.

4 Source-Grid-Load Flexible Interactive Operation Model

4.1 Objective Function

The objective of this paper is to minimize the overall system operation cost, so that the desalination would be more economic competitive. The comprehensive costs include unit operation and maintenance costs, demand response costs, environmental pollutant penalty costs, and interaction indicator costs. The objective function can be calculated according to the equation:

$$C = \min (C_m + C_x + C_e + C_\gamma) \tag{3}$$

where C is the comprehensive system operation cost, C_m is the unit operation and maintenance cost, C_x is the demand side response cost, C_e the is environmental pollutant penalty cost, and C_γ is the interaction indicator costs.

Unit Operation and Maintenance Costs. Unit operation and maintenance costs mainly include energy consumption costs and unit operation and maintenance costs.

$$C_m = C_{m1} + C_{m2} \tag{4}$$

where C_{m1} is the unit operation and maintenance cost, and C_{m2} is the unit energy consumption cost.

$$C_{m1} = \sum_{i=1}^{n} O_i C_i M_i \tag{5}$$

where n is the number of units, O_i is annual maintenance cost factor for the i-th device, C_i is capacity of the i-th device, and M_i is construction price of the i-th device.

Energy consumption cost mainly refers to the cost incurred by the micro-power generator unit during the power generation process. The wind turbine does not consume fuel, and the thermal turbine is the major object. Energy consumption cost can be calculated according to the equation:

$$C_{m2} = \sum_{t=1}^{T} \sum_{i=1}^{n} C_i P_i(t) \tag{6}$$

where C_{m2} is the unit energy consumption cost, C_i is the unit operation cost of i-th type distributed power, $P_i(t)$ is the output power of i-th type distributed power at t hour, and T is the system operating cycle.

Demand Response Cost. The implementation of demand response shall provide compensation measures for responding users according to the contract or agreement. This paper takes the desalination load as an interruptible load as an example, and controls the interruption of the desalination load by controlling the start and stop of the unit's pumps. The calculation method of compensation costs is as follows:

$$C_c = \sum_{i=1}^{N_1} m_c P_{1,i} T_{1,i} \tag{7}$$

where C_c is the demand response compensation cost, m_c is the compensation price per unit of electricity, N_1 is the number of interruptible loads, $P_{1,i}$ is the load interruption capacity of the i-th interruptible load, and $T_{1,i}$ is the interruption time.

The mathematical model of the direct income of demand response participating in the peak-shaving and valley-filling of distribution network is as follows:

$$C_{in} = \sum_{1}^{24} [P^+(t) - P^-(t)] C_{pu}(t) \tag{8}$$

where $P^+(t)$ is the amount of electricity transferred into the demand side at time t, $P^-(t)$ is the amount of electricity transferred out of the demand side at time t, $C_{pu}(t)$ is the hourly electricity price at time t, and C_{in} is the direct benefits of peak-shaving.

The demand response cost can be calculated according to the equation:

$$C_x = C_c - C_{in} \tag{9}$$

Environmental Pollutant Penalty Costs. The environmental impact is mainly reflected in the emissions of pollutants, mainly considering the emissions of carbon dioxide and sulphur dioxide. The calculation method of pollutant emissions is as followed.

$$E_s = \sum_{j=1}^{T} \sum_{i=1}^{n} (\alpha_{si} + \beta_{si} P_i(t) + \gamma_{si} P_i^2(t)) \tag{10}$$

$$E_c = \sum_{j=1}^{T} \sum_{i=1}^{n} (\alpha_{ci} + \beta_{ci} P_i(t) + \gamma_{ci} P_i^2(t)) \tag{11}$$

where E_s is the sulphur dioxide emission, E_c is the carbon dioxide emission, α_{si}, β_{si}, γ_{si} are the coefficients of the quadratic function of the unit's sulphur dioxide emissions with respect to the unit outputs, respectively, and α_{ci}, β_{ci}, γ_{ci} are the coefficients of the quadratic function of the unit's carbon dioxide emissions with respect to the unit output, respectively.

The calculation method for the total penalty cost of environmental pollutants is as followed.

$$C_e = F_s + F_c \tag{12}$$

$$F_s = \rho_s E_s \tag{13}$$

$$F_c = \rho_c E_c \tag{14}$$

where F_s, F_c are the penalty costs for sulphur dioxide and carbon dioxide, respectively. ρ_s, ρ_c are the penalties for sulphur dioxide and carbon dioxide, respectively.

Penalty Costs of Interaction Indicators. When the flexible load demand response participates in the source-grid-load flexible interaction, the interaction indicators are mainly determined from the power supply side and the grid side. If an accident occurs at the power supply side or the grid side, the cost of power outages and the cost of system backup capacity will increase.

$$F_c = \rho_c E_c \tag{15}$$

where C_y is the penalty cost of interaction indicators, C_1 is the outage cost, and C_2 is the system spare capacity cost.

The calculation formula of the outage cost is as follows.

$$C_1 = A(1 - e^{(1-r)/\tau}) \tag{16}$$

where A is the outage cost stable value. When the power outage continues for a period of time, the cost of unit power outages will gradually approach this stable value. τ is outage loss time constant, and τ determines how long it takes for the outage cost to reach a stable value after a power outage. Taking the seawater desalination load as an example, the value of A is 150, and the value of τ is 4.

The calculation formula of the system reserve capacity cost is as follows.

$$C_2 = \sum_{t=1}^{24} \left\{ [Q_t(1 + \vartheta) - \frac{P^{min}}{1 - q}] \times C_m + (Q_t' - \frac{P^{min}}{1 - q}) \times C_{rv} \right\} \tag{17}$$

where Q_t is the forecasting load demand at time t, Q_t' is the actual load demand, ϑ is the standby rate of grid load, P^{min} is the minimum output of the system, and C_{rv} is the variable cost once the unit reserve capacity is called by the system operator.

4.2 Constraints

The system is mainly constrained by the power balance constraints, the unit rotation reserve constraints, and the unit output constraints. The equation of the system power balance constraint is as follows.

$$\sum_{i=1}^{n} P_i(t) + P_w(t) + P_s(t) + P_x(t) = P_d(t) \tag{18}$$

where $P_i(t)$ is the output value of the thermal power units, $P_w(t)$ is the output value of the wind turbines, $P_s(t)$ is the output value of the photovoltaic units, $P_x(t)$ is the demand response value, and $P_d(t)$ is the value of load forecast.

The formula of the unit spinning reserve constraint is as follows.

$$\sum_{i=1}^{n} P_i^{\max} - P_d(t) \geq P_R(t) \tag{19}$$

where $P_i^{\max}$ is the maximum output of unit i, and $P_R(t)$ is the spinning reserve capacity of the system.

The formula of unit output constraint is as follows.

$$P_i^{\min} \leq P_i(t) \leq P_i^{\max} \tag{20}$$

where $P_i^{\min}$ is the minimum output of unit i.

5 Source-Grid-Load Flexible Interactive Operation Model

This paper takes a certain area in an eastern city in China as an example. The 24-h photovoltaic output in this area is shown in Fig. 3, and the 24-h wind power output is shown in Fig. 4.

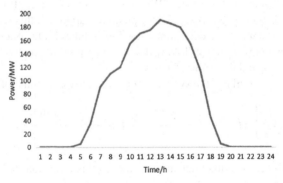

Fig. 3. General trend of PV.

It can be seen from Fig. 3 that the maximum output value of the photovoltaic unit is 185 MW, which appears at 14 h. The photovoltaic output occurs from 4:00–20:00. It can be seen from Fig. 4 that the peak output of the wind turbine appears at 3 h and 21 h, the maximum output value of the wind turbine is 470 MW; the minimum output value of the wind turbine is 120 MW, which appears at 14 h. The wind output is larger in the evening and lower in the daytime.

Due to the diversity of flexible loads such as seawater desalination loads, a particle swarm optimization algorithm with a simpler principle is adopted. The optimization results are shown in Fig. 5.

From Fig. 5, it can be seen that with the introduction of demand response, the peak load is 3200 MW, the valley load is 2610 MW, and the peak-to-valley difference is 0.184, which shows that the addition of demand response is beneficial to balance the daily load.

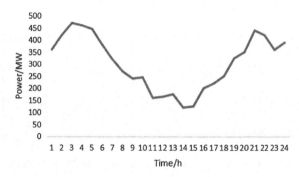

Fig. 4. General trend of wind generation.

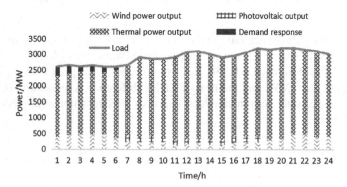

Fig. 5. Optimization Results.

At the same time, the lower the peak-to-valley ratio, the lower the cost of system reserve capacity. The capacity plays the role of peak-shaving and valley-filling, reducing the peak-to-valley difference of system power and improving the interaction effect.

In order to evaluate the effectiveness of the model, the cost and interaction indicators of several different models are compared. Among them, case 1 is a model considering the output of wind turbines and photovoltaics, but without the desalination load; case 2 is a model considering only the output of wind turbines to reduce the output of new energy; case 3 is the interaction model proposed in this paper.

Figure 6 is a comparison of the new energy output in Case 1 and Case 3, reflecting the impact of introducing the desalination load to increase the source-grid-load interaction capability on the new energy output in the system.

From Fig. 6, it can be seen that when considering the desalination load, the output of new energy is higher than the value of the new energy when the desalination load is not considered. The interactive effect can also reduce the phenomenon of wind and solar curtailment, thus maximizing energy efficiency and realizing the nature of source-grid-load interaction.

Figure 7 is a comparison of load curves for Case 1 and Case 3. After the seawater desalination load is introduced into the source-grid-load interactive system, the system

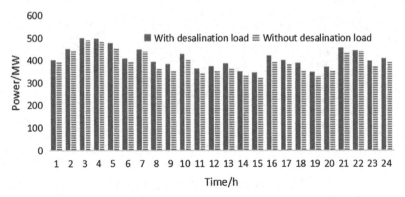

Fig. 6. Comparison of the impact of desalination load on new energy consumption.

load peak-to-valley difference ratio in the source-network-load interactive index is significantly reduced. In Case 1, the peak load is 3490 MW, the valley load 2310 MW, and the peak-to-valley difference is 0.338. However, in Case 3, the peak load is 3200 MW, the valley load 2610 MW, and the peak-to-valley difference is reduced to 0.184. The introduction of seawater desalination load improves the interaction ability between the source-grid-loads.

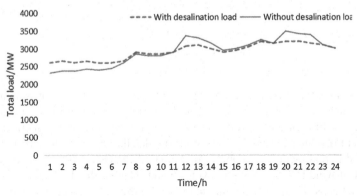

Fig. 7. Comparison of load curve for case 1 and case 3.

It can be seen from Fig. 7 that the seawater desalination load uses electricity during the low period of the load, and the valley of the load rises; the seawater desalination load stops using electricity during the peak period of the load, so that the peak load is reduced. Through optimization and adjustment, the load curve is basically stable. The introduction of seawater desalination load significantly reduces the peak-to-valley difference of the system load. The peak-shaving effect is significant, and the interaction capability between the source-grid-load is improved.

Figure 8 shows the comparison of the costs of interaction in the three scenarios; Table 1 shows the comparison of the indicators for evaluating the interaction capacity of the source network load in the three scenarios.

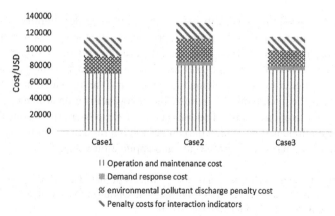

Fig. 8. Costs of interaction in three cases.

Table 1. Reduction in maximum temperature of different locations.

Case	Unit power supply reliability	Peak-to-valley difference
Case #1	0.692	0.338
Case #2	0.735	0.257
Case #3	0.821	0.184

As can be seen from Fig. 8 and Table 1, the total cost of interaction in case 1 is 113793 USD, the total cost of interaction in case 2 is 131999 USD, and the total cost of interaction in case 3 is 115461 USD. The comparison between Case 1 and Case 3 shows that due to the addition of seawater desalination load, although the unit's operating cost has increased in Case 3, the introduction of seawater desalination load is conducive to the dispatch of new energy sources, and it also plays a role of peak-shaving and valley-filling, reducing the wind and solar curtailment, improving the utilization rate of energy, and reflecting the essence of source-grid-load interactive operation. It can be seen from the comparison between Case 2 and Case 3 that although the addition of new energy has raised the start-stop cost and demand response call cost of the unit to varying degrees, it has reduced the operating costs of thermal power units and the penalty cost of pollutant emissions. With the increase of new energy output, the overall system cost has decreased. Through the comparison of the above three cases, the superiority of the source-grid-load interaction model is further verified. Meanwhile, the saved costs would be beneficial to welcome and circulate the desalination load in the corresponding areas.

6   Conclusions

This paper establishes the source-grid-load interaction indicators for distribution power networks from the power source side and the grid side, to evaluate the flexible interaction effect. The unit power supply reliability rate is used to evaluate the reliability of the power supply during the interaction of the power supply units, and the system power peak-to-valley difference ratio is used to evaluate the effect of the grid side participation in the interaction. The essence of the source-grid-load interaction is to maximize the utilization of energy. This paper considers the demand response and the introduction of distributed energy to balance the load, so as to maximize the energy utilization as much as possible. By introducing the proposed optimal operation model, the power grid operates in a more economical and efficient state, which maximizes the interaction capability between the source-grid-load and has certain promotion and application value.

Acknowledgments. This work was supported by Projects of State Grid Corporation of China., "Study on Multi-source and Multi-load Coordination and Optimization Technology Considering Desalination of Sea Water" (SGTJDK00DWJS1800011).

References

1. Siddaiah, R., Saini, R.P.: A review on planning, configurations, modeling and optimization techniques of hybrid renewable energy systems for off grid applications. Renew. Sustain. Energy Rev. **58**, 376–396 (2016)
2. Parida, A., Chatterjee, D.: An improved control scheme for grid connected doubly fed induction generator considering wind-solar hybrid system. Int. J. Electr. Power Energy Syst. **77**, 112–122 (2016)
3. Carta, J.A., González, J., Cabrera, P.l.: Preliminary experimental analysis of a small-scale prototype SWRO desalination plant, designed for continuous adjustment of its energy consumption to the widely varying power generated by a stand-alone wind turbine. Appl. Energy **137**, 222–239 (2015)
4. Astofi, M., Mazzola, S., Silva, P.: A synergic integration of desalination and solar energy systems in stand-alone microgrids. Desalination **419**, 169–180 (2017)
5. Kyriakarakos, G., Dounis, A.I., Arvanitis, K.G.: Design of a fuzzy cognitive maps variable-load energy management system for autonomous PV-reverse osmosis desalination systems: a simulation survey. Appl. Energy **187**, 575–584 (2017)
6. Karavas, C.S., Arvanitis, K.G., Kyriakarakos, G.: A novel autonomous PV powered desalination system based on a DC microgrid concept incorporating short-term energy storage. Sol. Energy **159**, 947–961 (2018)
7. Amir-Hamed, M., Vincent, W.S., Robert, S.: Autonomous demand-side management based on game-theoretic energy consumption scheduling for the future smart grid. IEEE Trans. Smart Grid **1**(3), 320–331 (2010)
8. Shao, C.C., Wang, X.F.: Cooperative dispatch of wind generation and electric vehicles with battery storage capacity constraints in SCUC. IEEE Trans. Smart Grid **5**(5), 2219–2226 (2014)
9. Martinez-Mares, A., Fuerte-Esquivel, R.: A robust optimization approach for the interdependency analysis of integrated energy systems considering wind power uncertainty. IEEE Trans. Pow. Syst. **28**(4), 3964–3976 (2010)
10. Lin, J., Sun, Y.Z., Song, Y.H.: Wind power fluctuation smoothing controller based on risk assessment of grid frequency deviation in an isolated system. IEEE Trans. Sustain. Energy **4**(2), 379–392 (2013)
11. Li, C., Savaghebi, M., Guerrero, J.M.: Operation cost minimization of droop- controlled AC microgrids using multiagent-based distributed control. Energies **9**(9), 717–736 (2016)

12. Maheshwari, P., Tambawala, Y., Nunna, H.: A review on plug-in electric vehicles charging: Standards and impact on distribution system. In: IEEE International Conference, pp. 1–6 (2014)
13. Fernandez, L.P., Roman, T.G., Cossent, R.: Assessment of the impact of plug-in electric vehicles on distribution networks. IEEE Trans. Pow. Syst. **26**(1), 206–213 (2011)
14. Zhou, B., Liu, B., Yang, D., Cao, J., Littler, T.: Multi-objective optimal operation of coastal hydro-electrical energy system with seawater reverse osmosis desalination based on constrained NSGA-III. Energy Convers. Manage. **207**, 112533 (2020)
15. Gray, S.: Seawater use and desalination technology. Treatise Water Sci. 73–109 (2011)
16. Maleki, A., Pourfayaz, F., Ahmadi, M.H.: Design of a cost-effective wind/photovoltaic/hydrogen energy system for supplying a desalination unit by a heuristic approach. Sol. Energy **139**, 666–675 (2016)
17. Maleki, A., Khajeh, M.G., Rosen, M.A.: Weather forecasting for optimization of a hybrid solar-wind-powered reverse osmosis water desalination system using a novel optimizer approach. Energy **114**, 1120–1134 (2016)
18. Liu, B.Y., Zhou, B.W., Yang, D.S.: Optimal capacity planning of combined renewable energy source-pumped storage and seawater desalination systems. Global Energy Interconnect. **2**, 310–317 (2019)
19. Zhong, M., Jin, L., Xia, J., Cheng, L., Chen, P., Zeng, R.: Coordinated control of coastal multi-source multi-load system with desalination load: a review. Global Energy Interconnect. **2**, 300–309 (2019)

The Optimal Dispatching Strategy of Microgrid Group Based on Distributed Cooperative Architecture

Xue Li, Yunpeng Zhang$^{(\boxtimes)}$, Dajun Du, Zhourong Zhang, and Meng Xia

Shanghai Key Laboratory of Power Station Automation Technology,
Shanghai University, Shanghai 200444, China
`lixue@i.shu.edu.cn, zypyh@shu.edu.cn`

Abstract. This paper establishes a distributed optimization operation model of microgrid group to solve the problem of privacy security and economic conflict among different stakeholders. Firstly, the factors affecting the operation cost are analyzed from the perspective of the internal equipment and external active/reactive power transactions of the microgrid. Then, each node in the microgrid group takes the overall operating cost as the optimal objective, and particle swarm optimization (PSO) algorithm is used to optimize the model. Furthermore, the collaborative optimization algorithm is adopted to reach an agreement of the respective optimization results through the exchange of key information among different node agents, and the optimal dispatching strategy is obtained independently. Finally, the model is applied to the improved 30-node microgrid group system. The results show that the model and algorithm can reduce the operating cost of the microgrid group, ensure the maximum benefit of the individual subject, and achieve the optimal operation of the multi-microgrid system.

Keywords: Microgrid group · Collaborative optimization algorithm · Particle swarm optimization (PSO) · Distributed model

1 Introduction

With the continuous development of microgrid, multiple microgrids close to each other form a microgrid group, which is conducive to improving the overall operation stability and economy of the multi-microgrid system. Compared with a single microgrid, the interest subjects of the microgrid group are diversified, so the multi-microgrid system needs to coordinate the interests of all the sub-microgrid. The traditional optimal dispatching adopts centralized optimization, which requires the collection of operating parameters and prediction data of all electrical equipment. The resulting optimization problem is of large scale, has high requirements on the communication system and control system, and has the risk of disclosure of privacy information of the microgrid [1,2]. Therefore,

© Springer Nature Singapore Pte Ltd. 2020
M. Fei et al. (Eds.): LSMS 2020/ICSEE 2020 Workshops, CCIS 1303, pp. 18–35, 2020.
https://doi.org/10.1007/978-981-33-6378-6_2

it is very important to study the collaborative optimization algorithm of the microgrid group and ensure the optimal overall economy of the microgrid group.

The operation optimization of interconnected micro-grid group in the ubiquitous power Internet of things is a complex energy management problem, which involves the cooperative game in the microgrid group with multiple stakeholders, the optimization of reactive power and energy storage equipment inside the micro-grid, and the coordination and scheduling between micro-grid groups. In [3–5], an optimization and coordination scheme of individual micro grid and micro grid group was proposed for the energy management problem of interconnected microgrids. An optimization strategy of power exchange between micro grids based on model predictive control is proposed in [6] for balance the benefits of energy storage and energy exchange between micro grids. Meanwhile, the power input cost of external distribution network is optimized, while the internal cost of micro grids is ignored. In [7], a two-step optimization method is adopted. Firstly, the power of each microgrid tie line is obtained through overall optimization. The microgrid is absolutely subject to the day-ahead dispatching results of the upper layer. Aiming at the multi-objective optimization problem related to the transaction cost and internal cost of the external power grid, a multi-objective framework to optimize the micro-grid group is proposed [8]. In [9], a two-stage energy management model is proposed, in which the upper layer optimizes the objectives related to the transaction of the distribution network, and the lower layer optimizes the operation cost and pollutant emission of the micro grid. By means of the improved particle swarm optimization algorithm for centralized solution, the uncertainty optimization scheduling model of off-grid island multi-microgrid is established [10]. In the above study, all the model data of the microgrid group were collected step by step to form a complete model of the whole network. However, when the network size of the microgrid group was large, this hierarchical centralized scheduling and computing method would result in an excessive burden of data and computation at the final convergence point, which required a high demand on the communication system and the control system.

In order to avoid the shortcomings of the centralized algorithm, a layered energy management strategy is proposed to optimize the internal cost of the micro-grid and the overall cost of the multi-micro-grid system [11,12]. A general distributed secondary control scheme is proposed to solve the problems of complex communication topology of centralized algorithm and high performance requirements of CPU, and global coordination of micro network group was carried out [13]. A two-layer energy management method is proposed, in which the energy management system of each microgrid optimizes itself and then coordinates and optimizes the multi-microgrid system with a central EMS to reduce the information communication between the microgrid and the distribution network [14]. The multi-agent two-layer game interaction model is established for the comprehensive energy system, and the multi-energy complementarity in cooperative game is simulated in the energy market bidding process [15]. A distributed nested energy management architecture is proposed to provide layered privacy for users of micro grid to solve the problem of privacy disclosure during multi-micro grid optimization [16]. A distributed convex optimization framework for energy exchange between micro-grids is proposed, which does not consider the internal cost of microgrids [17].

The above work ignores the underlying equipment associated with the distribution of the microgrid, and the algorithm only involves the level of the microgrid, not the equipment level of the microgrid, which further increases the overall computing burden. Therefore, how to establish the distributed optimal operation model in the microgrid group and achieve the overall optimal operation of the microgrid group on the basis of ensuring the optimal operation of the sub-microgrid is the main research purpose of this paper (Fig. 1).

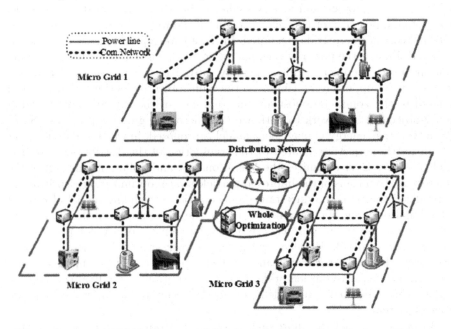

Fig. 1. A schematic block diagram of distributed optimization of microgrid group.

(1) How to obtain the optimal strategy independently through the communication between the distributed devices in the microgrid and the adjacent devices, and how to achieve the overall optimal operation of the microgrid group without violating the privacy of other microgrids is the primary challenge.

(2) How to ensure the optimal operation of the distributed micro grid group on the basis of the optimal operation of the sub-micro network cost is the second challenge of the overall optimal operation of the distributed micro network group.

In order to solve these problems, the optimal operation strategy based on distributed model is studied in this paper. The main contributions of this paper include:

(1) A distributed architecture microgrid optimization model is proposed, which takes into account the protection of individual microgrid privacy and reduces the amount of computation and communication in the microgrid through the mutual communication of adjacent devices in the microgrid, so as to obtain a coordinated optimization strategy at the microgrid equipment level.

(2) In view of the conflict between the optimal operation of a single micro network and the optimal operation of multiple microgrids, the optimal cost constraint of a single microgrid in multiple microgrids is increased, so as to improve the global optimal operation of microgrid groups under the condition that each microgrid meets the optimal operation of an individual.

2 Description of Cost Model of Microgrid Group System

The microgrid group is a network composed of multiple microgrids, each of which is also connected to the distribution network. Power electronic converters are used to unify different rated voltages among the microgrid groups, and power is transferred through a combined power transmission line. In addition, multiple microgrids in a microgrid group can belong to the same company or different companies. A single microgrid consists of a set of renewable distributed energy, diesel generators and energy storage devices. As shown in Fig. 2, a node in the microgrid can connect multiple devices, such as renewable distributed energy, diesel generators, and so on, and also overwrite a communication agent node. The node agent can only exchange information with neighboring agents and carry out complete distributed control over the devices under its control.

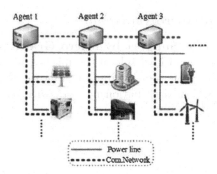

Fig. 2. Microgrid connection device diagram.

2.1 External Costs of Microgrid Groups

Cost of Electricity Purchased from the Distribution Network. After making full use of renewable energy, the microgrid will purchase the power that is lacking from the high-voltage distribution network or sell the remaining power to the distribution network at the right price [18].

$$C_1 = f_1 c_a \left\{ P_{net} + \sum_{i=1}^{n} \sum_{k=1}^{n} V_i V_k G_{ik} \cos \delta_{i,k} \right\} \tag{1}$$

$$P_{net} = \sum_{i=1}^{n} (P_i + P_{bat,i} + P_{link,i}) \tag{2}$$

c_a is the real-time electricity selling price, P_{net} is the net injected power of the microgrid, V_i is the voltage amplitude of node i, G_{ik} is the admittance of the line between node i and k, $\delta_{i,k}$ is the voltage Angle difference between node i and k. P_i is the injected power of the distributed power supply of node i, and $P_{bat,i}$ is the active power of the energy storage device in node i. $P_{link,i}$ is the active power of selling electricity between other microgrids and node i. f_1 is the price factor, $f_1 = 1$ means that the micro grid purchases electricity from the distribution network, $0 < f_1 < 1$ represents the micro network to distribution network electricity sales.

Electricity Cost from Other Micro Shopping. In a microgrid group, a single micro grid can also purchase or sell active and reactive power from other microgrids through connection lines in the micro grid group.

$$C_2 = \sum_{i=1}^{n} (c_p P_{link,i} + c_q Q_{link,i}) \tag{3}$$

c_p and c_q are respectively the negotiated prices of active and reactive power units traded between microgrids.

2.2 Internal Cost of Microgrid Group

Cost of Reactive Power Support by Micro Grid Equipment. The converters on the power side and load side of the micro grid can provide a certain amount of reactive power for the distribution network and other micro grids. Therefore, the micro grid or distribution network with reactive power support needs to compensate for it.

$$C_3 = -c_b \sum_{i=1}^{n} |Q_i| \tag{4}$$

c_b is the compensation price per unit of reactive power.

Cost Relating to Unfair Utilization of Converters. If the converter in the micro grid works at a state close to the rated capacity for a long time, internal components will be damaged due to high temperature, conduction loss, switching loss and other problems, which will increase the additional cost.

$$\alpha_i = \frac{\text{sqrt}\left(S_{i,\max}^2 - P_i^2\right)}{S_{i,\max}} \tag{5}$$

$$C_4 = c_\alpha \left\{ \sum_{i=1}^{n} \alpha_i |Q_i| \right\} \tag{6}$$

α_i is the unreasonable utilization of the converter of node i, and c_α is the cost factor for the unreasonable operation of the inverter.

Cost of Power Loss by Charging and Discharging of Energy Storage Equipment. The charging and discharging process of energy storage equipment will lead to power loss of the device, and additional cost will be added. There is a linear relationship between the efficiency of energy storage equipment and charging/discharging power [19].

$$C_5 = c_p \sum_{i=1}^{n} [1 - \eta_i] |P_{bat,i}| \tag{7}$$

$$\eta_i = \begin{cases} a_i - b_i |P_{bat,i}|, & if \cdot P_{bat,i} > 0 \\ 1/(a_i - b_i |P_{bat,i}|), & if \cdot P_{bat,i} < 0 \end{cases} \tag{8}$$

η_i is energy storage device charge-discharge efficiency, a_i and b_i is charging and discharging efficiency coefficient.

The Market Cost of Microgrid Transactions. The power transaction between microgrids will borrow the power lines of the public grid, so for each successful power transaction between microgrids, a part of the borrowing line fee will be paid to the public grid, which will be paid by the seller's microgrid.

$$C_6 = \beta \sum_{i=1}^{n} (P_{\text{link},i} + Q_{\text{link},i}) \tag{9}$$

β is the coefficient of the payment.

2.3   Constraints on Multi-microgrid Systems

Voltage constraint: the voltage amplitude of each node should be within a certain range.

$$V_{\min} < V_i < V_{\max} \tag{10}$$

In the optimization process, the reactive power emitted by the converter of each device should be within the effective range, that is $|Q_i| \leq |Q_{i,\max}|$. In the distributed microgrid model, the variation of reactive power output of the same converter in two consecutive iterations should be within a reasonable range, and should not exceed the reactive power limit that the converter can provide in a short time [20].

$$|Q_i[z] - Q_i[z-1]| \leq Q_{i,dr} \tag{11}$$

$Q_i[Z]$ is the output reactive power value of node i in the z iteration, and $Q_{i,dr}$ is the limit of reactive power increase of the converter at node i in a short time. The charging and discharging power of the energy storage equipment is determined by the equipment manufacturer, so the charging and discharging power of the energy storage equipment cannot exceed the rated power.

$$|P_{\text{bat},i}| \leq P_{\text{bat},i,\max} \tag{12}$$

3   A Distributed Cooperative Algorithm Based on PSO

The third section describes the algorithm is based on micro group of system overall cost as the objective function, the particle swarm optimization (PSO) algorithm with the distributed voltage state estimation and the consensus of the node agent negotiation more, use to ensure that both algorithms of each agent in the privacy, and ensure the multi-agent reach a consensus in the process of optimization, implementation piconets group of distributed optimization system.

3.1   Distributed Voltage Estimation

In the optimization process of the microgrid group, the reactive power injected by the device converter into the microgrid is taken as the control variable, while the voltage at each node is taken as the state variable. The corresponding relationship between the two is non-linear.

$$Q_i = V_i \sum_{k=1}^{n} V_k \left(G_{ik} \sin \delta_{ik} - B_{ik} \cos \delta_{ik} \right) \tag{13}$$

According to formula 12, approximate voltage estimation equations for active and reactive power and voltage states can be obtained:

$$
\begin{bmatrix} V_2 \\ V_3 \\ \vdots \\ V_{n-1} \\ V_n \end{bmatrix} = V_0 1_{n-1} + \frac{[Y(2:n,2:n)]^{-1}}{V_0} \begin{bmatrix} P_2 - jQ_2 \\ P_3 - jQ_3 \\ \vdots \\ P_{n-1} - jQ_{n-1} \\ P_n - jQ_n \end{bmatrix} \tag{14}
$$

1_{n-1} is a column vector with all one. Y is the nodal admittance matrix of a microgrid group, V_0 is the voltage of the balanced node, $P_i - jQ_i$ is the injected power of the node.

3.2   Multi-agency Consensus Negotiation

The connected graph G contains N nodes and E edges, each of which is under the control of the agent. A pair of nodes with a common edge is called "adjacent node", and each agent only communicates and exchanges information with its adjacent node. $\xi_k[z]$ is the estimate of state variables by agent k at an iteration

z, After the iteration times, if $|\xi_k[z] - \xi_l[z]| < \mu, (k \neq l)$, where μ is a very small positive number, the $\xi_k[z]$ might be more agent finally reached consensus. According to the consensus algorithm, each agent updates its estimates using the following discrete time model [21]:

$$\xi_k[z+1] = \sum_{l=1}^{N} C_{k,l}\xi_l[z] \qquad (15)$$

$C_{k,l}$ represents the communication coefficient between agent k and l, if k and l are not adjacent, then $C_{k,l} = 0$. Further, the estimated equation of multiple agents is as follows:

$$\bar{\xi}_k[z+1] = C * \bar{\xi}_l[z] \qquad (16)$$

C is a matrix composed of elements $C_{k,l}(k, l \in 1 \cdots n)$, and $\bar{\xi}_k$ is a column vector including all node state estimates. If C matrix is doubly stochastic matrix and the eigenvalues of C are less than or equal to 1, all the agent's state estimations will independently converge to the same value in the proof [18]. In order to improve the privacy of the agent node information in the negotiation process and to accurately its results, the dual random matrix C is introduced as follows [19]

$$c_{i,k} = \begin{cases} \frac{2}{n_i+n_k+1} & k = N_i \\ 1 - \sum_{i,l=1(i\neq l)}^{n} C_{i,l} & k = i \\ 0 & otherwise \end{cases} \qquad (17)$$

N_i is the set of all agents adjacent to agent i, n_i refers to the number of agents connected to agent i (Fig. 3).

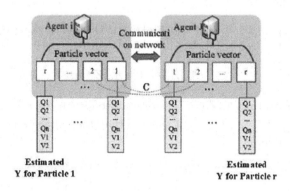

Fig. 3. Information exchange between adjacent agents.

3.3   Process of Distributed Cost Optimization Algorithm Based on PSO

This algorithm aims at the overall cost of the microgrid group, including the internal and external costs of the microgrid. However, due to the constraints of the micro grid group's voltage range, reactive power output and the game problem between self-optimization and global optimization, inequality constraints with penalty function factors should be added to the overall cost of the micro grid group.

$$f(x) = \sum_{i=1}^{6} C_i + \sum_{i=1}^{n} a_1 |V_i - V_0| + a_2 (Q_{i,dr} - |Q_{i,z} - Q_{i,z-1}|) \\ + a_3 (P_{\text{bat},i,\text{max}} - |P_{\text{bat},i}|) \tag{18}$$

where a_1, a_2 and a_3 are penalty function factors. In the distributed cost optimization algorithm based on particle swarm optimization, each agent is a separate particle swarm, which can calculate the cost of the entire microgrid group through its own information and information of adjacent nodes, and then independently iterate and optimize to obtain the optimal cost agreed with other nodes [22].

Step 0. Each agent randomly determines a particle matrix A_i, $A_i(m, r)$ corresponding to the matrix element $Q_{i,m}^r$, which refers to the rth reactive power output assumed by node-agent i to agent m.

$$A_i[Z] = \begin{bmatrix} Q_{i,1}^1 & \cdots & Q_{i,1}^r & \cdots & Q_{i,1}^p \\ Q_{i,2}^1 & \cdots & Q_{i,2}^r & \cdots & Q_{i,2}^p \\ \cdots & \cdots & \cdots & \cdots & \cdots \\ Q_{i,n}^1 & \cdots & Q_{i,n}^r & \cdots & Q_{i,n}^p \end{bmatrix} \tag{19}$$

Step 1. After the consensus processing of all agent nodes, each agent will update the particle matrix. The row vector update of the particle matrix is calculated as follows:

$$\begin{bmatrix} A_i[Z](m,1) \\ A_i[Z](m,2) \\ \cdots \\ A_i[Z](m,p) \end{bmatrix} = C(i,:) * \begin{bmatrix} A_1[Z](m,1) \cdot A_1[Z](m,2) \ldots A_1[Z](m,p) \\ A_2[Z](m,1) \cdot A_2[Z](m,2) \ldots A_2[Z](m,p) \\ \cdots \\ A_n[Z](m,1) \cdot A_n[Z](m,2) \ldots A_n[Z](m,p) \end{bmatrix} \tag{20}$$

Step 2. Substituting $A_i[Z](:, r)$ into Eq. 14 to obtain the rth voltage estimation vector $\left[V_{i,1}^r \cdot V_{i,2}^r \ldots V_{i,n}^r \right]^T$ of agent i, then the voltage estimation matrix of agent i is as follows:

$$V_i[Z] = \begin{bmatrix} V_{i,1}^1 \cdots V_{i,1}^r \cdots V_{i,1}^n \\ V_{i,2}^1 \cdots V_{i,2}^r \cdots V_{i,2}^n \\ \cdots \\ V_{i,n}^1 \cdots V_{i,n}^r \cdots V_{i,n}^n \end{bmatrix} \tag{21}$$

Step 3. Substituting $A_i[Z](:, r)$ and $\left[V_{i,1}^r \cdot V_{i,2}^r \ldots V_{i,n}^r \right]^T$ into the cost function 1–14,18 to obtain the rth particle of target function $f_i^r \left(Q_{i,1}^r \ldots Q_{i,n}^r \cdot V_{i,1}^r \ldots V_{i,n}^r \right)$ for agents i

Step 4. Repeat steps 1 to 3 to obtain the objective function $F[Z]$ of all agent particles for the Zth iteration, and compare to obtain the optimal target value of each agent $[f_{1,\min}[Z] \cdot \ldots \cdot f_{n,\min}[Z]]^T$. The contribution should contain no more than four levels of headings. Table 1 gives a summary of all heading levels.

$$F[Z] = \begin{bmatrix} F_1[Z] \\ F_2[Z] \\ \ldots \\ F_n[Z] \end{bmatrix} = \begin{bmatrix} f_1^1 \cdots f_1^r \cdots f_1^p \\ f_2^1 \cdots \cdot f_2^r \cdots \cdot f_2^p \\ \ldots \\ f_n^1 \cdots \cdot f_n^r \cdots f_n^p \end{bmatrix} \tag{22}$$

$$\begin{bmatrix} f_{1,\min}[Z] \\ f_{2,\min}[Z] \\ \ldots \\ f_{n,\min}[Z] \end{bmatrix} = \begin{bmatrix} \min\left(f_1^1 \cdots \cdot f_1^r \cdots \cdot f_1^p\right) \\ \min\left(f_2^1 \cdots \cdot f_2^r \cdots \cdot f_2^p\right) \\ \min\left(f_n^1 \cdots \cdot f_n^r \cdots \cdot f_n^p\right) \end{bmatrix} \tag{23}$$

Step 5. The agent based on iterative once get $A_i[Z]$ of PSO, if $\forall |f_{i,\min}[Z] - f_{j,\min}[Z]| < i, j = 1, 2 \ldots n(i \neq j)$, then outputs the global optimal target $F_{i,\min}[Z]$ and $\chi[Z]$; otherwise return step 1 (Fig. 4).

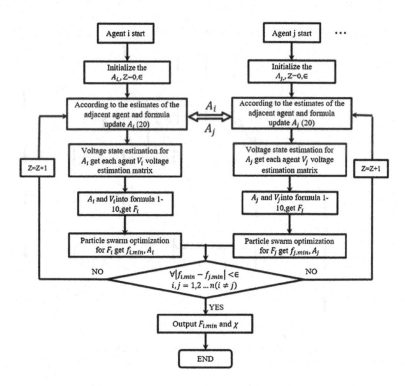

Fig. 4. Block diagram of algorithm operation.

4   Simulation Results and Discussion

In this section, three microgrids containing renewable energy, energy storage systems and loads are connected to each other. The parameters of each microgrid are shown in Table 1, and the structure is shown in Fig. 5. Each micro grid has its own power generation and load. The electricity price of the micro grid group system and the superior distribution network adopts the time-of-use electricity price mechanism. As shown in the Table 1, the peak-valley electricity price table adopted in this calculation example is adopted. The price factor f_1 in formula (9–15) is 0.35, C_q is 0.05*C_p, C_{q1} and C_α are both 0.03. The system has a reference capacity of 100 MVA and a reference voltage of 10 Kv. There are four types of converters for renewable energy and load, with rated power of 5 KvA, 10 KvA, 20 KvA and 50 KvA respectively, and rated power of converters for energy storage system is 20 KvA.

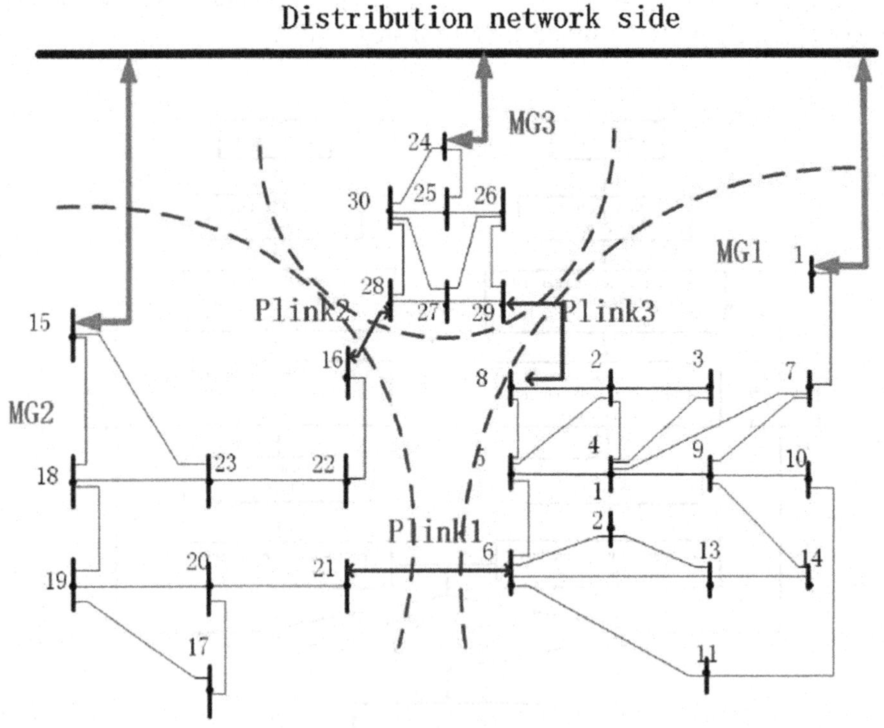

Fig. 5. Block diagram of algorithm operation.

The electricity price of the operating entity of the micro-grid group and the superior distribution network adopts the time-of-use electricity price mechanism [23]. As shown in Table 2, the calculation example of this paper adopts peak-

Table 1. Multi - microgrid system parameters.

Microgrid number	1	2	3
Microgrid location	Residential	Industrial	Commercial
Nodes in the microgrid	14	9	7
Nominal voltage	10KV	10 KV	10 KV
Nodes with RDG-BESSs	1, 7, 9, 13, 14	15, 16, 23	25, 27, 30
Nodes with links	1, 14	15, 18	24, 26
Type of RDGs	Solar PV	Wind	Solar PV

Table 2. Peak and valley electricity price table.

Price($/KWh)	Peak period: 14:00–17:00 19:00–22:00	Flat period: 08:00–14:00 17:00–19:00, 22:00–24:00	Valley period: 00:00–08:00
Power purchasing price	0.83	0.48	0.13
Power selling price	0.65	0.38	0.17

valley flat electricity price table. Table 3 shows the state parameters of the energy storage system in this case.

In this section, the microgrid swarm system is optimized for 24 cycles, the particle number of the algorithm is 30, the convergence accuracy is 0.0001, and the maximum iteration is 200 times. Each node agent uses this algorithm and independently realizes the solution of overall optimal operation of the multi-micro network system through different case studies. Case study 1 demonstrates the advantages of energy trading between microgrids; Case study 2 introduces the advantages of the optimization scheme of reactive trading. Case study 3 studies the conflict between single microgrid optimization and multi-microgrid system optimization.

Case study 1: effectiveness of distributed collaborative algorithm based on particle swarm optimization.

Table 3. State parameters of the energy storage system.

Energy storage system	Initial SOC	The lowest SOC	The highest SOC	Efficiency coefficient a_i b_i
MG1	40%	10%	90%	1.1 10.2
MG2	40%	10%	90%	1.4 10.5
MG3	40%	10%	90%	1.3 11.2

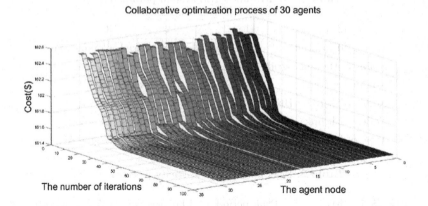

Fig. 6. The particle swarm optimization process of 30 agents.

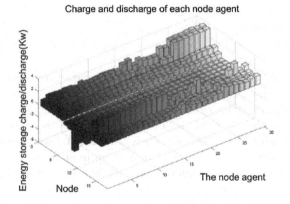

Fig. 7. Each agent reactive power output strategy.

Table 4. Optimization results of island mode and connection mode in MGS.

	Parameter	MG1	MG2	MG3	total
IASLAND	COST($)	356.5	−452.9	1052.1	955.7
	LINK_1(KVA)	–	–	–	–
	LINK_2(KVA)	–	–	–	–
	LINK_3(KVA)	–	–	–	–
	BESS(KVA)	40.3	36.4	33.7	110.4
LINK	COST($)	214.9	−655.5	889.1	448.5
	LINK_1(KVA)	–	−121	−21	−142
	LINK_2(KVA)	121	–	56	177
	LINK_3(KVA)	21	−56	–	−35
	BESS(KVA)	57.5	50.4	42.1	150

During more than 30 nodes of the network system, a set of 30 node agent, the thirty agent communication exchange information with neighboring agents, only with more micro network system as a whole running cost as the optimization goal, for power generation and load in the system according to the rated capacity of inverter randomly given, based on the final convergence situation of 30 agents to determine the effectiveness of the algorithm.

After 62 iterations of the optimization objective function of the 30 agents in the Fig. 6, all agents not only independently achieve the optimal goal, but also reach a consensus among themselves and obtain a unified optimal scheduling strategy. The reactive power output strategy shown in the Fig. 7 below is basically the same.

Case study 2: advantages of collaborative optimization of multi-micro network connection systems In the simulation of case 2, the distributed collaborative algorithm based on particle swarm optimization is used to carry out unconnected island optimization and overall optimization for the multi-microgrid system respectively. The optimization results are shown in Table 4. Each of these parameters are system in 24 h for the sum of value, COST represents the total operation COST of micro grid 24 cycle, cycle LOAD refers to the microgrid 24 total LOAD, LINK_i stands for micro grid I sell electricity (greater than zero representative to sell, buy less than zero represents), BESS represents the total energy storage power generation of the micro grid in 24 cycles (greater than zero represents power generation, less than zero represents charging).

Figure 8 is microgrid 1 in an island and connection mode of the cost for 24 h, can be found in the optimized algorithm of two kinds of mode, compared with the island model, coupling model can bring the micro grid and multiple piconets system more economic advantage, because the excess electricity compared with public power grid is more favorable price to sell to adjacent grid, so the adjacent grid can buy more preferential than public power grid electricity.

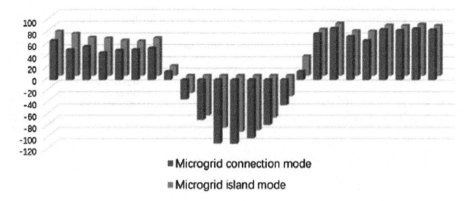

Fig. 8. Each agent reactive power outut strategy.

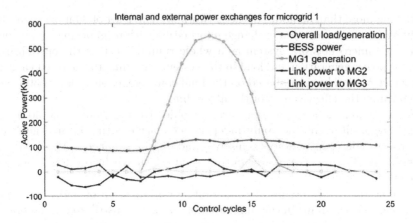

Fig. 9. Internal and external power exchanges for microgrid 1.

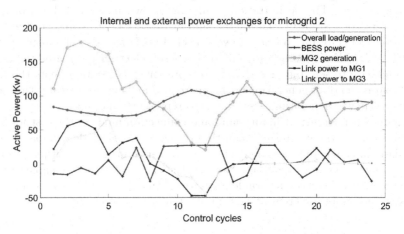

Fig. 10. Internal and external power exchanges for microgrid 2.

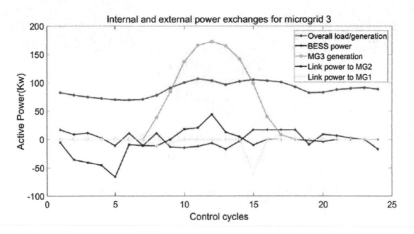

Fig. 11. Internal and external power exchanges for microgrid 3.

Figure 9, 10 and Fig. 11 show the optimization results of the multi-microgrid system group in the connected mode. The microgrid actively calls the energy storage system and renewable energy and trades the surplus electricity with the adjacent microgrid, thus reducing the operation cost of the microgrid. For microgrid 1, in the ninth to eleventh control cycle, the power generation is higher than the local load, so the electricity is sold to other microgrids and the local energy storage system is charged. Similarly, 2 in the third control cycle of the piconets, generating a surplus, and the micro network and micro network power shortage, reduced more power sharing between piconets micro nets system operation cost of each piconets, under such a mechanism of sharing power, micro network voltage deviation within 3%, visible this distributed algorithm can be applied to actual multiple microgrid system (Fig. 12).

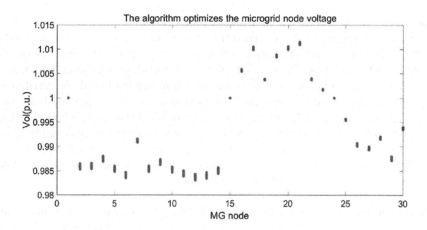

Fig. 12. 24 cycle microgrid node voltage.

Through the Shared power cooperation between microgrids, the power grid purchased by the multi-microgrid system in the connection mode is reduced by 24.6% compared with the island mode. This is because more of the extra power between the microgrids is sold to the adjacent microgrids with insufficient electricity through the connection lines, reducing the electricity trade with the public grid. If there is no connection line, the extra power will be sold to the high-voltage grid at the Internet price. The results also show that the energy storage charge-discharge power used in the island mode is less than that used in the connection mode. In island mode, the power needed to be purchased by the microgrid can only come from the public grid at a high price, so the microgrid must reduce the cost by reducing the utilization of the energy storage system. However, when the multi-microgrid system actively trades electrical energy with each other at a lower cost in the connection mode, the sub-microgrid energy storage system charges from other microgrids at a lower price or discharges to other microgrids at a higher price, which improves the utilization rate of the energy storage system and realizes the cost optimization of the microgrid group system (Table 5).

Table 5. Utilization rate of microgrid energy storage system.

Energy storage charge-discharge utilization	IASLAND	LINK
MG1	56%	63%
MG2	62%	69%
MG3	48%	55%

5   Conclusion

This paper mainly studies the multiple piconets system optimal operation strategy, internal operation and external transaction cost in system as economic indicators, and combined with the limit risk, inverter voltage value of reactive power output, which establishes distributed micro power grid optimization model, and put forward a kind of collaborative optimization algorithm based on distributed architecture. Based on the modified 30-node system, this paper analyzes the advantages of multi-micro grid energy transaction, and the results show that the operation cost of the multi-micro grid system can be reduced by reasonably planning the inter-micro grid energy transaction and energy storage charge-discharge, and the optimal operation of the multi-micro grid system can be obtained while ensuring the maximum benefits of the sub-micro grid.

Acknowledgement. Supported by Natural Science Foundation of China (No. 61773253, 61533010), Key Project of Science and Technology Commission of Shanghai Municipality under Grant No. 16010500300.

References

1. Chen, Q., Keyou, W., Guojie, L., Bei, H., Shaojun, X., Wei, Z.: Hierarchical and distributed optimal scheduling of AC/DC hybrid active distribution network. Proc. CSEE **37**(7), 1909–1917 (2017)
2. Antoniadou-Plytaria, K.E., Kouveliotis-Lysikatos, I.N., Georgilakis, P.S., Hatziargyriou, N.D.: Distributed and decentralized voltage control of smart distribution networks: models, methods, and future research. IEEE Trans. Smart Grid **8**(6), 2999–3008 (2017)
3. Arefifar, S.A., Ordonez, M., Mohamed, Y.A.R.I.: Energy management in multi-microgrid systems-development and assessment. IEEE Trans. Power Syst. **3**(2), 910–922 (2017)
4. Li, J., Liu, Y., Wu, L.: Optimal operation for community based multi-party microgrid in grid-connected and islanded modes. IEEE Trans Smart Grid **9**(2), 756–765 (2018)
5. Nikmehr, N., Najafi Ravadanegh, S.: Optimal power dispatch of multi-microgrids at future smart distribution grids. IEEE Trans Smart Grid **6**(4), 1648–1657 (2015)

6. Rahbar, K., Chai, C.C., Zhang, R.: Energy cooperation optimization in microgrids with renewable energy integration. IEEE Trans. Smart Grid 9(2), 1482–1497 (2018)
7. Yanqian, Z., Xiaodong, D., Yang, Z.: Research optimally coordinating dispatching of multi microgrid based on bi-level optimization theory. Shaanxi Electric Power 44(12), 1–5+28 (2016)
8. Chiu, W.Y., Sun, H., Poor, H.V.: A multiobjective approach to multimicrogrid system design. IEEE Trans. Smart Grid 6(5), 2263–2272 (2015)
9. Lu, T., Wang, Z., Ai, Q., Lee, W.J.: Interactive model for energy management of clustered microgrids. IEEE Trans. Ind. Appl. 53(3), 1739–1750 (2017)
10. Yonzhi, Z., Hao, W., Yining, L.: Dynamic dispatch of multi-microgrid for neighboring Islands based on MCS-PSO algorithm. Autom. Electric Power Syst. 38(9), 204–210 (2014)
11. Bui, V.H., Hussain, A., Kim, H.M.: A multiagent-based hierarchical energy management strategy for multi-microgrids considering adjustable power and demand response. IEEE Trans. Smart Grid 9(2), 1323–1333 (2018)
12. Wang, Y., Mao, S., Nelms, R.M.: On hierarchical power scheduling for the Macrogrid and cooperative microgrids. IEEE Trans. Ind. Inform. 11(6), 1574–1584 (2015)
13. Liu, W., Gu, W., Xu, Y., Wang, Y., Zhang, K.: General distributed secondary control for multi-microgrids with both PQ-controlled and droop-controlled distributed generators. IET Gen. Tran. Dist. 31(2), 707–718 (2017)
14. Marzband, M., Parhizi, N., Savaghebi, M., Guerrero, J.M.: Distributed smart decision-making for a multimicrogrid system based on a hierarchical interactive architecture. IEEE Trans. Energy Conv. 31(2), 637–648 (2016)
15. Ran, H., Qian, A., Ziqing, J.: Bi-level game strategy for multi-agent with incomplete information in regional integrated energy system. Autom. Electric Power Syst. 42(4), 194–201 (2018)
16. Hong, L., Jifeng, L., Shaoyun, G.: Coordinated scheduling of grid-connected integrated energy microgrid based on multi-agent game and reinforcement learning. Autom. Electr. Power Syst. 43(1), 40–50 (2019)
17. Gregoratti, D., Matamoros, J.: Distributed energy trading: the multiple-microgrid case. IEEE Trans. Ind. Electr. 62(4), 2551–2559 (2015)
18. Wang, T., O'Neill, D., Kamath, H.: Dynamic control and optimization of distributed energy resources in a microgrid. IEEE Trans. Smart Grid 6(6), 2884–2894 (2015)
19. Xu, Y., Zhang, W., Hug, G., Kar, S., Li, Z.: Cooperative control of distributed energy storage systems in a microgrid. IEEE Trans. Smart Grid 6(1), 238–248 (2015)
20. Zhou, D., Blaabjerg, F.: Bandwidth oriented proportional-integral controller design for back-to-back power converters in DFIG wind turbine system. IET Renew. Power Gen. 11(7), 941–951 (2017)
21. Olfati-Saber, R., Fax, J.A., Murray, R.M.: Consensus and cooperation in networked multi-agent systems. Proc. IEEE 95(1), 215–233 (2007)
22. Utkarsh, K., Trivedi, A., Srinivasan, D., Reindl, T.: A consensus-based distributed computational intelligence technique for real-time optimal control in smart distribution grids. IEEE Trans. Merging Topics Comput. Intell. 1(1), 51–60 (2017)
23. Qiming, F., Jichun, L., Yangfang, Y., Qi, W., Shichao, N., Lifeng, K.: Multi-local grid optimization economic dispatch with different types of energy. Power Syst. Technol. 43(2), 452–461 (2019)

An Electric Power Emergency Repair System: Structure, Navigation, and Software Design

Jisheng Cui[1], Hong Gang[1], Peng Qiu[1], and Boyu Liu[2(✉)]

[1] State Grid Liaoning Electric Power Co. Ltd Jinzhou Power Supply Company, Jinzhou 121001, Liaoning, China
{cjs_jz,gh_jz,qiup_jz}@ln.sgcc.com.cn
[2] College of Information Science and Engineering, Northeastern University, Shenyang 110000, Liaoning, China
liuboyu@stumail.neu.edu.cn

Abstract. The development of power grids gradually raises the demand for their emergency repair capabilities and efficiency. This paper focuses on the enhancement of the power emergency repair system from three aspects: 1) the structure of an improved electric power emergency repair system is proposed; 2) the navigation method for the repairing is established based on the application of BPNN-Dijkstra algorithm; 3) the software structure of the electric power emergency repair system is proposed. The proposed electric power emergency repair system has the advantages of real-time situation monitoring and instant repair unit dispatch, which could be considered as the enhancement to the state-of-the-art electric power emergency repair techniques.

Keywords: Electric power emergency repair · Repair path·BPNN-Dijkstra·software design

1 Introduction

As electric power networks continue to develop around the world, the practical significance of their emergency repair systems becomes gradually apparent. However, the complexity and the number of the equipments and devices in the power systems becomes higher and higher, thus, the enhancement of the emergency repair systems draws more and more attentions than ever before. To maintain ordinary power supply to power consumers and scientifically resolve the efficiency problem of decision-making and repair scheduling in electric power emergency events, the emergency repair system should be established with appropriate system structure, otherwise the retardation would cause great economic losses [1].

The concept of electric power emergency repair system has been proposed for a period of time [2]. Tian et al. [3] discussed the concept and framework of the electric power emergency management platform. In their later research [4], an electric power repair system based on user-service platform is introduced. However, among these researches, the data transmission systems are based on telephone, fax, and other

© Springer Nature Singapore Pte Ltd. 2020
M. Fei et al. (Eds.): LSMS 2020/ICSEE 2020 Workshops, CCIS 1303, pp. 36–46, 2020.
https://doi.org/10.1007/978-981-33-6378-6_3

approaches, which is not efficient enough to help establishing a fast-respond emergency repair system.

One of the problems of electric power emergency repairing is to determine the optimal path for repair unit navigation. Broumi et al. [5] applied improved Dijkstra algorithm to determine the optimal repair path. Lin et al. [6] proposed a bilevel algorithm which the first stage is Dijkstra algorithm, and the second stage is simulated annealing algorithm. Zhang et al. [7] took vehicular parameters and other driving parameters into consideration, they proposed the approach to determine optimal repair path based on Dijkstra algorithm but additionally assessed the petrol consumption by vehicles. Xing et al. [8] proposed a multi-point model which considered the spatiotemporal characteristics of the power system emergency repairing, then the radial-based artificial neural network is introduced. Although the existed literature has revealed that Dijkstra-based methods for optimal path determining is matured and has achieved some results, there are still problems exists. In traditional dispatch methods, traffic information is not implemented, thus, there is no real-time feedback for a given path during driving. In this paper, a BPNN-Dijkstra-based optimal repair path navigation method is introduced and implemented in the proposed emergency repair system to function as a novel aspect.

Inspired by the aforementioned works and driven by the motivation to improve the electric power emergency repair system, the main contributions of the paper can be summarized as: 1) the structure of an improved electric power emergency repair system is proposed; 2) the navigation method for the repairing is established based on the application of BPNN-Dijkstra algorithm; 3) the software structure of the electric power emergency repair system is proposed.

The remainder of this paper is organized as follows. Section 2 introduces the structure of the electric power emergency repair system, the software design of the electric power emergency repair system, and the BPNN-Dijkstra method to determine the optimal path of emergency repairing. Section 4 proposes a case study to validate the BPNN-Dijkstra method. Section 5 concludes the paper.

2   Electric Power Emergency Repair System Structure

The proposed electric power emergency repair system should be consist of following units: 1) emergency repair intelligent dispatch center; 2) vehicle-mounted emergency repair system; 3) expert GPS intelligent terminal system; 4) monitoring management information system. Figure 1 is the schematic diagram of the electric power emergency repair system structure.

The electric power emergency repair system above has the following features: 1) GPS system modules are applied to locate and monitor repair units; 2) GIS system modules are attached to the management system in order to graphically demonstrate repair situations; 3) GPRS system modules are implemented to realize electric power data transmission. All these features meet the requirement for an electric power emergency repair system to achieve real-time situation monitoring and instant repair unit dispatch [9].

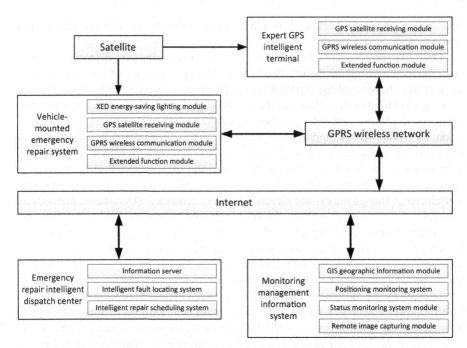

Fig. 1. Schematic diagram of the electric power emergency repair system

3   Software Implementation of the Emergency Repair System

3.1   General Design of Software Structure

The design of the software structure of the emergency repair system includes the selection of the system network structure, the design of the software structure, and the design of the software functions.

For system network structure, the Browser/Server structure (B/S structure), is adopted in the proposed design, for it is the most widely applied structure type at present. This structure consists of two sides: browser side and server side. The server side is the platform which enables the emergency repair system realize its functions by the web systems, thus, almost all calculations, queries and are commenced on this side. On the other hand, the browser is used to display the results of the server side and issue commands, thereby becomes the most important application on the browser side [10]. By this way, the functions of the client side under the client/server mode are concentrated on the server side. Only browser-related functions are retained. This reduces the requirement of the client-side performance and simplifies system developing and maintenance [11]. Based on the advantages mentioned above, the multi-layer B/S structure is selected as the network structure of the emergency repair system [12].

For system software structure, since the system adopts multi-layer B/S network structure, the application, user interface and other operation modules are separately established. Meanwhile, all layers of the system perform bidirectional data transmission on the system support platform, which could ensure the confidentiality of the documents.

WWW browser is applied as the user interface, to realize management and operation of the system, which could greatly relief the client workload and save the total cost [13]. The software structure of the system is shown in Fig. 2.

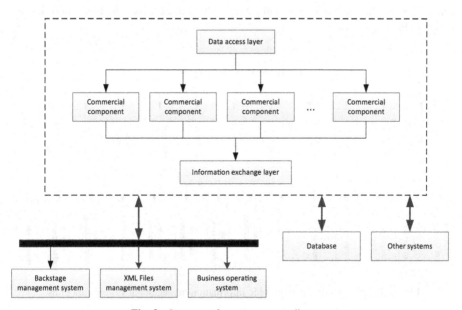

Fig. 2. System software structure diagram

3.2 Basic Information Management Sub-system

The basic information management system mainly maintains the basic information of the system, including querying, editing, and other functions for the basic information of emergency vehicles, emergency personnel, and emergency experts. Its function module is shown in Fig. 3.

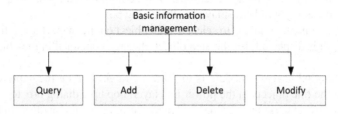

Fig. 3. Block diagram of basic information management system functions

The query module can query and display the basic information in the system; the add, delete and modify modules can edit the basic information.

3.3   Geographic Information Management Sub-system

The geographic information management program maintains the geographic information data used in the emergency repair system, mainly including power grid facilities, transportation road networks and other geographic entities. Its functional modules are shown in Fig. 4.

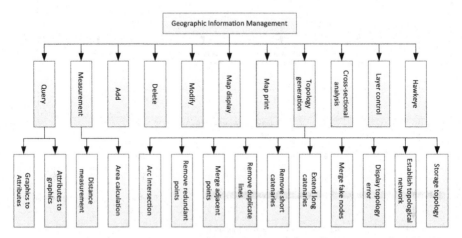

Fig. 4. Block diagram of geographic information management system functions

The add, delete, and modify modules provide relevant editing functions for geographic information data records and map entities; the map display module is used for the map display of geographic information and zooming of the map.

The query module is divided into graphics-to-attributes query and attributes-to-graphic query. Graphic-to-attributes query: Query the attribute information from the entity graphics on the map. Click the mouse to select the object, it will be highlighted, and a dialog box will be popped up to display the relevant data of the objects' attributes. Attributes-to-graphic query is divided into precise query and fuzzy query. Enter the unique attribute value of the feature entity, such as coordinates, place name, etc. in the dialog box during precise query. The object will be highlighted on the map and other attributes of the entity will be displayed in the dialog box.

The measurement module is to select a solid object on the map. If it is a line object, the total length is displayed; for the area object, the area contained in the object can be displayed.

Cross-sectional analysis module is to draw a straight line on the capture layer of the map, cut the line objects in the power line layer, pop up a dialog box to display the relevant attributes of the cut power line, and make a cross-sectional analysis of the power line with the depth attribute.

Layer control is module used to select the desired layer to be displayed on the interface. It can display a certain layer separately or display multiple layers in combination.

The topology generation module is used to generate the topology required for network analysis. The related operations include arc intersection, removing redundant points, merging adjacent points, removing duplicate lines, removing short catenaries, extending long catenaries, merging fake nodes, establishing topological networks and storage topologies. Before establishing the topology structure, the program will pop up a dialog box to display the abnormal situation during the topology generation process, and the abnormalities could be manually handled according to the prompt information. The process is shown in Fig. 5.

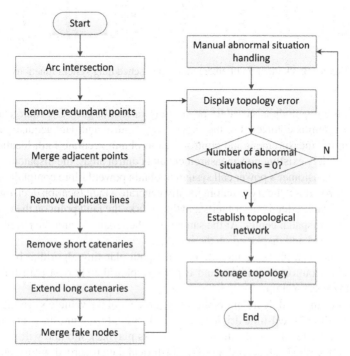

Fig. 5. Flow chart of topology generation

Among them, the abnormal situation check mainly checks the hanging nodes and merge operations in the topology. The program highlights the line segments with dangling nodes and the marked places, number them, select a certain sequence number, the program will automatically navigate to that place. After that, whether the merge operation of dangling nodes and marks conforms to the real situation of the network can be artificially judged. If it is not, it can be modified manually; if it is, the dangling node is removed from the anomaly.

3.4 Emergency Repair Monitoring and Scheduling Sub-system

In the event of a sudden power failure, the emergency repair scheduling program is responsible for locating, scheduling, and monitoring the emergency vehicles, personnel, and materials. Its functional modules are shown in Fig. 6.

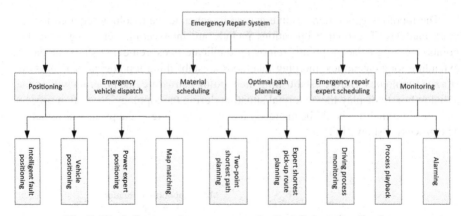

Fig. 6. Block diagram of emergency repair scheduling system functions

The positioning module is used for positioning various targets. The emergency repair scheduling system is connected to the supervisory control and data acquisition system (SCADA) in the measurement and control area to obtain accurate fault location information. For widely existing non-measurement and control areas, the dispatching system connects to the customer's power call system to obtain power failure complaint information, and uses rough set theory to automatically generate a fault location decision table to intelligently locate sudden power faults and then marks the fault point on the monitoring interface of the dispatch center. At the same time, the repair car and the required repair experts are located and displayed on the interface.

The optimal path planning module uses the path planning algorithm based on the geographic distribution of vehicles and experts to plan the shortest path for repaired vehicles and expert pickup.

The monitoring module monitors the whole process of the vehicle's driving process, displays the vehicle's real-time driving route on the monitoring interface, and records it in the table of the repair log database. The process playback module will perform the monitoring route playback according to the historical data recorded in this table. If the vehicle is in a special situation such as a failure during driving (such as traffic jams on the road), it will send an alarm message to the dispatch center. The dispatch center can get in touch with the vehicle personnel to deal with the alarm accordingly (such as rearranging the drive route), or rearrange other emergency repair vehicles to the emergency repair site), record the alarm to the emergency repair log database table.

3.5 BPNN-Dijkstra-Based Navigation Method

BP neural network (BPNN) is one of the most widely used artificial neural networks, it has been studied abundantly and achieve satisfying results in many fields of research. BPNN neural network prediction model consists of the input layer, the implicit layer and the output layer. BPNN has the following advantages:

1) Non-linear mapping capability: The BPNN neural network essentially realizes a mapping function from input to output. Mathematical theory proves that the three-layer neural network can approximate any nonlinear continuous function with arbitrary precision. This makes it particularly suitable for solving problems with complex internal mechanisms.

2) Self-learning and self-adaptive ability: BPNN neural network can automatically extract "reasonable rules" between output and output data through learning during learning, and adaptively memorize the learning content in the weights of the network.

3) Generalization ability: The generalization ability means that when designing the pattern classifier, it is necessary to consider the network to ensure the correct classification of the required classification objects, and also care about whether the network has been able to deal with unseen ones after training. Modes or modes with noise pollution are classified correctly.

4) Fault tolerance: BPNN neural network will not have a great impact on the global training results after its local or partial neurons are destroyed, that is to say, the system can still work normally even if the system is damaged locally.

Dijkstra algorithm is a classic shortest path planning algorithm. It can be properly used in dense graph. The BPNN-Dijkstra method is to combine both algorithms into one flow. The flowchart of the BPNN-Dijkstra method is shown in Fig. 7.

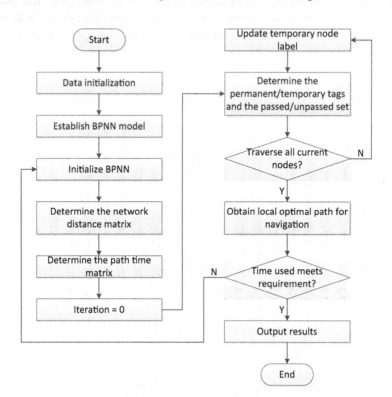

Fig. 7. Flowchart of the proposed BPNN-Dijkstra method

4   Case Study

In this section, a brief case study is introduced to validate the methodology. Assume an undirected graph of a power emergency repair system is obtained as Fig. 8 shows.

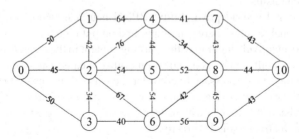

Fig. 8. Undirected graph of a power emergency repair system

Now, assume that at time step i, an accidental failure occurred in the power system at Node 10. The relevant information has been submitted to the emergency repair system through GIS and GPRS modules mentioned in Sect. 2. The scheduling unit arranges the available repair vehicles and calculate the time t required for various paths. The vehicle with the minimum of t will be determined as the vehicle to repair the fault. Also, assume that the time limit to update vehicle speed of the roads is 2 min. The velocity at the roads in the graph are forecasted. The prediction of vehicle speed at each road is shown in Fig. 9.

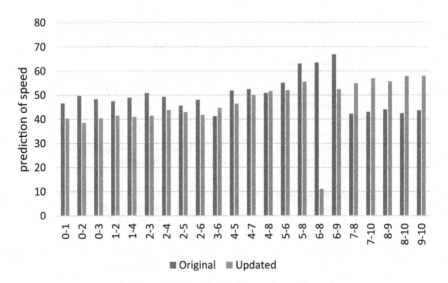

Fig. 9.   prediction of vehicle speed at each road

The results of permanent tags, calculated by BPNN-Dijkstra method, under this situation, are presented in Table 1.

Table 1. Results of permanent tags

	Original	Updated		Original	Updated
v_0	0	0	v_6	2.0146	1.6176
v_1	1.0771	1.0725	v_7	3.1748	2.5725
v_2	0.9070	0.9025	v_8	2.6796	1.9993
v_3	1.0431	1.0450	v_9	2.8534	2.1494
v_4	2.3881	1.8840	v_{10}	3.7216	2.6350
v_5	2.0884	1.8151			

When the vehicle speeds of the roads stay original and the path is selected as $\{v_0, v_3, v_6, v_8, v_{10}\}$. The time of arriving at Node 6 is 2.0146 min, which has exceeded the time limit, thus the speed is updated, and the time of each path might change. In this case, after the time is changed, the path is re-planed by the method. Thus, the final path that the method navigates is: $\{v_0, v_3, v_6, v_9, v_{10}\}$, the total time spend is 3.8297 min.

It can be seen that the BPNN-Dijkstra method can ensure the traffic information can be taken into account in a proper and in-time way. The method reveals its advantage when the roads in the original optimal path have accidently and drastic changes.

5   Conclusion

This paper proposed the enhancement of electric power emergency repair system from perspective of structure, navigation and software design. The proposed system comprehensively uses GPS technology, GPRS technology and GIS technology. Each technology can make use of its own advantages to accurately and quickly complete the positioning and real-time monitoring of repaired vehicles, and at the same time it can accurately, safely send, receive and transmit information. The BPNN-Dijkstra method is used for the intelligent scheduling of repairing vehicles, and the optimal path of the repaired vehicles is intelligently determined, which makes the navigation of the repair scheduling work faster and more efficient. The comprehensive application of these advanced technologies can greatly improve the efficiency of power system repairs, and increase the satisfaction of power users with repair work, providing a more friendly and efficient platform for power companies to better serve society and service users. Meanwhile, it can reduce the economic loss caused by the fault and save the cost of power supply.

Acknowledgement. The authors would like to grant the support from State Grid Liaoning Electric Power Corporation under the project "The Electric Power Emergency Repair System Based on a Novel Energy-Saving Light Sources".

References

1. Zhou, X.: Application of satellite communication technology in emergency repair of electric power emergency communication. Commun. World **08**, 170–171 (2018)

2. Tian, S., Zhu, C., Chen, X., et al.: Theory of electric power emergency management and its technological countermeasures. Pow. Syst. Technol. **31**(24), 22–27 (2007)
3. Tian, S., Chen, X., Zhu, C., et al.: Study on electric power emergency management platform. Pow. Syst. Technol. **32**(1), 26–30 (2008)
4. Xu, J.: Design and Implementation of Power Failure Repair Information System Based on Mobile Terminal. University of Electronic Science and Technology (2014)
5. Broumi, S., Bakali, A., Talea, M., Smarandache, F.: Applying Dijkstra algorithm for solving neutrosophic shortest path problem. In: IEEE Proceedings of The 2016 International Conference on Advanced Mechatronic Systems, 30 November–3 December 2016, Melbourne, Australia, pp. 412–416 (2016)
6. Lin, X., Geng, F., Jin, Z., Xu, Q.: The optimization of delivery vehicle scheduling considering the actual road network factors. In: 2016 International Conference on IEEE Logistics, Informatics and Service Sciences (LISS), 24–27 July 2016, Sydney, NSW, Australia (2016)
7. Zhang, J., et al.: Vehicle routing in urban areas based on the Oil Consumption Weight-Dijkstra algorithm. In: IEEE IET Intelligent Transport Systems, 15 August 2016 (2016)
8. Xing, S., Gu, Y., Shen, L., Yang, X., Zhuang, G.: A traffic congestion prediction model based on vehicle speed of urban expressways. Traffic Inf. Secur. **34**(2), 48–50 (2016). https://doi.org/10.3963/j.issn1674-4861.2016.02.00
9. Zhong, A.: Research on Intelligent Dispatching System of Electric Power Repair Based on GIS. Chongqing University (2008)
10. Ding, L.: Design and Implementation of ERP System Sales Management Subsystem. Beijing University of Posts and Telecommunications (2010)
11. Lu, H.: Design and Implementation of Oilfield Development Information System. University of Electronic Science and Technology (2010)
12. Xu, Z.: Research on Original Image Database System Based on WEB. People's Liberation Army Information Engineering University (2004)
13. Zheng, Q.: Design and Implementation of Electric Emergency Command System. Jilin University (2014)

Research on Information Interaction Technology Between Multi-service Application Systems of Power Grid

Hao Chen[1], Yuan Li[1], Huang Tan[1], Ling Zhao[2], Qiuyun Mao[3(✉)], Jing Li[3(✉)], and Qian Guo[3]

[1] China Electric Power Research Institute, Beijing 100000, China
`{chenhao2010,liyuan3,tanhuang}@epri.sgcc.com.cn`
[2] College of Mechanical and Electrical Engineering, China Jiliang University, Hangzhou 310018, China
`lxh7400@126.com`
[3] Electric Power Research Institute of State Grid Zhejiang Electric Power Co Ltd., Hangzhou 310000, China
`1456419666@qq.com, ljagu@163.com, 16a0102140@cjlu.edu.cn`

Abstract. This paper analyzes information interaction and sharing technology of power information collection system for power supply service command system and 95598 business support system. Through the analysis of information interaction characteristics and business information exchange topology among the three platforms, the business requirements of supporting the other two platforms with power information collection system are summarized, and the issues in the interaction of multi-business application systems are investigated. This paper presents an information interaction model for multi-service application system structure and designs a data interface platform based on the proposed information interaction model and improved schema. The scheme is applied in the information transmission process of low voltage distribution network repair in a region. The superiority of this scheme is verified by the comparison of related function data after the new interface design comes online. It provides a research basis for the future business development and practical promotion of smart grid.

Keywords: Information interaction · Data interface design · Power users electricity information acquisition system · Multi-service platform · Power grid business

1 Introduction

With the improvement of scientific and technological level and the growth of the national economy, the coverage of the power grid becomes more extensive, the service level requirements for the power grid are upgraded at the same time. In addition to ensuring the regular development of daily work and daily life electricity business for power users, it is very important to ensure the reliability of power supply. According to incomplete statistics, 80% of power outage faults in China are caused by distribution network

© Springer Nature Singapore Pte Ltd. 2020
M. Fei et al. (Eds.): LSMS 2020/ICSEE 2020 Workshops, CCIS 1303, pp. 47–60, 2020.
https://doi.org/10.1007/978-981-33-6378-6_4

faults. Distribution network is an important part of power system to supply power to users, which requires high reliability and quality of power supply. In the process of emergency repair of distribution network, the timeliness and integrity of the information transmission of emergency repair plays a significant part in improving the efficiency of emergency repairment [1–3]. This requires improvement of the collaborative operation capability among various power network business systems, optimization of the information interaction between platforms, and improvement of the efficiency of information transmission.

At present, there are three main business systems of electric power enterprises including power user electric energy data acquire system, power supply service command system and 95598 business support system. Power user electric energy data acquire system is mainly responsible for the user terminal information acquisition and centralized management [4]. The power supply service command system is a supporting system for the collaborative command of distribution operation centered on reliable power supply [5]. The 95598 system is mainly responsible for docking with customers and is the main service window of State Grid for power customers [6]. Power enterprises need to establish a unified interface platform for multi-service application system to achieve the resource sharing of user electricity information collection data and to support the application of more business functions. The existing platform information interaction framework has some defaults in data transmission for power grid. Only a few provincial companies integrate the relevant functions into marketing business application system (SG186) at the moment. Besides, the business application system calls the corresponding functions of the power consumption information collection system through the web service interface, or directly opens the relevant account to the business department, but its efficiency and security cannot be guaranteed [7]. The existing general data transmission framework built by using the message middleware ActiveMQ has a problem that it is hard to guarantee the quality of service while accessing high concurrent data [8]. In this paper, an information interaction model for multi-service application system structure is proposed. Moreover, a data interface platform with high concurrency is built on the basis of power user electric energy data acquire system. The pressure of high access to the database of the collection system is released, and at the same time, with the achievement of the efficient and accurate transmission of data and the promotion of the efficiency of the distribution network repair work, a more secure and efficient power environment for social production and life will be provided.

2   Introduction to Multi-service System Platform of Power Grid

Business data interaction needs of power information collection system, power supply service command platform and 95598 business support system can be grouped into two categories: data support requirements and business support requirements. Data support class requirements are mainly for applications that are not time-efficient and obtain data in the frequency of day intervals. It supports applications related to external business systems such as data query, work order classification, etc. Business support needs mainly focus on the application requirements of time-sensitive, real-time or quasi-real-time frequency to obtain data. It supports external business systems for fast fault location, emergency rescue, online problem solving and other businesses.

At present, the information provided by the user information collection system to the power supply service command platform and 95598 business support system includes real-time data, frozen data, field data and operation results. The four types of data are as below:

Real-time data: Power-off records of smart meters, power outage events of collection terminals, etc.

Frozen data: Indicators of energy meters, voltage/current curves, etc.

Field data: Transmit the relay status of the power meter; Transmit the current three-phase voltage and current of the power meter.

Operation result: Query the execution result of remote fee control instruction.

To sum up, the power information collection system obtains the user's power data and then accurately calculates and manages the storage. 95598 Business Support System is a bridge between electric power enterprises and customers for business acceptance and power information communication. It receives the current power indication value, relay status and power-off record of the power meter provided by the electric energy information acquisition system, and operates on this information to users. For example, to judge if a user is in arrears or not will be based on the record of the execution result of fee control, to determine the cause of customer outage as a assistance will be based on the relay status of meters, etc. At the same time, the power information collection system not only provides intelligent meter real-time power outage events for the power supply service command platform and combines the topological relationship between the platform and users, but also accurately locates the impact range of power outage and provides support for quick repair of power outage faults in low voltage distribution network. Data fusion flowchart for three platforms is shown in Fig. 1.

3 Research on Multi-service Information Interaction Technology in Power Network

The vast variety of power network data determines that various platforms need to use a variety of interfaces and information transmission methods while interacting. This results in greater workload and difficulty and will put greater pressure on the system when a large number of interfaces are running at the same time. Therefore, in combination with the actual needs of the modern society, it is particularly important to select an appropriate way of information interaction on the basis of the combination of different kinds of technologies.

3.1 Research and Improvement on Efficiency of Data Sharing

Data sharing is mainly divided into two aspects: data acquisition and data transmission on the network. The relational database is often utilized in the power information collection system to store the collected data. When business applications and interfaces request shared data from the database, it is necessary to solve the problem of sharing the same data multiple times to improve the efficiency of data flow. Data acquired from the power information collection system is supposed to be transferred to other business systems. Applying RMI technology and KafKa to data transfer can improve data transfer efficiency.

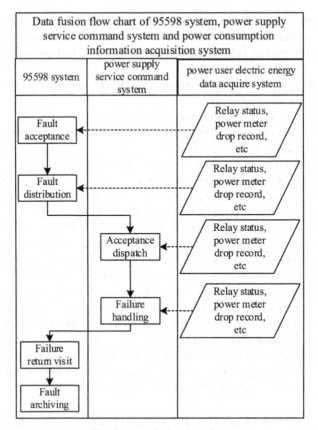

Fig. 1. Data fusion flowchart

Caching Technology. When information exchange between multi-service application systems, other business platforms need to acquire data from the power information collection system. In the case of large concurrent access to the database and large amount of data to be shared, caching technology can reduce direct interaction with the database and reduce the pressure on the database. In the meanwhile, more recorded data, static data, non-real-time data, and so on, can be pre-loaded into the cache to improve the overall response speed of the service [9, 10]. For information exchange in multi-service applications, caching technology can be applied to internal local data caching and distributed data caching for the entire cluster. Common distributed caches involves Redis, Memcache, and so on.

With the addition of local caching technology, when data needs to be shared externally or to be requested actively by third-party clients, services can determine whether to access the database by checking whether there is data in the local cache. This can effectively reduce the interaction with the database and reduce concurrent access to the database to a certain extent. It improves the response speed and reduces the burden on the database. In practical development, cached data is stored in memory and a policy for persistence to disk is specified, which not only makes the local cache read efficiently, but also ensures the stability of the data. Moreover, developers can configure the cache emptying policy to empty the corresponding cache in time after database-related data is written to ensure data consistency.

Data Connection Pool. The application of data connection pools can improve the efficiency of data acquisition. As the application starts, a batch of connections that interact with the database is created based on the configuration of the connection pool and saved in the pool. When a request for data is received, the database will be accessed directly using the created connection, omitting the process of creating and destroying the connection [11, 12]. Apply the startup process as shown in Fig. 2.

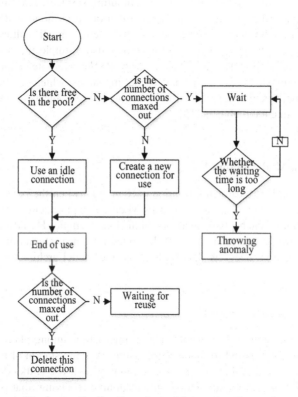

Fig. 2. Application flow chart of data connection pool.

When a connection failure occurs, the connection policy or data source can be switched quickly by using the fault-tolerant mechanism provided by the connection pool, enabling the application to resume data sharing in a short time. While query efficiency decreases, developers can quickly locate locations that may affect data acquisition performance by viewing the connection and performance monitoring logs and resolve them.

Kafka. Kafka is a distributed, partitionable, replicable messaging system. The biggest feature of it is the ability to process large amounts of data in real time to meet a variety of demand scenarios [13, 14]. Kafka summarizes the categorization of messages, each of which is referred as a topic - that can be subscribed to and processed by multiple objects. The Kafka message queue enables the middleware to transfer the message to the processing module for real-time processing, and then transfers the message queue to the data center for uniform storage in the large data platform. This reduces the time loss from the successful collection to the acquisition by the external system and meets the requirements of the external system for data timeliness. At the same time, it copes with the growing demand of various business systems for data collection applications.

RMI Technology. In addition, the power grid business platform can be divided into several nodes (such as interface services, static computing services, real-time computing services, etc.). According to the business and function, each node is deployed independently. Based on the interaction between nodes, there is a call within the platform. Hence, the interaction between nodes is implemented using simple and easy-to-use Java RMI (Java Remote Method Invocation) technology. RMI is a remote method call, which is used to communicate between different virtual machines, which can be on different hosts, enabling objects in client Java virtual machines to call objects in server-side Java virtual machines like local objects [15].

3.2 Interaction Model Design I

Some methods can be used for information exchange between the power information collection system and the other two platforms when the requirement of real-time information interaction of each platform is not high. For example, HTTP multiple handshake authentication mode interaction, WebService interface interaction, DBlink data synchronization and Oracle_fcw data replication, the model architecture is shown in Fig. 3.

3.3 Interaction Model Design II

When the real-time requirement of information interaction among platforms becomes high, Kafka push can be used for data transmission. IV main station of Electric power collection system and distribution network automation will use Kafka double authentication method to achieve single-emission and double-receiving, that is, distribution network automation IV main station receives transformations data, while the collection system can receive transformations data. The blackout data of public transformer and line fault signals are published through the unified data publishing service. The logical structure of the interface is shown in Fig. 4.

4 Design of Multi-business System Interactive Interface Platform

Based on the above information interaction model and data efficiency improvement technology, a data interface scheme is developed. In terms of technical implementation, the unified interface platform is divided into interface service layer, background computing

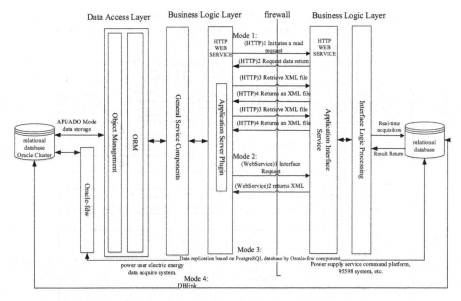

Fig. 3. Interaction model design 1.

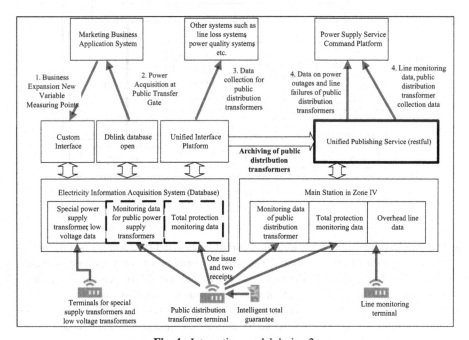

Fig. 4. Interaction model design 2.

layer, data source layer, file service layer and communication layer. Web Service + file and MQ (message queue) are used to achieve data interaction. This supports XML, JSON and other data formats. It can meet different real-time requirements of data interaction at the same time. The technical implementation architecture is shown in Fig. 5.

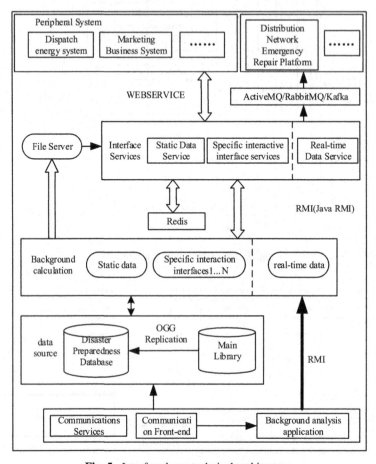

Fig. 5. Interface layer technical architecture.

- The interface service layer is a unified export of external data and is responsible for interaction with external systems. The services provided mainly include: WebService service for peripheral system access; HTTP file service for peripheral system to download data files; and interaction with background computing layer.
- The background computing layer focuses on the business logic processing of the interface platform and is responsible for parsing the data requests forwarded by the interface layer. This provides data collection support for interface service layer through interaction with data layer.
- The data source layer refers to the disaster recovery database or historical database of the acquisition system, which is the main data source of static data and statistical data of the interface platform.

- The communication service layer is composed of the front-end computer and the back-end analysis application, which provides the interface platform with real-time data of the collection system through the mechanism of one transmitter and two receivers or the technology of stream processing.
- File service layer is a server for storing platform data files, which provides support for interface service layer to download data files.

The platform adopts appropriate data to transfer technologies for different data types and adds data connection pools and Redis technologies to ease the pressure on the database. The comprehensive use of multiple technologies improves the efficiency of information exchange between multi-service application systems.

First, the communication service layer obtains data from the main station of the power information collection system through the technology of single-emission and double-receiving or stream processing and transfers the data to the main library of the data source layer. Within the data source layer, the primary database data is copied to the interface library through OGG or ETL (Extract-Transform-Load) technology. The interface library is the data source of static data, which can alleviate the pressure of data transmission by the main library. In addition, the communication service layer transmits real-time data to the background computing layer through RMI technology. The background computing layer receives data from both the data source layer and the communication service layer and processes the received data. It provides corresponding data based on data requests from the interface service layer and transfers the processed data to the file server. The interface service layer stores the acquired data in a distributed cache, which makes it easy to share the same data multiple times without having to repeatedly access the database. 95598 Business Support System invokes static data through WebService technology to the interface service layer. Power supply service command system has a high requirement for real-time data transmission because of its power supply emergency repair characteristics. It invokes real-time data through ActiveMQ or Kafka technology to interface service layer. At this point, the information exchange between the power information collection system, 95598 business support system and power supply service command system are completed.

5 System Application and Analysis

Taking a regional power network in Zhejiang as an example, this data interface design is applied to external data release and to actual low voltage distribution network emergency repair. The emergency repair data of low voltage distribution network in the original interactive mode in 2018 and the new interface scheme in 2019 are compared and analyzed. As the new data interface officially launched in January 2019, part of the data was selected for comparative analysis from January to September in 2018 and in 2019 respectively. The application analysis is divided into two parts: the application function analysis inside the power grid system and the customer-oriented rush repair quality analysis.

5.1   Analysis of the Internal Application Function of Power Network System

During the pilot operation of the interface platform on-line, it publishes data such as daily frozen energy capacity and load curve to the line loss system for more than 27 million power users in the province. The data acquisition time consumption information of the Unified Interface Platform and the original external data publishing interface platform is shown in Table 1.

Table 1. Time consuming information of data acquisition before and after new interface application.

Data item	Number of data acquired	Original time consuming	Now time-consuming
Daily frozen	27652655	13 min	6 min
Electric energy	51439121	16 min and 40 s	9 min and 10 s
Load curve	222629	30 s	10 s
PV user profile	196297	20 s	8 s

According to the time consumption comparison of obtaining data items before and after using the new interface platform in the table, it can be observed that for the same type of data items, the new interface platform takes much less time to transmit than the original. Especially for the daily frozen energy capacity with more than 27 million data bars, the time consumed by the new interface platform is only half of the original time consumed. Visually reflecting the application of the new interface platform greatly improves the efficiency of information exchange.

Figure 6 and Fig. 7 shows the time consumption comparison curve of fault acceptance and handling during the same period in the distribution network emergency repair activities in 2018 and in 2019.

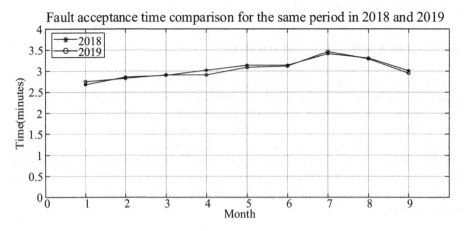

Fig. 6. Fault acceptance time comparison for the same period in 2018 and 2019.

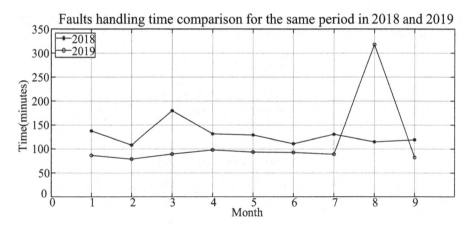

Fig. 7. Fault handling time comparison for the same period in 2018 and 2019.

As can be seen from Fig. 6, the time curves of the failure acceptance in 2018 and 2019 are very similar, with little change in the same period. It can be discovered from the same period of the fault handling in Fig. 7 that except for August, the time taken from occurrence to completion of the fault in 2019 is significantly lower than that in 2018. From the analysis of long-term data changes on curve trend, the trend of curve development in July 2019 and before is relatively flat. Therefore, August 2019 data can be regarded as a special case, not included in the comparison, and the time of failure handling in 2019 is generally lower than that in 2018. It can be observed that this interface design can effectively enhance the timeliness of distribution network repair activities after being applied in the information exchange of various platforms.

5.2   Analysis of the Internal Application Function of Power Network System

Table 2 is a comparison of the number of power network refunds and the satisfaction rate of return visits for the same period in 2018 and in 2019. It shows that although the number of returns in 2019 has increased than that in 2018, it still remains within a reasonable range, which may be related to the increase of power grid user base in 2019. In 2019, the satisfaction rate of return visits to work orders was basically maintained above 99%. It indicates that the user's satisfaction is still high after interface replacement.

Figure 8 shows the total complaint curve for the same period of emergency repair of low voltage distribution network in 2018 and in 2019. As can be demonstrated from the graph, the total number of complaints decreased significantly in 2019, and the trend is the same as that of last year, which verifies the reliability of the data. This implies that the new interface platform improves the efficiency of emergency repair and improves the quality of service.

Figure 9 shows the accuracy of outage information publishing in 2018 and in 2019. It can be considered that the accuracy of outage information publishing in these two years has been maintained at 100%, which has remained at a high level for a long time.

Through comparing the time-consuming data of information released through the above interfaces, analyzing related data such as fault handling efficiency, work order

Table 2. Comparison table of number of power grid chargebacks and work order return visit satisfaction rate in the same period of 2018 and 2019.

Contrast Item	Return odds for national and provincial networks		Work Order Return Satisfaction Rate	
Month	Year			
	2018	2019	2018	2019
1	0	0	–	99.3
2	0	0	–	99.4
3	0	0	–	99.2
4	1	0	–	99.29
5	0	0	99.12	99.17
6	1	4	99.17	98.91
7	0	1	99.19	99.24
8	0	1	99.12	99.38
9	0	3	98.95	98.77

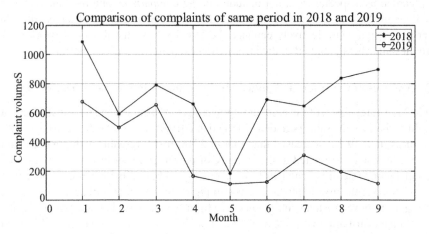

Fig. 8. A comparison chart of the total number of complaints in 2018 and 2019.

quality and service evaluation in low voltage distribution network emergency repair, this comprehensive analysis shows that after the new interface platform becomes online, data transmission time is reduced, and the speed of emergency repair of distribution network is improved, the satisfaction rate of grid customers is improved. This further demonstrates that the information interaction interface design proposed in this paper can effectively improve the information interaction efficiency between power information

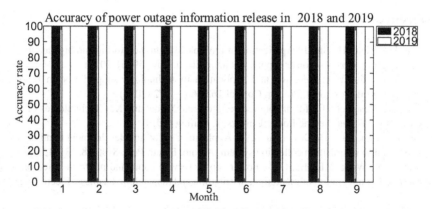

Fig. 9. Accuracy of blackout information release in 2018 and 2019.

collection system and other business platforms, besides, it can improve the efficiency of distribution network repair, the level and quality of service, and provide users with a better service experience.

6   Conclusion

By analyzing the information interaction modes of the telecommunication information collection system, the power supply service command system and 95598 business support system, this paper researches the multi-platform information exchange technology and improvement methods, puts forward the information interaction models among the three platforms, and designs a new data interface platform scheme. This interface can simultaneously meet the data transmission requirements of different timeliness, and it has high concurrency and reduces database pressure. Through investigating the functional data obtained from the application of this scheme in the emergency repair activities of a low voltage distribution network in a region, it is regarded that the information interaction design proposed in this paper is feasible and effective and can optimize the information interaction efficiency among platforms. Power failure information can be conveyed accurately and quickly, which can shorten the time of emergency repair for low voltage distribution network, improve the efficiency of emergency repair and the quality of power supply and user service. The proposed design is an expression of the progress of the comprehensive application of information technology, which can be extended to other aspects of multi-platform interaction and can provide a research basis for future business development and practical promotion of smart grid.

Nevertheless, there are still space for improvement in the design. Along with the development of power business, the types and quantities of power user data are increasing, some of which are sensitive. Enhanced encryption and protection of these data can be considered as a future research direction.

Foundation. Basic   public   welfare   research   project   of   Zhejiang   Province (LGG20E070003).

References

1. Pan, J.Y., Zhao, H., Shi, J.: Application of multiple system monitoring and big data analysis in emergency repair of power supply. Zhejiang Electr. Pow. 59–61(2015)
2. Wu, G.P., Liu, Y.Q.: Research and application of technology support system for smart distribute grid. Pow. Syst. Protect. Control **162**,166 + 172 (2015)
3. Chen, F.: Research of Information Secure Exchange Model and Key Technologies of Smart Grid. North China Electric Power University, Beijing (2014)
4. Hu, J.Y., Zhu, E.G., Du, X.G., Du, S.W.: Application status and development trend of power consumption information collection system. Autom. Electr. Pow. Syst. **38**, 131–135 (2014)
5. Yang, F.Y., Wang, Y.L., Ba, G.N.: Power supply service command platform based on deep information integration of marketing, distribution and dispatching business. Electr. Eng. 35–37 (2019)
6. Li, M.D.: Construction idea and realization method of provincial 95598 customer service system. Electr. Pow. IT 45–49 (2011)
7. Lin, H.Y., Zhang, J., Xu, K.P., Pi, X.J.: Design of interactive service platform for smart power consumption. Pow. Syst. Technol. **36**, 255–259 (2012)
8. Dai, J., Zhu, X.M.: Design and implementation of an asynchronous message bus based on Active MQ. Comput. Syst. Appl. **19**, 254–257, 215 (2010)
9. Zhan, L.Q., Chen, H.L., Ren, X.W.: Optimized application of redis caching technology in automatic weather station data call. Comput. Syst. Appl. 77–83 (2019)
10. Banday, M.T., Khan, M.: A study of recent advances in cache memories. In: 2014 International Conference on Contemporary Computing and Informatics (IC3I), Mysore, pp. 398–403 (2014)
11. Liu, F.: A method of design and optimization of database connection pool. In: 2012 4th International Conference on Intelligent Human-Machine Systems and Cybernetics, Nanchang, Jiangxi, pp. 272–274 (2012)
12. Gupta, K., Mathuria, M.: Improving performance of web application approaches using connection pooling. In: 2017 International conference of Electronics, Communication and Aerospace Technology (ICECA), Coimbatore, pp. 355–358 (2017)
13. Shree, R., Choudhury, T., Gupta, S.C., Kumar, P.: KAFKA: the modern platform for data management and analysis in big data domain. In: 2017 2nd International Conference on Telecommunication and Networks (TEL-NET), Noida, pp. 1–5 (2017)
14. Hiraman, B.R., Viresh, M.C., Abhijeet, C.K.: A study of apache kafka in big data stream processing. In: 2018 International Conference on Information, Communication, Engineering and Technology (ICICET), Pune, pp. 1–3 (2018)
15. Meesala, S.K.: Parallel processing implementation on clusters of terminals using Java RMI. In: 2012 International Conference on Computing, Communication and Applications, Dindigul, Tamilnadu, pp. 1–6 (2012)

Impact Analysis of False Data Injection Attack on Smart Grid State Estimation Under Random Packet Losses

Meng Xia[1], Dajun Du[1], Minrui Fei[1(✉)], and Kang Li[2]

[1] Shanghai Key Laboratory of Power Station Automation Technology, School
of Mechatronic Engineering and Automation, Shanghai University, Shanghai, China
mrfei@staff.shu.edu.cn
[2] School of Electronic and Electrical Engineering, University of Leeds,
Leeds LS2 9JT, UK

Abstract. Supervisory control and data acquisition (SCADA) system
has been widely used in traditional power systems for operation and control. As increasingly more ICT technologies are deployed to improve the
smartness of the power grid, cyber security is becoming an important
issue in the development of smart grids, for example, false data injection
attack (FDIA) poses a serious threat. The paper analyzes the impact
of false data injection attack on smart grid state estimation under random packet losses. First, a measurement model of power grids under
random packet loss is established, and an attack vector range that can
fool the attack detector is acquired. Then, a mean square error matrix
of weighted least squares estimation is proposed, taking into account
potential false data injection attacks. A IEEE-14 nodes system is used
to evaluate the performance of the weighted least squares state estimation under three different scenarios, namely false data injection attack
only, random packet loss only, and under both random packet loss and
false data injection attack.

Keywords: False data injection attack · Random packet losses ·
Weighted least squares estimation · Smart grid

1 Introduction

Modern power systems transmit electricity from generators to users via large-
scale transmission and distribution networks. To ensure safe and reliable operation of the system, increasingly more ICT technologies are introduced into the
power systems to improve the smartness [1]. However, the introduction of modern communication networks not only facilitates information interaction and
wide-area system monitoring, protection and control of power grids but also
makes it vulnerable to network invasion [2,3]. In recent years, cyber-attacks on
power grids around the world have been viewed as a principal threat, not just a
conceptual one.

© Springer Nature Singapore Pte Ltd. 2020
M. Fei et al. (Eds.): LSMS 2020/ICSEE 2020 Workshops, CCIS 1303, pp. 61–75, 2020.
https://doi.org/10.1007/978-981-33-6378-6_5

For example, Iran's Blushehr nuclear power plant was attacked by the Stuxnet virus in 2010, which caused the delay of power generation and seriously damaged Iran's industrial facilities [4]. The transmission lines in Ukraine were continuously tripped in 2015, while the information system was implanted with malicious software, which blocked the system restart [5]. In 2019, several cities in Venezuela including its capital city Caracas plunged into darkness, and power outages affected 21 of the country's 23 states. According to the media reports, the direct cause of the power failure was a cyber-attack on the country's largest hydropower station. Soon after, several transformer explosions occurred in the federal district of Caracas, causing another power failure [6].

The power system control center collects measurement data from different power devices and components through the supervisory control and provide instructions back to the system [7]. State estimation is a key functionality in real-time power system monitoring and supervisory control. By analyzing the data collected by the SCADA systems, the current operating state of the power grids can be estimated while bad data and anomalies in the collected measurements can be eliminated.

However, state estimation can be vulnerable to cyber-attack in the open network environment. The false data injection attack (FDIA) against the state estimator in the SCADA system was investigated by Liu et al. in 2009 [8], and it was found that existing bad data detection methods relying on Chi-square detector may not work in response to some false data injection attacks. An experienced attacker can deliberately design the attack vector such that these attacks can bypass the Chi-square detector. Once the sensor is successfully hacked, the tampered measurement will spread in the network, resulting in system performance degradation or even instability [9].

In the research area of false data injection attack, some researchers aim to identify the vulnerability of the system and build the attack models [10–13], and this helps to improve the understanding of the attack mechanism in order to design a better defense system. For example, a linear spoofing attack strategy and the corresponding feasibility constraints are demonstrated where fake data can be effectively designed to cause system failure [10]. In [11], the potential impact of unobservable attacks is investigated, and the least measurable attack strategy is proposed. Under the fully measurable model and partially measurable model, the existence conditions of unobservable subspace attacks are derived, based on which two attack strategies are proposed in [12]. The first strategy directly affects the system state by hiding attack vectors in the system subspace, and the second strategy misleads the bad data detection mechanism. Meanwhile, other researchers focus on the detection and defense of the system in the presence of attacks [14–19]. For example, both active detection and estimation-based detection are proposed in [14]. In the active detection method, a reasonable excitation signal is designed to be superimposed on the control signals, which improves the detectability of attacks on the actuator attack. The other method estimates the value of the attack by using the unknown input observer. In [15], a FDIA attack detection mechanism based on the increments of analytic measurements in the micro-grid environment was proposed.

Most existing researches are based on the analysis of the acquired measurements, but the impact of data communication is not considered. The FDIA in smart grid applications is an attack that reduces the integrity of data acquired by the system. In the existing communication technology, data transmitted through the network is often in the form of packets [20]. Most existing approaches construct the attack model and detect the attack using acquired measurements and the estimation of measurements [10–18]. However, in addition to potential FDIA, the data transmitted through the network may also be affected by network characteristics such as data losses during the transmission phase. This paper investigates the data injection attack on power system state estimation considering data losses in communication. The main contributions are as follows:

- A DC (direct current) model of the system under data injection attack is deduced, taking into account the random packet losses.
- The mechanism of weighted least squares state estimation and bad data detection are analyzed and an undetected range of attack vectors is derived.
- Based on the established DC measurement model, the mean square error matrix of state estimation under the FDIA is analyzed.

The remainder of the paper is organized as follows. The transmission model of sensor measurements in the power grids under random packet loss is discussed in Sect. 2. Section 3 analyses the effects of random packet loss and data injection attacks on weighted least squares estimation, and the range of attack vectors is also studied. Simulation results are presented in Sect. 4, and the weighted least squares state estimation results under three different cases are compared.

2   Problem Formulation

2.1   Data Transmission Model

The SCADA system in the power grids collects sampled measurements from sensors through the communication network. However, due to limitations of the communication technology, data may get lost during the transmission. Figure 1 illustrates the whole process from data sampling and transmission to state estimation.

As shown in Fig. 1, at time instant t_{k-1}, the measurement device samples and transmit the sensor measurements to the network in the form of packets. Due to network induced delays, after the transmission delay d_{k-1}, the SCADA system will receive the sampled measurements at time instant $t_{k-1} + d_{k-1}$. Further, some data may be lost during the transmission process, such as the data at time instant t_k shown in Fig. 1. Once the SCADA system obtains the measurements, the estimator can receive the data after the computing time delay of c_{k-1}. Power grids are typical complex cyber-physical systems with numerous sensors, and all sensor data will go through the similar process as shown in Fig. 1 when they are transmitted to the SCADA system.

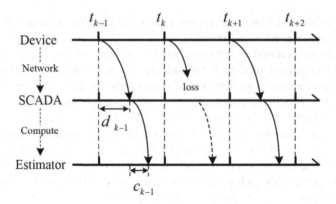

Fig. 1. The data sampling, transmission and state estimation process.

Define the measurements received by the SCADA system at sampling instant k as z_k, $z_k \in R^m$, and if there exists data packet losses, two popular compensated methods are often adopted. One is to directly replace the lost data with 0 [21]. Another is to replace the lost data with the previous sampled data. This paper adopts the first method, i.e., the loss packet is set as 0. For random packet losses, the received measurements can be expressed by

$$z_{lk} = \lambda_k z_k, \tag{1}$$

where $\lambda_k \in R^{m \times m}$ is a diagonal matrix whose diagonal elements are either 1 or 0. When a measurement is lost, its corresponding value is set to 0.

2.2 Power Grid Measurement Model

When the system is subject to a false data injection attack, the measurement process of the grids is shown in Fig. 2. When a sensor device samples measurements, it may be invaded by an attacker by deception, and false data are injected. Next, the sensor transmits the corrupted data to the SCADA over the network. When random packet loss is not considered at the sampling instant k, the AC measurement model can be described as

$$z_k = h(x_k) + v_k, \tag{2}$$

where z_k is denoted as the measurement vector, x_k is the system state vector, v_k is the Gaussian measurement noise, and $h(x_k)$ is the functional dependency between measurements and state variables.

If the ground admittance and branch conductance are ignored and assume that the voltage phase difference between two nodes is negligible, the voltage amplitude of the nodes is close to unit quantity 1. The DC measurement model can be used to approximate AC measurement model. The DC measurement model can be expressed as

$$z_k = Hx_k + v_k, \tag{3}$$

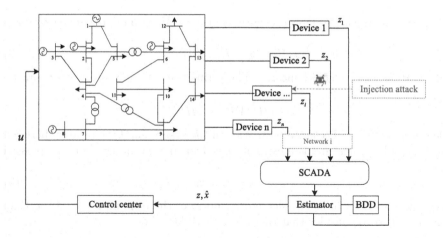

Fig. 2. The power grid measurement process subject to false data injection attack.

where H is the steady-state functional dependency between measurements and state variables.

When only random packet loss is considered, at the sampling instant k, the DC measurement model can be expressed as

$$z_{lk} = \lambda_k(Hx_k + v_k). \tag{4}$$

When only a data injection attack is considered and assume that the injected value is a_k and $a_k \in R^m$. If a_k is nonzero, the corresponding measurement is tampered. Then the measurement contains the attack vector a_k, which can be expressed as

$$z_{ak} = z_k + a_k, \tag{5}$$

where a_k is the attack vector injected to measurement.

When random packet loss is considered, the measurement function can be expressed as

$$z_{lak} = \lambda_k z_{ak}. \tag{6}$$

Equation (6) is the measurement model under the false data injection attack which considers both the influence of random packet loss and data injection attack on the measurements of the grid.

3   Analysis of Weighted Least Squares Estimation

State estimation is used for monitoring the operating state of the grid and remove bad data, and the weighted least square method is a popular state estimation method. The false data injection attack aims to mislead the state estimation, and it is necessary to have a detailed analysis of the state estimator. According to the weighted least squares estimation, the objective function can be expressed as

$$\min J(x_k) = (z_k - Hx_k)^T W(z_k - Hx_k), \tag{7}$$

where W is the weighted matrix. The estimation of the system state can be expressed as

$$\hat{x}_k = (H^T W H^T)^{-1} H^T W z_k. \tag{8}$$

Define $\hat{z}_k = H\hat{x}_k$ as the state estimation of the system, and the residual between the real and the measurement estimation is defined as r_k, and r_k can be expressed as

$$r_k = z_k - \hat{z}_k. \tag{9}$$

According to the Chi-square detector, 2-norm of the residual must be less than the threshold to consider that there is no bad data, i.e.,

$$\|r_k\|_2 \leq \tau, \tag{10}$$

where τ is the threshold of the Chi-square detector, which can be obtained by checking the Chi-square distribution table. When there is only a false data injection attack, the injected increment must meet certain conditions in order not to be detected. According to (8), for a given a_k, the state estimation can be expressed as

$$\hat{x}_{ak} = (H^T W H^T)^{-1} H^T W z_{ak}, \tag{11}$$

where $\hat{x}_{ak}$ the corrupted estimation due to FDIA. The estimate of the measurement is $\hat{z}_{ak} = H\hat{x}_{ak}$, and the residuals can be expressed as

$$\begin{aligned} r_{ak} &= z_{ak} - \hat{z}_{ak} = z_k + a_k - (H\hat{x}_k + H(H^T W H)^{-1} H^T W a_k) \\ &= (I - H(H^T W H)^{-1} H^T W)(z_k + a_k). \end{aligned} \tag{12}$$

To evade the detector, Eq. (13) must be satisfied, that is

$$\|r_{ak}\|_2 \leq \tau. \tag{13}$$

Let $B = (I - H(H^T W H)^{-1} H^T W)$, Eq. (13) can be re-written as

$$\|B(z_k + a_k)\|_2 \leq \tau. \tag{14}$$

According to the compatibility

$$\|B(z_k + a_k)\|_2 \leq \|B\|_2 \|(z_k + a_k)\|_2, \tag{15}$$

when $\|B\|_2 \|(z_k + a_k)\|_2 \leq \tau$ hold, the Eq. (14) will be hold, where $\|B\|_2 = \sqrt{\eta_{\max}(B^T B)}$ is the induced norm and $\eta_{\max}(B^T B)$ is the maximum eigenvalue of the matrix $B^T B$.

Therefore,

$$\|(z_k + a_k)\|_2 \leq \frac{\tau}{\|B\|_2}. \tag{16}$$

Remark 1. Inequality (16) represents a subset of the attack vector which will not trigger an alarm from the bad data detector.

Corollary 1. *Equation (17) is the non-detectable spoofing range of the attack vector.*

$$\|a_k\|_2 \leq \frac{\tau}{\|B\|_2} - \|z_k\|_2. \tag{17}$$

According to the triangle inequality, it's easy to prove Corollary 1 is true. The specific derivation is given as follows.

According to the triangle inequality of vector 2-norm,

$$\|(z_k + a_k)\|_2 \leq \|z_k\|_2 + \|a_k\|_2, \tag{18}$$

when $\|z_k\|_2 + \|a_k\|_2 \leq \frac{\tau}{\|B\|_2}$ hold, the Eq. (16) will be hold. So Eq. (17) is a safe range of the attack vector.

When packets are randomly lost, the integrity of the collected data by SCADA is destroyed. However, due to the redundancy of data in data acquisition of the power grids, the effect of the loss of a small number of measurements may small. To study the effect of data injection attack on the performance of state estimation under random packet losses, the mean square error (MSE) of weighted least squares state estimation under random packet losses is derived.

Suppose that the state vector x_k, the attack vector a_k, and the noise v_k obey the Gaussian distribution where the mean value is $\mu_{x_k} = 0$, and the variance is R_{x_k}, R_{a_k}, R_v. When there is random packet loss, the measurement model of the system is shown by Eq. (6). Combined Eq. (11) with Eq. (6), the state estimation of the system can be expressed as

$$\begin{aligned}\hat{x}_{lak} &= ((\lambda_k H)^T W \lambda_k H)^{-1} (\lambda_k H)^T W (z_k + a_k) \\ &= (H^T \lambda_k W H)^{-1} H^T \lambda_k W (z_k + a_k).\end{aligned} \tag{19}$$

When the system state estimation residual is defined as $\varepsilon_{x_k} = \hat{x}_{lak} - x_k$, ε_{x_k} can be expressed as

$$\begin{aligned}\varepsilon_{x_k} &= (H^T \lambda_k W H)^{-1} H^T \lambda_k W (z_k + a_k) - x_k \\ &= (H^T \lambda_k W H)^{-1} H^T \lambda_k W (H x_k + v_k + a_k) - x_k \\ &= (H^T \lambda_k W H)^{-1} H^T \lambda_k W (v_k + a_k)\end{aligned} \tag{20}$$

When there is random packet losses and data injection attack, the mean square error matrix of system state estimation is

$$\begin{aligned}R_{\varepsilon_{x_k}} &= E\{\varepsilon_{x_k} \varepsilon_{x_k}^T\} = (H^T \lambda_k W H)^{-1} \\ &+ (H^T \lambda_k W H)^{-1} H^T \lambda_k W R_{a_k} \lambda_k W H (H^T \lambda_k W H)^{-1}.\end{aligned} \tag{21}$$

Let $B_k = (H^T \lambda_k W H)$, $R_{\varepsilon_{x_k}}$ can be expressed as

$$R_{\varepsilon_{x_k}} = B_k^{-1} + B_k^{-1} H^T \lambda_k W R_{a_k} \lambda_k W H B_k^{-1}. \tag{22}$$

Ideally, when there is no packet losses and data injection attacks, $\lambda_k = I$, $a_k = 0$. Then the mean square error matrix of the weighted least squares state estimation is

$$R_{\varepsilon_{x_k}} = (H^T W H)^{-1}. \tag{23}$$

Comparing Eq. (22) and (23), it can be found that the existence of random packet losses will not only affect the state estimation, but also affect the effect of data injection attack.

4 Simulation Study

To assess the impact of data injection attack under random packet losses on smart grid state estimation, IEEE-14 node system is used in the simulation experiments, as shown in Fig. 3. IEEE-14 node system has 54 measurements, where 1–14 are the measurements of the active power of the bus, 15–34 are the measurements of branch power of the incoming node, and 35–54 are the measurements of branch power of the outgoing node. Assuming that the noise of each measurement obeys the Gaussian distribution, i.e., $v_i \tilde{} N(0, 0.02^2)$, where $i = 1, 2, \cdots, 54$. Considering the phase angle of the reference bus $\delta_1 = 0$, it is only necessary to estimate the state quantity of the other 13 nodes, and $H \in R^{54 \times 13}$.

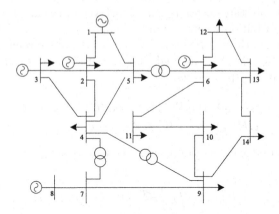

Fig. 3. The power grids measurement process.

Firstly, node 1 is selected as the reference node, and the state truth value and the measurement truth value are obtained by 100 power flow calculations. It is assumed that the white noise obeys the Gaussian distribution $(0, 0.02^2)$ and the measurement error covariance matrix is constant.

Performance Index: From Eq. (23) under ideal conditions, when there is no data injection attack and transmission packet losses, the mean square error matrix of system state estimation is $R_{\varepsilon_{x_k}} = (H^T W H)^{-1}$. In order to measure the state estimation performance, Eq. (24) is used as the performance index.

$$\text{Performance} = \frac{\|(xreal - \hat{x})(xreal - \hat{x})'\|_F}{\left\|(H^T W H)^{-1}\right\|_F},\tag{24}$$

where $xreal$ is system status truth value, $\hat{x}$ is the estimation, and $\|\ \|_F$ is frobenius norm of matrix.

When there only exist data injection attacks, while Eq. (17) is satisfied, three different attacks are randomly selected, where one measurement is tampered in a_{88}, five measurements are tampered in a_{41}, and ten measurements tampered in a_{55}. The dimension of each non-zero in the attack vector was randomly selected in $[-(\frac{\tau}{\|B\|_2} - \|z_k\|_2)/p, (\frac{\tau}{\|B\|_2} - \|z_k\|_2)/p]$, where p is the number of the tampered devices. The details of the attack vector are shown in Table 1. The estimated results are also illustrated in Fig. 4.

Table 1. Details of the attack vector in data injection attack only

Attack		Details								
a_{88}	Index	52								
	Value	37.97								
a_{41}	Index	4		15		43		44		45
	Value	0.84		−6.05		9.47		8.44		−8.57
a_{55}	Index	11	12	16	25	29	47	48	50	51	54
	Value	−5.06	−1.18	−1.01	−3.22	−3.59	−0.1	0.17	−0.04	−2.17	−5.25

According to Fig. 4, the data injection attack has a great impact on the survivability of system state estimation. However, with the increase of attack dimensions, the impact of the attack on the estimation decreases gradually if the attack vector remains non-detectable by satisfying Eq. (16).

The performance index is also illustrated in Fig. 5. This is a result from the attack vector limited by Eq. (17). The more dimensions of the attack, the lower the amplitude of each dimension in the attack vector will become.

In the packet loss only scenario, three packet loss rates are randomly selected, which are 2%, 5% and 10% respectively. The specific information of random packet losses is shown in Table 2, and the estimation results are illustrated in Fig. 6.

As show in Figs. 6 and 7, a small amount of random data packet loss in the data transmission of the sensor does not have significant impact on the system state estimation. This is due to the existence of the measurement redundancy of the power system, which guarantees the safety and reliability of power system state estimation.

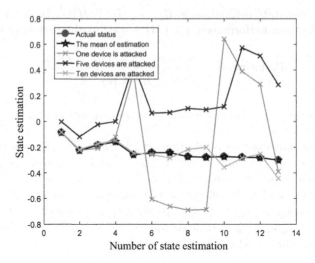

Fig. 4. State estimation under only data injection attack.

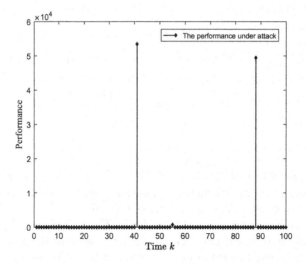

Fig. 5. Estimation performance under only data injection attack.

Table 2. Details of the packet loss due to random packet loss only

Probability	Time	Index
2%	66	2
5%	99	16,28,48
10%	62	3,5,7,8,39

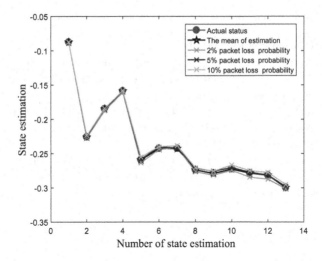

Fig. 6. State estimation under random packet loss scenario.

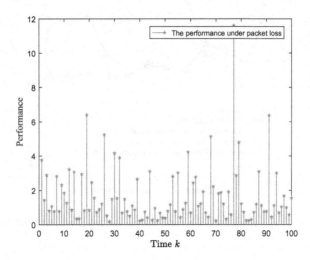

Fig. 7. Estimation performance in random packet loss only scenario.

Furthermore, comparing Fig. 4 with Fig. 6, it is clear that the data injection attack has a greater impact on the system state estimation. Again, the performance indexes as shown in Figs. 5 and 7 are not in the same order of magnitude.

When both packet losses and data injection attacks are presented, 5% packet loss rate and 5 dimensions attacked are simulated.

Table 3. The details of the packet loss and attack vectors

Time	Attack and packet loss		Details					Common index
44	a_{44}	Index	9	15	24	30	49	No
		Value	−9.33	−7.28	−6.88	−3.79	−9.67	
	Packet loss index		7,20,22					
92	a_{92}	Index	2	14	30	44	46	30,44
		Value	6.37	−6.40	0.94	9.84	−6.14	
	Packet loss index		21,30,44					
53	a_{53}	Index	12	24	28	48	53	12,48,53
		Value	−2.19	−8.03	5.65	8.68	−1.45	
	Packet loss index		12,48,53					

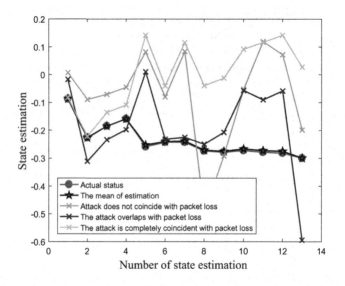

Fig. 8. State estimation under both random packet loss and data injection attack.

Three scenarios, including random packet losses and data injection attack are not coincidences, some coincident, and all occurred coincidently are analyzed. The specific information of random packet loss and attack vectors are listed in Table 3, and the estimation results are illustrated in Fig. 8. It can be seen that notification of data injection attack and random packet loss will have a great impact on system state estimation results.

As shown in Fig. 9, when the random packet losses occur coincidently with the attack, and the estimation performance is better than the non-overlap, but the impact of the attack vector itself is greater.

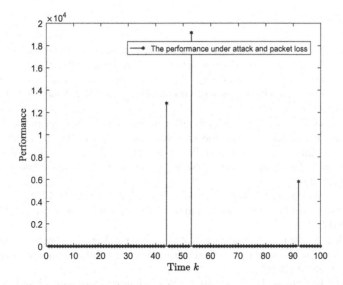

Fig. 9. Estimation performance under both random packet loss and data injection attack.

5 Conclusions

This paper has analyzed the impact of false data injection attacks on smart grid state estimation under random packet losses. Firstly, the measurement model of power grid under random packet losses is established, and an attack vector range that can escape the detector is derived. Then, the weighted least squares estimation is analyzed, and a non-detectable range of attack vectors in the data injection attack is derived. It is proved that as long as the attack vectors are selected in the derived range, the existing "bad data" detection device will not respond. Further, considering the false data injection attack, the mean square error matrix of the weighted least squares estimation is provided. Finally, simulation experiments on a IEEE-14 node system is used to compare the effects of data injection attack, random packet loss, and simultaneous random packet loss and data injection attack on the system state estimation.

Acknowledgement. Supported by Natural Science Foundation of China (No. 61633016, 61533010), Key Project of Science and Technology Commission of Shanghai Municipality (No. 19510750300, 19500712300, 16010500300), Industrial Internet Innovation and Development Project (TC190H3WL).

References

1. Yan, J., Guo, F., Wen, C.: False data injection against state estimation in power systems with multiple cooperative attackers. ISA Trans. **101**(10), 225–233 (2020)
2. Sahoo, S., Dragicevic, T., Blaabjerg, F.: Cyber security in control of grid-tied power electronic converterschallenges and vulnerabilities. IEEE J. Emer. Sel. Top. Power Electr. **15**, 1–15 (2019)
3. Shu, J., Guo, Z., Han, B.: A bilevel optimization model for power network spurious data injection attack. Autom. Electr. Power Syst. **43**(10), 95–101 (2019)
4. Liang, G., Zhao, J., Luo, F., Weller, S.R., Dong, Z.Y.: A review of false data injection attacks against modern power systems. IEEE Trans. Smart Grid **8**(4), 1630–1638 (2017)
5. Liang, G., Weller, S.R., Zhao, J., Luo, F., Dong, Z.Y.: The 2015 Ukraine blackout: implications for false data injection attacks. IEEE Trans. Power Syst. **32**(4), 3317–3318 (2017)
6. Gong, X.: Analysis of the situation of the power outage in Venezuela and recommendations for the safety of critical infrastructure. J. Inf. Technol. Network Secur. **38**(04), 1–2+14 (2019)
7. Gong, X.: Upadhyay, D., Sampalli, S.: Scada (supervisory control and data acquisition) systems: Vulnerability assessment and security recommendations. Comput. Secur. 89, 101666 (2020)
8. Liu, Y., Ning, P., Reiter, M.K.: False data injection attacks against state estimation in electric power grids. ACM Trans. Inf. Syst. Secur. **14**(1), 1–33 (2011)
9. Li, L., Yang, H., Xia, Y., Yang, H.: Event-based distributed state estimation for linear systems under unknown input and false data injection attack. Signal Process. **170**, 107423 (2020)
10. Guo, Z., Shi, D., Johansson, K., Shi, L.: Optimal linear cyber-attack on remote state estimation. IEEE Trans. Control Network Syst. **4**, 4–13 (2016)
11. Zhao, Y., Goldsmith, A., Vincent Poor, H.: Minimum sparsity of unobservable power network attacks. IEEE Trans. Autom. Control **62**(7), 3354–3368 (2017)
12. Kim, J., Lang, T., Thomas, R.J.: Subspace methods for data attack on state estimation: a data driven approach. IEEE Trans. Signal Process. **63**(5), 1102–1114 (2015)
13. Zhong, H., Du, D., Li, C., Li, X.: A novel sparse false data injection attack method in smart grids with incomplete power network information. Complexity 1–16 (2018)
14. Muniraj, D., Farhood, M.: Detection and mitigation of actuator attacks on small unmanned aircraft systems. Control Eng. Pract. **83**, 188–202 (2019)
15. Huaye, P., Chen, P., Hongtao, S., Mingjin, Y.: Incremental detection mechanism of microgrid under false data injection attack. Inf. Control **48**(5), 522–527 (2019)
16. Chen, R., Li, X., Zhong, H., Fei, M.: A novel online detection method of data injection attack against dynamic state estimation in smart grid. Neurocomputing **344**, 73–81 (2019)
17. Du, D., Chen, R., Li, X., Wu, L., Zhou, P., Fei, M.: Malicious data deception attacks against power systems: a new case and its detection method. Trans. Inst. Measur. Control **41**(6), 1590–1599 (2019)

18. Du, D., Li, X., Li, W., Chen, R., Fei, M., Wu, L.: ADMM-based distributed state estimation of smart grid under data deception and denial of service attacks. IEEE Trans. Syst. Man Cybernet.-Syst. **49**(8), 1698–1711 (2019)
19. Xia, M., Du, D., Fei, M., Li, X., Yang, T.: A novel sparse attack vector construction method for false data injection in smart grids. Energies **13**(11) (2020)
20. Aghanoori, N., Masoum, M.A., Abu-Siada, A., Islam, S.: Enhancement of microgrid operation by considering the cascaded impact of communication delay on system stability and power management. Int. J. Electr. Power Energy Syst. **120**, 105964 (2020)
21. Ding, D., Han, Q.L., Xiang, Y., Ge, X., Zhang, X.M.: A survey on security control and attack detection for industrial cyber-physical systems. Neurocomputing **275**, 1674–1683 (2018)

Research on Home Energy Optimization Strategy Based on Micro-sized Combined Heat and Power

Qingchun Li[1](✉), Ye Zhang[1](✉), Shanglai Li[2](✉), Jian Zhang[1](✉), and Pengyu Liu[1](✉)

[1] State Grid Liaoning Electric Power Supply Co. Ltd., Shenyang 110003, China
150070699@qq.com, 953332684@qq.com, stonezhj@163.com,
lpy567@yeah.net
[2] Department of Electrical Engineering, College of Information Science and Engineering,
Northeastern University, Shenyang 110819, China
liaoningshiyoulai@qq.com

Abstract. Under the background of the energy Internet, the demand of residential users for electricity and heat has shown different degrees of growth. In this paper, a household energy optimization strategy with micro cogeneration is proposed. Firstly, the composition of household energy microgrid and the operation mode of mCHP are introduced. Then, the load and user comfort of household energy microgrid are modeled. Finally, the optimal scheduling model of household energy was established with the goal of low energy cost, which provides the optimized scheme for the comprehensive utilization of natural gas and electric energy for household users.

Keywords: Home energy microgrid · Micro-sized combined heat and power · Optimized dispatching

1 Introduction

With the development and construction of cities, people's demand for electricity and heat has been increased to different degrees [1, 2]. Nowadays, the world is faced with the problem of energy decreasing and natural environment deteriorating. Combined heat and power [3–6] (CHP) may be the answer to the problem of how to use the existing primary energy to efficiently and cleanly solve the problem of increasing residents' demand for electricity and heat.

The CHP system is divided into regional type, household type and building type according to its size [7]. The cogeneration involved in this paper is household micro cogeneration, it is according to the principle of energy cascade utilization [8], at first it uses oil, coal, natural gas for heating, and then uses the steam generated during heating to generate electricity [9].

© Springer Nature Singapore Pte Ltd. 2020
M. Fei et al. (Eds.): LSMS 2020/ICSEE 2020 Workshops, CCIS 1303, pp. 76–86, 2020.
https://doi.org/10.1007/978-981-33-6378-6_6

2 Home Energy Microgrid

The energy internet was proposed in the third industrial revolution. It is based on Internet technology and closely combined with multiple energy networks to realize the new concept of multiple energy input on the energy input side and optimal utilization of load demand on the energy output side [10, 11]. The home energy microgrid is a miniature of the energy Internet on the home side. It is a port model [12] to realize multi-energy input and multi-energy output through the energy converter. The specific model is shown in Fig. 1.

Fig. 1. Energy Input-Output Port model

In this paper, a home energy microgrid based on micro cogeneration (mCHP) is constructed [13], the household energy micro network includes gas generator, heat recovery unit, heat storage tank, storage battery, heat exchanger, hot water tank and other equipment, and the household energy micro network model is shown in Fig. 2.

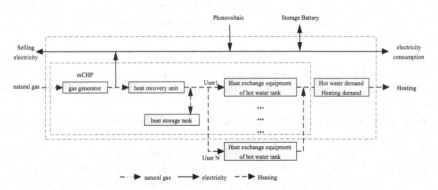

Fig. 2. Home energy micro grid model

The source of heat energy is only provided by micro-cogeneration of heat and power, and cannot be generated by conversion of electric equipment. The electricity of the user is mainly obtained from micro-cogeneration, storage battery and power purchase from the power grid company. When the power demand of the user is greater than the power output from micro-cogeneration, the user needs to purchase power from the power grid company and discharge through the battery. When the user's demand for electricity is less than the output of the micro combined heat and power generation, the user can sell the remaining energy to the power company through the power feedback or through the storage battery for storage.

3   Household Energy Microgrid Load Model and Comfort Model

3.1   Load Model

The load of home energy management center mainly includes electric load and thermal load. The classification and energy sources of various loads are shown in Fig. 3.

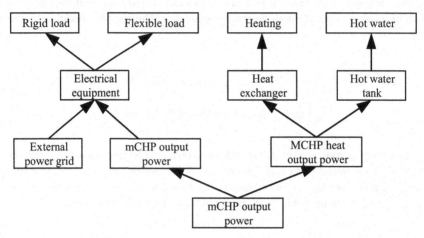

Fig. 3. Classification of load and energy source

Electrical Load Model. The household electric equipment can be divided into rigid load and flexible load according to its power consumption time and whether the power consumption can be adjusted.

Rigid load is an unadjustable electrical equipment. this kind of load is mainly to meet the basic needs of residential users.

Flexible load is an electrical equipment that can adjust the electricity consumption time and power consumption. This kind of load only needs to complete its own electricity consumption task within the allowed working period to meet the electricity demand of users.

Thermal Load Model. The heat load mainly consists of two parts: the heating load and the hot water tank.

Hot Water Tank. Micro-cogeneration of heat and power uses natural gas to generate heat energy for heating, so that the temperature of the hot water tank is always in the temperature range set by the user, to ensure that the user's demand for hot water is met at any time.

The Heating Load. There is always heat exchange between indoor and outdoor, and the rate of exchange per unit time depends on the thermal resistance of building materials. The greater the thermal resistance R is, the less heat the indoor absorbs from the outdoor, and the greater the reverse.

Convertible Load Model Between Electric and Natural Gas. The source of electric energy can be obtained through either micro-cogeneration or purchase power from grid. Therefore, the household energy management system can choose the source of electricity.

3.2 Modeling of Indoor Temperature Comfort

Household energy management should not only optimize the use of household energy, but also consider the energy comfort of users.

In order to establish a relationship between the indoor temperature and the residents' temperature comfort, the author introduced the Predicted Mean Vote (PMV) to quantify the temperature comfort. If the quantization value is 0, it means that the indoor temperature is considered to be the most comfortable state. The corresponding relationship between specific PMV and comfort is shown in Table 1.

Table 1. Correspondence between PMV and comfort

PMV	3	2	1	0	−1	−2	−3
Feeling	Hot	Warm	Slightly warm	Neutral	Slightly cool	Cool	Cold

The relationship between temperature and PMV is shown in Eq. (1):

$$I_{PMV} = \begin{cases} 0.3895(x - 26)\, x \geq 26 \\ 0.4065(26 - x)\, x \leq 26 \end{cases} \tag{1}$$

According to equation, when the temperature is 26 °C, the PMV value is 0, the human body's temperature comfort is the best. According to the standard given by ISO7730, when PMV is between −0.5 and 0.5, it is considered to be an acceptable temperature range for human body. The corresponding temperature range of PMV is 24.8 °C to 27.3 °C. Therefore, this temperature range is taken as the indoor temperature comfort range.

4 Optimal Operation Model of Household Energy Considering User Comfort

4.1 Establishment of the Objective Function of Home Energy Optimization

Considering the indoor temperature comfort, the optimal scheduling model of household energy management system is established, which takes the minimum energy consumption of household energy microgrid as the objective function. The objective function is shown in the following equation:

$$\min C_{pay} = \sum_{t=1}^{T} C_t = \sum_{t=1}^{T} \left(\alpha_t^g f_t^{CHP} \Delta T + \alpha_t^e p_t^e \Delta T \right) \tag{2}$$

ΔT represents the sub-period of optimization decision of household energy management system. If the selected working sub period is too short, the action of optimal control will be too frequent, which will seriously damage the service life of the equipment; If the selected working time is too long, the process of optimal control will be rough and the optimal effect will not be achieved. Considering that the electrical equipment is a household appliance, the selection range of working sub-time segment is 10 min to 60 min.

It is assumed that the adjustable loads of user U participating in the optimal scheduling are $L_1, L_2 \ldots L_N$, the working hours of each load are $t_{L1}, t_{L2} \ldots t_{LN}$, the number of work subsegments to be worked for each load is $m_1, m_2 \ldots m_N$, The work redundancy time of each load is $T_1, T_2 \ldots T_N$, the length of the work subtime segment optimized by user U is t_U, the redundant time of a single electrical device n is calculated as shown in Eqs. (3) and (4):

$$m_n = \left[\frac{t_{Ln}}{t_U} \right] + 1 \tag{3}$$

$$T_n = m_n \times t_U - t_{Ln} \tag{4}$$

The total redundancy time T of all adjustable loads participating in the optimal scheduling is shown in Eq. (5):

$$T = \sum_{n=1}^{N} T_n \tag{5}$$

4.2 Constraint Condition

Electrical Load Power Balance Constraints. The electric power balance of the family should be maintained at every moment, as shown in Eq. (6):

$$p_t^{CHP} + p_t^e = p_\Sigma \tag{6}$$

p_Σ represents the total electrical load power required by the home user at time t.

Operating Constraints for Micro-cogeneration. Thermal power balance should be maintained at each moment, as shown in Eq. (7):

$$h^t + h_w^t = h_t^{CHP} \tag{7}$$

The upper and lower limits of the output power of micro-cogeneration are shown in Eq. (8):

$$h_{min}^{CHP} \le h_t^{CHP} \le h_{max}^{CHP} \tag{8}$$

Constraints on Indoor Temperature. According to the standard of PMV, the comfortable temperature of human body has a range, and it is necessary to control the indoor temperature, as shown in Eq. (9):

$$T_{in.min} \leq T_{in} \leq T_{in.max} \tag{9}$$

$T_{in.min}$, $T_{in.max}$ represent the upper and lower limits of the body's comfortable temperature respectively.

Constraints on the Temperature of the Hot Water Tank. According to the user's hot water usage and requirements, it is necessary to control the temperature of the water heater within the range required by the user, as shown in Eq. (10):

$$T_{w.min} \leq T_w \leq T_{w.max} \tag{10}$$

$T_{w.min}$, $T_{w.max}$ represent the minimum and maximum temperatures required by the user for the water heater respectively.

Reliability Index Constraints Condition. On the basis of the initial constraints, the reliability index set by the user is replaced by the new constraints or directly added to the problem of intelligent power consumption optimization.

4.3 Optimization Process of Household Energy

Considering the user's electricity comfort and the operation mode of micro-cogeneration, the specific process is shown in Fig. 4.

5 Example Analysis

5.1 Basic Parameter Setting

In order to verify the effectiveness of energy optimization strategy, the energy usage of a certain day in winter was analyzed. In order to simplify the verification process, photovoltaic and battery are not involved in this example.

In this example, the day is divided into 96 time periods, each period is 15 min, and each period is the minimum operation period for the optimization of the household energy management system. It is assumed that the user adopts peak electricity price. The specific electricity price information is shown in Fig. 5. Natural gas is a fixed price of 2.7 yuan per cubic meter; The consumption curve of user's hot water on a certain day is shown in Fig. 6. The outdoor temperature curve and rigid load curve on a certain day in winter are shown in Fig. 7 and Fig. 8. The parameters of flexible load and micro-cogeneration are shown in Table 2 and Table 3. The exchange parameters of hot water tank and room temperature are shown in Table 4 and Table 5.

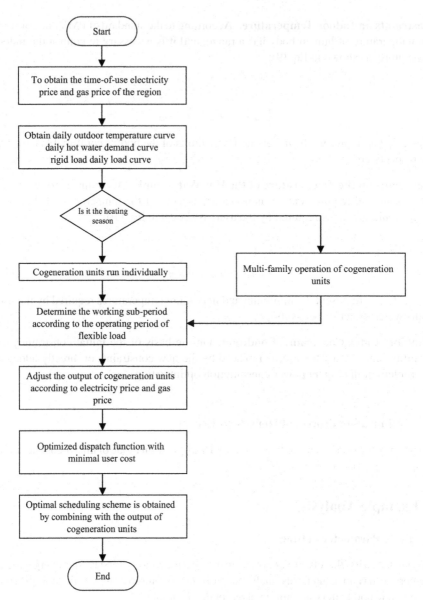

Fig. 4. Flow chart of home energy optimization based on mCHP multi-time scale

5.2 Plan of Flexible Load

The household energy management system controls the changes of household air heating power, hot water tank heating power, heat storage tank heat storage power and total heat power on a certain day in winter, as shown in Fig. 9, and the indoor temperature and water temperature of the hot water tank under control are shown in Fig. 10.

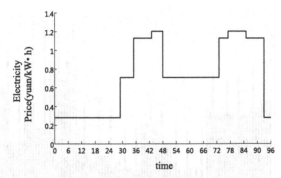

Fig. 5. Time of use tariffs in a day

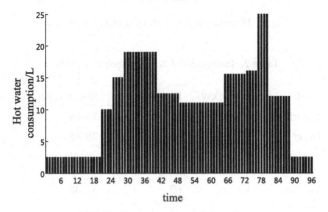

Fig. 6. Hot water demands in a day

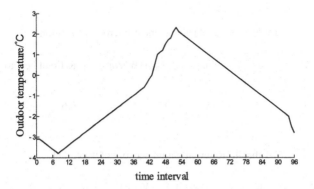

Fig. 7. Temperature curve of an extra day in winter

When the natural gas price is fixed, the thermal power output of micro-cogeneration is directly related to the electricity price, combined with the example of natural gas price and peak electricity price information, it can be concluded that when the price is less than 0.742 yuan, use natural gas to generate power, this will make the cost of the family goes

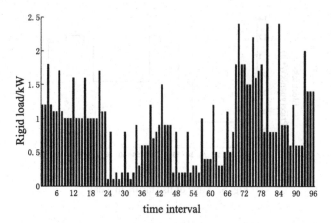

Fig. 8. Demand curve of inflexible electricity loads

Table 2. Parameters of flexible electricity loads

Load	Power(kW)	Run time(min)	Run time period
Electric iron	1.3	15	76–88
Dryer	0.9	45	28–48
Washing machine	0.5	45	28–44
Vacuum cleaner	1.1	60	32–48
Dishwasher	1.28	75	48–72

Table 3. Related parameters of micro-cogeneration

Parameter	h_{min}^{CHP} (kW)	h_{max}^{CHP} (kW)	η_h (%)	$Q_{gas}(kWh/m^3)$	Ratio of heat to electricity
Value	0.3	3	50	9.7	0.6

Table 4. Relevant parameters of hot water tank

Parameter	$T_{cw-min}(°C)$	$T_{cw-max}(°C)$	$V_w(L)$	$\rho_w(kg/m^3)$	$C_w(kWh/kg \cdot °C)$
Value	65	75	150	1000	1.1667×10^3

higher. When the electricity price is higher than 0.742 yuan, use natural gas to generate power, the cost of the family will be reduced. Finally, the energy cost of a household in

Table 5. Relevant parameters of room temperature

Parameter	$R(°C/kW)$	$C_{air}(kW/°C)$
Value	18	0.525

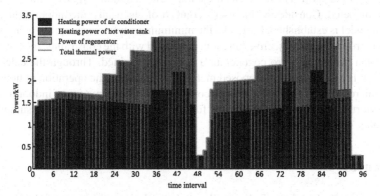

Fig. 9. Heat power change histogram and total thermal power curve of each device

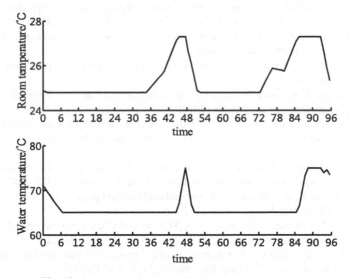

Fig. 10. Indoor temperature and water temperature curve

a certain day in winter is calculated to be 16.49 yuan, of which 22.55 yuan is needed to buy natural gas and −6.06 yuan is the comprehensive cost of purchasing and selling electricity.

6 Conclusion

In this paper, we optimize the household energy plan under the control of the household energy management system which based on household energy microgrid involving a variety of energy inputs. Firstly, this paper introduces and builds the model of household energy microgrid. Then, the electrical load, thermal load and electrical convertible load are modeled. Considering the user's comfort of electricity, the indoor temperature comfort model is established. Last, take the minimum energy consumption of household energy microgrid as the objective function, combined with the constraints, the household energy plan based on micro cogeneration is finally obtained. Through the calculation example, it is proved that the proposed method can optimize the operation of household energy microgrid and reduce the energy cost of users on the premise of indoor temperature comfort, so as to provide a solution for the comprehensive energy consumption of households.

References

1. Chenglong, Z., Xiandong, T., Yuyan, W., et al.: Analysis on the reasons for the growth of power consumption since the "13th five-year plan" and its medium and long-term prospects. China Electr. Pow. **52**(08), 149–156 (2019)
2. Jun, P., Jian, W., Zhong, M.: Effects of reducing atmospheric pollution by replacing centralized coal-fired heating with natural gas in China's cities. Chin. Environ. Sci. **35**(01), 55–61 (2015)
3. Xin, M., Kai, Z., Xi, C.: Development countermeasures of building intelligent power grid. Electr. Pow. Constr. **30**(6), 6–10 (2009)
4. Gellings, C.W., Lordan, R.J.: The power delivery system of the future. IEEE Pow. Eng. Rev. **17**(1), 70–80 (2002)
5. Li, Y.: Research on Power Structure Optimization in China Based on Sustainable Development. Harbin engineering university (2010)
6. Chao, M., Aoyang, H., Litao, Y.: Optimization model of combined heat and power generation household energy hub scheduling. J. Qingdao Univ. (Eng. Technol. Edn.) **33**(04), 51–55 + 68 (2018)
7. Lirong, D., Hongbin, S., Runze, C.: Research on node energy price of combined heat and power supply system for energy Internet. Pow. Grid Technol. **40**(11), 3375–3382 (2016)
8. Lian, J., Yajun, Z., Shangheng, Y.: Micro-combined heat and power generation for household use. Mod. Chem. Ind. **32**(05), 6–9 + 15 (2012)
9. Fadlullah, Z.M., Quan, D.M., Kato, N.: GTES: an optimized game-theoretic demand-side management scheme for smart grid. IEEE Syst. J. **8**(2), 588–597 (2014)
10. Qiuye, S., Fei, T., Huaguang, Z.: Construction of dynamic coordinated optimal control system for energy Internet. Chin. J. Electr. Eng. **35**(14), 3667–3677 (2015)
11. Geidl, M., Koeppel, G., Favreperrod, P.: Energy hubs for the future. Pow. Energy Mag. IEEE **5**(1), 24–30 (2007)
12. Huayi, Z., Fuzhuo, W., Can, Z.: Operation optimization model of household energy microgrid with comfort factor. Pow. Syst. Autom. **40**(20), 32–39 (2016)
13. Fang, L., Xiu, Y., Haitao, H.: Comprehensive optimization of microgrid energy under the operation mode of thermo electric decoupling including cogeneration. J. Pow. Syst. Autom. **28**(01), 51–57 (2016)

Filter-Based Fault Diagnosis of Heat Exchangers

Jianhua Zhang[1], Hongrui Li[2], Penghao Fan[2(⊠)], Guolian Hou[2], and Mifeng Ren[3]

[1] State Key Laboratory of Alternate Electrical Power System with Renewable Energy Sources,
North China Electric Power University, Beijing 102206, China
[2] School of Control and Computer Engineering, North China Electric Power University,
Beijing 102206, China
1216597720@qq.com
[3] College of Information Engineering,
Taiyuan University of Technology, Taiyuan 030024, China

Abstract. In this paper, a filter is proposed to diagnose fouling and sensor faults of heat exchangers. First, a typical heat exchanger system model is established, a filter based on the survival information potential is then designed for non-Gaussian stochastic systems. Consequently, the state estimation generated by the filter is used for fault detection, and the support vector machine is then employed to identify fouling and fault location of heat exchangers. The simulation results show that the fault diagnosis of heat exchangers can be effectively.

Keywords: Heat exchanger · Fault diagnosis · Survival information potential · Sensor fault · Support vector machine

1 Introduction

Heat exchanger is the main equipment in the heat transfer process. It is widely used in electric power production, petrochemical production, aerospace, mechanical processing, and other industrial industries. Therefore, safe and economic operation of heat exchangers is of great significance for industrial processes and production.

Some impurities of the working fluid will gradually be adsorbed by heat transfer during operation, hence, the inside surface of the device is gradually thickened to form foul [1]. Fouling will have a series of negative effects on the heat exchanger. More recent attention has focused on fouling of heat exchangers. In [2], the sound of the heat exchanger was collected and analyzed in order to detect the internal fouling; A small perturbation excitation of a fixed frequency was applied to the inlet temperature of the heat exchanger, the fouling characteristics inside the heat exchanger was then revealed in [3]; A physical model was established to investigate fouling of heat exchangers in [4]. However, the above methods for diagnosing the fouling fault of the heat exchanger depend on the information collected from sensors of heat exchangers. Once the sensor fails, it will lead to misjudgment of the state of heat exchangers.

The main aim of this study is to investigate filter based fault diagnosis method for heat exchangers. Due to the noises disturbed heat exchangers are not necessarily Gaussian, an extended minimized error entropy criterion will be used to design filters

© Springer Nature Singapore Pte Ltd. 2020
M. Fei et al. (Eds.): LSMS 2020/ICSEE 2020 Workshops, CCIS 1303, pp. 87–96, 2020.
https://doi.org/10.1007/978-981-33-6378-6_7

for heat exchangers. Following the improvements on generalized entropy [5], the survival Information Potential (SIP) is used to design filter to estimate states for heat exchangers instead of entropy. Faults occurred in heat exchanger can then be detected with the aid of the proposed filter. Afterwards, support vector machine (SVM) neural network is used to diagnose sensor faults and fouling.

The rest of the paper is organized as follows: In Sect. 2, a heat exchanger model with non-Gaussian noise and fouling faults is established. Section 3 utilizes a SIP-based filter to estimate state and detection faults. In Sect. 4, SVM is used to classify and locate the faults. The simulation results are given in Sect. 5, in which the effectiveness of the proposed fault diagnosis method is verified. The last section summarizes the full text.

2   Modeling of Heat Exchanger

Heat exchanger is the main equipment of heat transfer systems, which mainly realizes the function of exchanging heat between two or many kinds of working substances, it is widely existing in the chemical and electric power production industry, and tubular heat exchanger in the field of heat exchanger occupies a large proportion, applied in many occasions. The structure of the tubular heat exchanger is shown in Fig. 1, the heated fluid from the right direction to the left side through the outer tube, the cold fluid in the inner tube is left to the right.

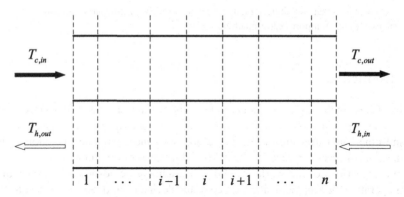

Fig. 1. The model of the heat exchanger

Using the lumped parameter method, the evaporator was divided into n segments, then i and $i + 1$ section are studied. In this model, the heat dissipation of the external tube wall is not considered and the inner tube wall is assumed to haven't heat storage capacity, while the working fluid does not undergo a phase change during operation. Thus, according to the conservation of mass and momentum of laws, the differential equations of the heat exchangers in sections i and $i + 1$ can be written as follows:

$$M_h c_h \frac{dT_{h,i}}{dt} = \dot{m}_h c_h (T_{h,i+1} - T_{h,i}) + A_h U_h \Delta T \tag{1}$$

$$M_c c_c \frac{dT_{c,i+1}}{dt} = \dot{m}_c c_c (T_{c,i} - T_{c,i+1}) + A_c U_c \Delta T \tag{2}$$

where $= \frac{(T_{h,i}+T_{h,i+1})-(T_{c,i}+T_{c,i+1})}{2}$, M and $\dot{m}$ stand for mass of fluid in one section and mass flow rate respectively. c and U represent specific heat and heat transfer coefficient respectively. T and A are temperature and surface area of heat transfer. The subscripts, c and h denote cold side and hot side respectively.

In this study, a simplified heat exchanger model is used, i.e. let $n = 2$, as shown in Fig. 2.

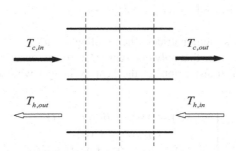

Fig. 2. The simplified model of the heat exchanger

The state space equation for the heat exchanger can be described as follows:

$$\begin{cases} \dot{x}(t) = Ax(t) + Bu(t) + v(t) \\ y(t) = Cx(t) + \omega(t) \end{cases} \tag{3}$$

where $x(t) = \begin{bmatrix} x_1 & x_2 & x_3 & x_4 \end{bmatrix}^T = \begin{bmatrix} T_{h,in} & T_{h,out} & T_{c,in} & T_{c,out} \end{bmatrix}^T$ and $u(t) = \begin{bmatrix} T_{h,in} & T_{c,in} \end{bmatrix}^T$ are the state vector and input vector, severally. $y(t) = \begin{bmatrix} T_{h,in} & T_{h,out} & T_{c,in} \end{bmatrix}^T$ is the system output vector of the heat exchanger. $v(t)$ and $\omega(t)$ are system noise and measurement noise respectively. After necessary assumption and simplification, the physical model of heat exchanger can be built as follows [6]:

$$A = \begin{bmatrix} \frac{-(1+\frac{\alpha}{2})}{\tau_h} & 0 & \frac{\alpha}{2\tau_h} & \frac{\alpha}{2\tau_h} \\ \frac{1-\frac{\alpha}{2}}{\tau_h} & \frac{-(1+\frac{\alpha}{2})}{\tau_h} & \frac{\alpha}{2\tau_h} & 0 \\ \frac{\beta}{2\tau_c} & \frac{\beta}{2\tau_c} & \frac{-(1+\frac{\beta}{2})}{\tau_c} & 0 \\ \frac{\beta}{2\tau_c} & 0 & \frac{1-\frac{\beta}{2}}{\tau_c} & \frac{-(1+\frac{\beta}{2})}{\tau_c} \end{bmatrix}, \tag{4}$$

$$B = \begin{bmatrix} \frac{1-\frac{\alpha}{2}}{\tau_h} & 0 \\ 0 & \frac{\alpha}{2\tau_h} \\ 0 & \frac{1-\frac{\beta}{2}}{\tau_c} \\ \frac{\beta}{2\tau_c} & 0 \end{bmatrix}, \quad C = \begin{bmatrix} 1 & 0 & 0 & 0 \\ 0 & 1 & 0 & 0 \\ 0 & 0 & 1 & 0 \end{bmatrix}, \tag{5}$$

where

$$\alpha(t) = \frac{A_h U_h(t)}{\dot{m}_h(t)c_h}; \quad \beta(t) = \frac{A_c U_c(t)}{\dot{m}_c(t)c_c}; \tag{6}$$

$$\tau_h(t) = \frac{M_h}{\dot{m}_h(t)}; \quad \tau_c(t) = \frac{M_c}{\dot{m}_c(t)}; \tag{7}$$

When fouling occurs, the model parameters of the heat exchanger in (3) will deviate from healthy heat exchanger without fouling, its model can be expressed by

$$\begin{cases} \dot{x}(t) = \tilde{A}x(t) + \tilde{B}u(t) + v(t) \\ y(t) = \tilde{C}x(t) + \omega(t) \end{cases} \tag{8}$$

where $\tilde{A}$, $\tilde{B}$, and $\tilde{C}$ are system matrices when fouling occurs.

In this paper, we assume that only one sensor fails once sensor failure occurs. When sensor failures of heat transfer appear, the heat exchanger with sensor faults can be formulated as follows:

$$\begin{cases} \dot{x}(t) = Ax(t) + Bu(t) + v(t) \\ y_{ideal}(t) = Cx(t) \\ y_{real}(t) = y_{ideal}(t) + f_s y_F(t) + \omega(t) \end{cases} \tag{9}$$

where $y_{ideal}(t)$ is the output vector of a healthy heat transfer, under the circumstance, all sensors are normal, moreover, all sensors are not disturbed by measurement noises. For actual sensors of the heat exchanger, they are usually disturbed by measurement noises, even these noises are not necessarily non-Gaussian. When sensor fault occurs, the output $y_{real}(t)$ should be supplemented by $f_s y_F(t)$. $y_F(t)$ is a scalar function that represents a sensor fault signal. f_s stands for a sensor fault event vector which corresponds to the ith sensor fault. In this work, let $f_s = \begin{cases} [1\ 0\ 0]^T, i = 1 \\ [0\ 1\ 0]^T, i = 2 \\ [0\ 0\ 1]^T, i = 3 \end{cases}$.

3   Design of SIP-Based Filter

Since there are stochastic disturbances in heat transfers, moreover, the disturbances are not necessarily Gaussian, a filter to estimate the state of a heat exchanger system is proposed with an improved criterion.

The heat exchanger can also be formulated by following discrete state-space model:

$$\begin{cases} x_{k+1} = Gx_k + Hu_k + v_k \\ y_k = Fx_k + \omega_k \end{cases} \tag{10}$$

where G, H, and F are discrete matrices of A, B, and C, respectively. And the proposed filter is designed as follows:

$$\begin{cases} \hat{x}_{k+1} = G\hat{x}_k + Hu_k + L_{k+1}e_k \\ \hat{y}_k = F\hat{x}_k \end{cases} \tag{11}$$

where $e_k = y_k - \hat{y}_k$ is the output estimation error, in this work, $e_k = \begin{bmatrix} e_k^1 & e_k^2 & e_k^3 \end{bmatrix}^T$.

In order to achieve fault diagnosis goal on the basis of the filter, the gain matrix L of the designed filter needs to be taken as an appropriate value, so that the magnitude and randomness of the output estimation error e_k are minimized. Due to the uncertainty and random nature of the heat exchanger system, the following survival information potential (SIP) [7] of the estimation error are used to obtain the gain matrix L

$$S_\alpha(e_k) = \int_{\mathbb{R}^m_+} \bar{F}^\alpha_{|e_k|}(\xi) d\xi \tag{12}$$

where $\bar{F}^\alpha_{|e_k|}$ is the survival function of the output estimation error e_k, α is the order of the survival function and $\alpha > 0$. $\mathbb{R}^m_+$ is the set of output estimation errors, m is the dimension of e_k, in this work, let m $= 3$.

In this section, the 'sliding window' is used to calculate the SIP of the output estimation error of the nearest P samples $\vec{e}_{(k-P):(k-1)}$, where $\vec{e}_{a:b} = \{e_a, e_{a+1}, \ldots, e_b\}$ is the sampling sequence of the sliding window, and the window width $P = b - a + 1$. Therefore, the SIP of the output estimation error at time k is:

$$S_\alpha\left(e_k | \vec{e}_{(k-P):(k-1)}\right) = \int_{\mathbb{R}^m_+} \left(\frac{1}{P} \sum_{i=k-P}^{k-1} \min(e_k, e_i)\right)^\alpha d\xi$$

$$= \frac{1}{P^\alpha} \sum_{i=k-P}^{k-1} \left(\prod_{j=1}^m \min\left(\left|e_k^j\right|, \left|e_i^j\right|\right)\right) \tag{13}$$

Using the SIP constructed by Eq. (13) as the performance index (i.e. $\bar{J} = S_\alpha(e_k)$), in the filter design process, our task is to find a suitable gain matrix to minimize the performance index, which can be expressed as $L_k^* = \arg \min J_k$.

There are many classic methods to solve this optimization problem, such as Newton's method, gradient descent method, swarm intelligence optimization algorithm and so on. In this paper, the heat transfer search (HTS) algorithm is used to obtain the optimal gain matrix [8].

4 Detection and Location of the Faults

In this section, the SIP-based filter is used to detect and locate fouling and sensor faults of the heat exchanger. Figure 3 shows the diagram of the proposed fault diagnosis method.

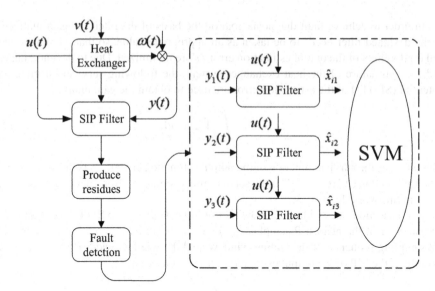

Fig. 3. Heat exchanger fault diagnosis diagram

4.1 Fault Detection

In Fig. 3, the residual generated by the filter is used to detect fouling and sensor faults. When fouling or sensor faults occur in the heat exchanger, fouling and sensor faults can be detected based on following decision rule:

$$E[e_y] \begin{cases} \leq \theta & \textit{if no fault occors} \\ > \theta & \textit{if fault has occered} \end{cases} \tag{14}$$

where $E[.]$ is mathematical expectation and θ the threshold.

4.2 Fault Location

In this work, SVM is used to identify fouling and sensor faults. SVM uses standard fault sets for training, which can diagnose the location and type of fault. In the heat exchanger system, three SIP-based filters are used to isolate fouling and sensor faults, four state variables are used as input to drive the state estimation of each SIP-based filter. The estimated state variables of each filter are extracted into the SVM for fault diagnosis.

At each sampling instant, the fault diagnosis method performs the following steps:

Step 1: Collect input u and output y_{real} from the heat exchanger system.
Step 2: The state estimate vector $\hat{x}$ is obtained from the SIP-based filter.
Step 3: Calculates residuals (11) and determines whether a fault occurs according to the fault detection law (14).
Step 4: If a failure occurs, a set of SIP-based filters is utilized for state estimation.
Step 5: fouling and sensor faults can be identified by SVM.

5 Simulation Results

In this section, the reliability and accuracy of the proposed fault diagnosis method are verified by simulation. It is generally considered that the sensor fault occurs at the exit of the hot fluid, the entrance of the hot fluid, or the entrance of the cold fluid, and fouling fault occurs inside the heat exchanger.

5.1 Healthy Heat Exchanger System

It can be seen that the designed SIP-based filter can achieve excellent performance when the system without fault, as shown in Fig. 4.

Figure 4(a) is an iterative curve for SIP-based filter performance index, it can be seen that with the increase of time, the performance index reaches the minimum value, and the convergence rate is very fast. Figure 4(b) and Fig. 4(c) are estimated errors of state and output respectively, and they tend to zero over time, which means that the designed SIP-based filter has good performance, and can accurately estimate the state variables and output variables of the system when there is no fault generation; in Fig. 4(d), it can be seen that the gain matrix of the filter is stable over time, and the optimal gain of the filter is obtained. Therefore, using the SIP-based filter has good adaptability to the heat exchanger system, it can effectively reduce the uncertainty of the output.

5.2 Unhealthy Heat Exchanger System

In consideration of the failure of the above-mentioned heat exchanger system, all failure types are listed in Table 1. Since there are ten types of faults, nine eigenvalues are used as input to classify the faults, that is, for each SIP-based filter, three estimates of the first state variable are taken out, namely:

$$X = [\hat{x}_{11,k-30}, \hat{x}_{11,k-20}, \hat{x}_{11,k-10}, \hat{x}_{12,k-30}, \hat{x}_{12,k-20}, \hat{x}_{12,k-10}, \hat{x}_{13,k-30}, \hat{x}_{13,k-20}, \hat{x}_{13,k-10}] \tag{15}$$

where $\hat{x}_{11}$, $\hat{x}_{12}$, $\hat{x}_{13}$ represent the estimated value of each filter for the first state, respectively.

Select 20 sets of data as a training sample for each type of failure, then normalized, input into the SVM for training, Fig. 5 shows the training results.

It can be seen from Fig. 5 that the accuracy of training is 100%, the false alarms rate is 0.00%, and the missed alarms rate is 0.00%. The fault type of training sample is identical to the fault type of SVM diagnosis. This shows that the SVM algorithm can be used to identify the fault type effectively, and a reasonable segmentation surface is established to separate the data with different characteristics.

SVM algorithm is then used for identify fouling and sensor faults. The test results are shown in Fig. 6. the accuracy of the test has reached 99.9%, the false alarms rate is 0.10%, and the missed alarms rate is 0.00%. Specifically, the fault of sensor 3 has one data diagnosed as Sensor 3's constant bias fault, the remaining fault types can be diagnosed accurately. This may be caused by the system adding random disturbances or data normalization errors. In this case, the accuracy of classification can be improved by increasing the training samples or the dimension of the feature vector.

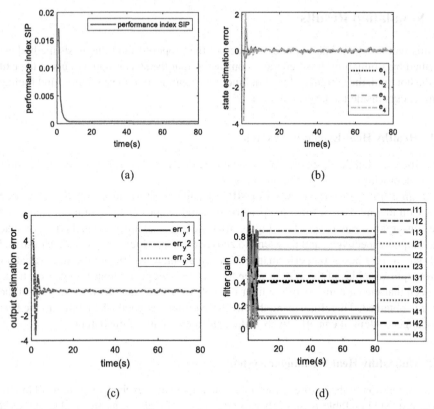

Fig. 4. (a) Performance index, (b) state estimation error, (c) output estimation errors, (d) filter gain.

Table 1. Fault-symptom table for the heat exchanger

Fault description	#
Constant bias fault occurred in the sensor 1	1
Constant bias fault occurred in the sensor 2	2
Constant bias fault occurred in the sensor 3	3
Constant gain fault occurred in the sensor 1	4
Constant gain fault occurred in the sensor 2	5
Constant gain fault occurred in the sensor 3	6
Fouling fault occurred in the sensor 1	7
Fouling fault occurred in the sensor 2	8
Fouling fault occurred in the sensor 3	9
Fouling fault occurred in the heat exchanger	10

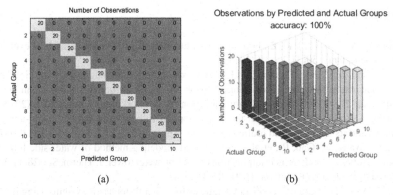

(a)   (b)

Fig. 5. (a) Number of observations, (b) Observations by predicted and actual group.

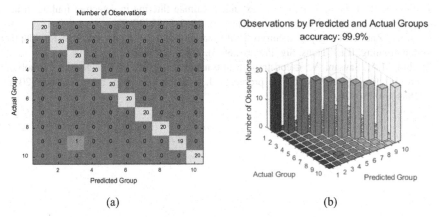

(a)   (b)

Fig. 6. (a) Number of observations, (b) Observations by predicted and actual group.

6 Conclusion

In this paper, the HTS algorithm is used to minimize the SIP to obtain the filter gain matrix, and a filter based on the SIP is designed. Based on this filter, a fault diagnosis method for heat exchangers is proposed. The simulation results show that this method can effectively achieve the fault diagnosis of the heat exchanger.

The fault diagnosis method designed in this paper can be used to identify fouling and sensor faults effectively. However, a limitation of this study is that the proposed method can not necessarily to carry out fault diagnosis when multiple sensors fail at the same time, both the false alarms rate and the missed alarms rate will increase. Therefore, in future work, the diagnosis method of multi-sensor simultaneous faults should be further studied.

Acknowledgements. This work was supported by China National Science Foundation under Grant (61973116) and State Key Laboratory of Alternate Electrical Power System with Renewable Energy Sources (LAPS2019-0415). These are gratefully acknowledged.

References

1. Huanliang, Y., Rui, C., Yao, Z.: Fouling corrosion analysis of hydrocracking heat exchanger. J. Petrochem. Univ. **30**(3), 15–19 (2017)
2. Zhu, X., Shu, L., Zhang, H., et al.: Preliminary exploration: fault diagnosis of the circulating-water heat exchangers based on sound sensor and non-destructive testing technique. In: International East Conference on Communications & Networking in China. IEEE (2013)
3. Lalot, S., Desmet, B.: The lock-in technique applied to heat exchangers: a semi-analytical approach and its application to fouling detection. Appl. Therm. Eng. **114**, 154–162 (2017)
4. Shen, C., Wang, Y., Gao, R., et al.: An improved modeling method of water-side fouling in enhanced tubes of condensers in application of cooling water tower. J. Therm. Sci. **28**(1), 30–39 (2019). https://doi.org/10.1007/s11630-018-1021-4
5. Guo, L., Wang, H.: Fault detection and diagnosis for general stochastic systems using B-spline expansions and nonlinear filters. IEEE Trans. Circ. Syst. I: Regul. Pap. **52**(8), 1644–1652 (2005)
6. Jonsson, G.R., Palsson, O.P.: Use of extended Kalman filtering in detecting fouling in heat exchangers. Int. J. Heat Mass Transf. **50**(13–14), 2643–2655 (2007)
7. Chen, B., Zhu, P., Principe, J.: Survival information potential: a new criterion for adaptive system training. IEEE Trans. Sig. Process. **60**(3), 1184–1194 (2012)
8. Tawhid, M.A., Savsani, V.: ϵ-constraint heat transfer search (ϵ -HTS) algorithm for solving multi-objective engineering design problems. J. Comput. Des. Eng. **5**(1), 104–119 (2017). S228843001730026X

Application of Logistic Regression in Identifying Power Empty Nested Users

Zhangchi Ying[1], Chengye Shu[1], Yue Xie[2], Shuo Jiang[1], Min Yu[3], Qiuyun Mao[2(✉)], Jing Li[2(✉)], and Qian Guo[2]

[1] Jinhua Power Supply Company State Grid Zhejiang Power Co., Ltd., Jinhua 321100, China
1105130383@qq.com, 408749634@qq.com, 1189524@qq.com
[2] College of Mechanical and Electrical Engineering, China Jiliang University, Hangzhou 310018, China
{xieyue,16a0102140}@cjlu.edu.cn, 1456419666@qq.com, ljagu@163.com
[3] Zhejiang Huayun Information Technology Co., Ltd., Hangzhou 310000, China
graceyu1231@hotmail.com

Abstract. Based on power users' electricity information, this paper mainly researches on the application effect of using logistic regression classification method to identify empty-nested users. In the paper, based on daily energy consumption of more than 3,000 electric power users in one year, a suitable classification feature quantity is selected and preprocessed. Logistic regression classification method is used for the second classification and the control method is Support Vector Machine method. Classification targets are empty-nested users and non-empty-nested users. Classification models are built through training and further test, besides, the results of multiple classifications are analyzed subsequently. The analysis results show that the Logistic Regression Classification method has a good application effect in power empty nest user identification. This research applies big data analysis of power to the identification of empty nest users, which provides reference for power enterprises to accurately identify empty nest users based on power information, saves human resources, and facilitates precise supporting work.

Keywords: Logistic regression · Support vector machine · Power user identification · Big data analysis

1 Introduction

Electric power enterprises have a large amount of user power data, including users' annual power consumption, daily power consumption, payment methods and the like. These data imply user's electricity characteristics information, which can reflect user's type and their consuming habits [1–3]. However, due to a large number of power users, the data are often very complex and diverse. Hence it is difficult to observe the user characteristics. Processing these data to extract valuable information has become a major research trend. The large data classification method can be used to analyze the power users' power consumption data. Users are divided into several categories according

© Springer Nature Singapore Pte Ltd. 2020
M. Fei et al. (Eds.): LSMS 2020/ICSEE 2020 Workshops, CCIS 1303, pp. 97–107, 2020.
https://doi.org/10.1007/978-981-33-6378-6_8

to their different usage habits, in this way, different user power classification models are obtained [4–6]. Reference [7] applies the Logistic Regression and Support Vector Machine classification methods to the sensitivity evaluation of an index and it obtains good classification results.

With the growth of life expectancy and the decrease of population fertility, the degree of aging in China is increasing nowadays [8]. According to the Statistical Bulletin of Social Services Development in 2019, as of the end of 2019, 2538.8 million people aged 60 years and above in the country, accounting for 18.1% of the total population, of which 17.6 million people aged 65 years and above, accounting for 12.6% of the total population, this data has increased since 2018. In current Chinese society, family size tends to be small and family structure is simple. The population structure of "four, two, one" families is widespread, which leads to an increasing number of the elderly living alone and in empty nests [9, 10]. At the same time, the government and society lack effective technical means to identify empty nesters. Reference [11] carries out a research of empty nester recognition through mobile communication data. Empty nester users are identified by neural network algorithm using mobile phone user's age information, call and SMS lists, and the amount of online meals. Nevertheless, such methods are limited to a large extent because of the potential to leak the privacy of users.

The research content of this paper is to classify power users based on their power information using logistic regression method, and to analyze and compare the classification results. This analysis can help power grid companies understand the power characteristics of empty nesters by locating empty nesters through classification. It has certain social value since it provide a research basis for accurate supporting work and improves work efficiency.

2   Introduction to Logical Regression

Logistic regression (LR) is a classical classification method and is often used for binary classification problems [12, 14]. When classifying data, the expected classification result is usually a fixed value, such as 0 or 1. The steps to classify data using logistic regression are as the followings.

2.1   Building Prediction Functions

When a prediction function is applied to a classification, the result can only be two values, either one or the other. This paper uses the Logistic function to construct the prediction function, which is in the form of the following.

$$g(z) = \frac{1}{1 + e^{-z}}. \tag{1}$$

When classifying data, it needs to establish a boundary to divide it.

$$z = \theta^T x = \theta_0 x_0 + \theta_1 x_1 + \cdots \theta_n x_n = \sum_{i=0}^{n} \theta_i x_i. \tag{2}$$

The prediction function is as follows.

$$h_\theta(x) = g\left(\theta^T x\right) = \frac{1}{1 + e^{-\theta^T x}}. \tag{3}$$

Here $h_\theta(x)$ is the probability that the data will be judged as 1, then the probability that it will be judged as 0 is $1 - h_\theta(x)$.

2.2 Build Loss Function

The size of loss function reflects the classification effect. When constructing the loss function, we need to use the characteristic quantity provided by multiple samples for parameter learning and modify parameter to minimize the loss function through constant iteration.

Take likelihood function.

$$l(\theta) = \prod_{i=1}^{m} (h_\theta(x_i))^{y_i} (1 - h_\theta(x_i))^{1-y_i}. \tag{4}$$

Loss function obtained from log likelihood function.

$$J(\theta) = -\frac{1}{m} \sum_{i=1}^{m} \cos t(h_\theta(x_i), y_i)$$

$$= -\frac{1}{m} \left[\sum_{i=1}^{m} (y_i \log h_\theta(x_i) + (1 - y_i) \log(1 - h_\theta(x_i))) \right]. \tag{5}$$

2.3 Regularization

In order to solve the problem of over fitting, we can regularize it by adding penalty terms to the cost function. The process of the regularized θj iterating through the gradient descent method is shown in (6), and finally θj converges to a set of the most appropriate values to minimize the value of $J(\theta)$.

$$\theta_j = \theta_j - \frac{\alpha}{m} \sum_{i=1}^{m} (h_\theta(x_i) - y_i) x_i^j - \frac{\lambda}{m} \theta_j. \tag{6}$$

In addition, Newton method can be used for parameter iteration. When Newton method is utilized to solve the optimal parameters, it is solved iteratively according to the Taylor expansion. The zero point of derivative function of likelihood function is found by Newton method, that is, the objective extremum. The iteration speed of this method is faster and the efficiency is higher. The iterative process is as follows.

$$\theta = \theta - \alpha(H(\theta)^{-1} \frac{\partial J(\theta)}{\partial(\theta)} + \lambda\theta). \tag{7}$$

3   Data Source and Processing

The data source of this paper is the daily power consumption of 3591 household users in a certain area of Zhejiang Province in 360 days of a year. That is to say, as a sample, each user has 3591 samples in total and each sample contains 360 power consumption values. The goal of classification is to identify the users in the area with and without empty nests. The sample data types are shown in Table 1.

Table 1. Number of user type.

User type	Empty nest users	Non empty nest users
Sample size	961	2630

In the process of empty nest user identification, it is necessary to extract the effective features of the above user power consumption information. Then the extracted features are processed and adopted for user recognition.

Four power users are randomly selected from empty nest users and non-empty nest users. Draw the annual power consumption curve of four users as shown in Fig. 1.

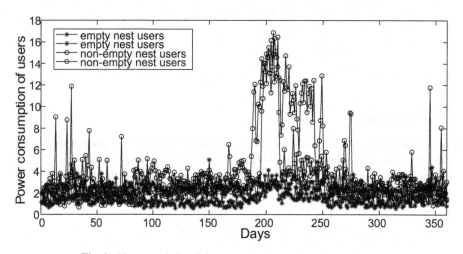

Fig. 1. The annual electricity consumption curve of power users.

It can be observed from Fig. 1 that the annual power consumption of non-empty nest users is higher than that of empty nest users. Because normally there is only elderly people at empty nest users' home, the average annual power consumption is usually lower than that of non-empty nest users. At the same time, empty nesters usually use household appliances less frequently. Therefore, the fluctuation range of this kind of users' electricity consumption will be smaller than that of non-empty nest users. Similarly, the corresponding annual extreme difference of electricity consumption will be smaller. In addition, the fluctuation of power consumption of empty nest users in daily power consumption activities will not be too large, which is more stable than that of non-empty nest users. Accordingly, the variance of annual power consumption will be smaller than that of non-empty nest users. Through many experiments, these three items are the characteristic quantities that can best reflect the characteristics of the data itself. Therefore, this paper uses the average value of annual power consumption, the extreme difference of annual power consumption and the variance of annual power consumption as the characteristic quantity of the classification algorithm.

The data processing of the above three features is slightly different. Before taking the average value of annual power consumption and the extreme difference of annual power consumption, the data needs to be differential processed. A user's electricity consumption will fluctuate in a year. Especially in winter and summer, there will be a kind of peak power consumption, and there will be a significant rise and fall before and after the waveform, with a large overall change. After the difference processing of data, it can reflect the change of data details more obviously. When the variance is taken, the annual power consumption needs to be filtered first. The filtering period is 15 days. The filtered data reduces the influence of noise and more obviously reflects the overall trend of data change.

Finally, in order to improve the data processing speed and classification accuracy, the three extracted features are normalized.

4   Result and Analysis of Empty Nested User Recognition

This paper uses logistic regression to identify and classify users based on the extracted electricity eigenvalues.

4.1   Sample Selection and Model Training

Logistic regression classification is considered as supervised learning, which determines the need to train a given sample to obtain a classification model. Therefore, 1700 user data from the samples are randomly selected as training sets for each classification. The proportion of empty-nested and non-empty-nested users is about 1:3, which is the same as the proportion of the total sample. Moreover, to avoid the impact of special circumstances, a total of 15 classifications were made and the result data were recorded. The data is divided into two non-overlapping parts involving the training set and the test set. In each classification, the training and test adopted by these two sets are exactly the same.

Newton's method is the iteration method used in the logistic regression classification. This method has fewer iterations than the gradient descent method and can improve the efficiency of finding the optimal parameters. The threshold value for classification decision is 0.5.

The recall rate and accuracy rate were used as the evaluation criteria for classification results. The classification results are defined as the followings.

Number: Order in 15 Classifications.

Actual users: The actual number of empty nested users in the test set for each classification.

Classified users: Number of empty nested users determined by classification.

Correct classification: Number of actual empty nested users in the number of classified users.

Recall Rate: Correct classification/Actual Users.

Accuracy: Correct classification/Classified users.

4.2 Result Analysis

Table 2 is the result of 15 user classifications using logistic regression.

Table 2. Logistic regression classification results.

Number	Actual users	Classified users	Correct classification	Recall rate	Accuracy
1	489	199	121	24.74%	60.80%
2	501	191	116	23.15%	60.73%
3	512	249	145	28.32%	58.23%
4	511	146	93	18.20%	63.70%
5	514	214	134	26.07%	62.62%
6	488	293	167	34.22%	57.00%
7	493	231	139	28.19%	60.17%
8	488	312	174	35.66%	55.77%
9	524	253	159	30.34%	62.85%
10	483	186	116	24.02%	62.37%
11	501	245	139	27.74%	56.73%
12	475	219	158	33.26%	72.15%
13	500	200	124	24.80%	62.00%
14	517	283	168	32.50%	59.36%
15	517	199	125	24.18%	62.81%

As can be seen from Table 2, the accuracy of logistic regression in the classification of empty nested users remains approximately 60%, with fluctuations. The highest accuracy can reach 72.15%, while the lowest accuracy can reach 55.77%. Recall rates range from 20% to 30%, possibly due to sample imbalance. That is, the ratio of empty-nested users to non-empty-nested users is about 1:3, which results in inadequate sample training for logistic regression model and the recall rate of classification results is low.

Support Vector Machine (SVM) is a binary classifier. Its essence is to find a hyperplane in the sample and separate the samples [15, 16]. In order to comprehend the application effect of logical regression in empty nest users' classification more comprehensively, Support Vector Machine is utilized as the contrast classification method and the two effects are compared and analyzed.

The SVM method was used to classify users 25 times. Fifteen of them yielded classification results. The other 10 times were overcategorized, meaning that all users were classified as non-empty nest users. The results of the 15 classifications are summarized in Table 3.

Table 3. Support vector machine classification results.

Number	Actual users	Classified users	Correct classification	Recall rate	Accuracy
1	489	207	132	26.99%	63.77%
2	501	244	145	28.94%	59.43%
3	512	225	135	26.37%	60.00%
4	511	252	164	32.09%	65.08%
5	514	199	132	25.68%	66.33%
6	488	258	153	31.35%	59.30%
7	493	226	146	29.61%	64.60%
8	488	259	159	32.58%	61.39%
9	524	226	145	27.67%	64.16%
10	483	187	123	25.47%	65.78%
11	501	239	144	28.74%	60.25%
12	475	220	134	28.21%	60.91%
13	500	281	174	34.80%	61.92%
14	517	231	135	26.11%	58.44%
15	517	265	168	32.50%	63.40%

As can be observed from Table 3, the accuracy of support vector machine for empty-nested user classification is maintained at approximately 60%. Recall rates are distributed between 20% and 30%.

Figure 2 is a comparison of the classification accuracy of Logistic Regression and Support Vector Machine.

In Fig. 2, the accuracy of empty nest user classification by logistic regression method is basically above or below 60%; and the trend is stable. As Table 2 indicates, we can see that when the number of classified users increases, the classification accuracy decreases accordingly. The classification accuracy of empty nest users by using Support Vector Machine method varies from 60% to 70%.

From the above analysis, it can be acknowledged that the classification accuracy of logistic regression is similar to that of support vector machine. However, at the same time, Support Vector Machine (SVM) produces a classification result that cannot distinguish empty-nested users, that is, all users are classified as non-empty-nested users. The probability of this result is about 40%. By contrast, the classification results of logistic regression are more stable. The reason for this result is related to the selection of training samples and the classification characteristics of Support Vector Machine (SVM). The

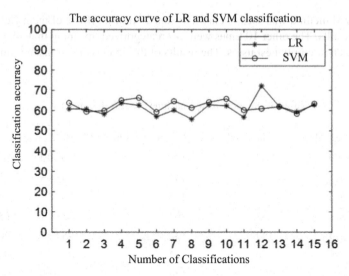

Fig. 2. The accuracy curve of logistic regression and support vector machine classification.

classification of Support Vector Machines relies on a few support vectors to delineate hyperplanes, which are subject to randomness each time the data is retrained and tested for disruption. Because the training samples are randomly selected for each classification, when the characteristic quantity of a training sample does not reflect the characteristics of this type of users, the appropriate classification parameters cannot be obtained, which results in poor classification results. Nonetheless, all sample points are considered in the logistic regression classification and the classification boundary is constructed based on the distribution of all samples, which is more comprehensive. Hence, the classification effect is better than SVM.

Figure 3 is a comparison of recall rates for logistic regression and Support Vector Machine classification.

Figure 3 shows that the recall rates for Logistic Regression Classification and Support Vector Machine Classification are similar and remain between 20% and 30%. Tables 2 and 3 imply that accuracy decreases when recall rates rise. The increase in the number of categorized users is the main reason for the increase in recall rates, which means that more non-empty-nested users are categorized as empty-nested users. A good classification model needs a suitable balance point between the two values. The classification model can be evaluated by drawing ROC curves.

The accuracy of the two classification models is evaluated by drawing ROC curves at any one time as shown in Fig. 4. At this time, the AUC value of logistic regression is 0.7658. The AUC value of the Support Vector Machine is 0.6944. It can be considered that the AUC value of logistic regression classification model is higher than that of Support Vector Machine classification model, that is, the model of logistic regression classification has higher accuracy. Meanwhile, according to the ROC curve evaluation criteria, it is known that the classification effect of Support Vector Machine is not ideal,

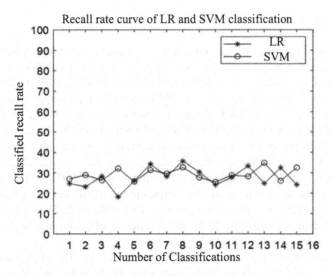

Fig. 3. Recall rate curve of logistic regression and support vector machine classification.

so the application of logistic regression method in power empty nest users' identification is better than that of Support Vector Machine method.

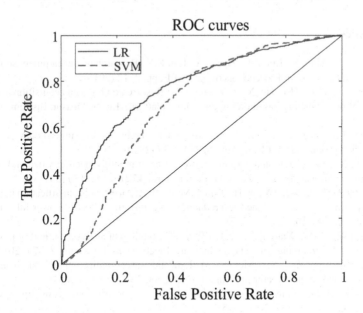

Fig. 4. ROC curves of logistic regression and support vector machines.

5 Conclusion

Based on the power user's electricity information, this paper researches on the application of logistic regression method in empty-nest users' identification and utilizes the Support Vector Machine method as a control. From the above data and analysis results, it can be concluded that although the accuracy of the two classification methods is similar, the Support Vector Machine method cannot identify empty nest users, which means it has over prediction. In the comparison of the two methods, the logic regression method is more stable in power empty nest user identification. At the same time, it can be seen from ROC curve that the accuracy of the logistic regression model established by Newton method is higher than that of SVM classification model. The accuracy of logistic regression classification is stable and the corresponding recall rate is low. It is necessary to ensure the stability of power empty nest user identification. Therefore, the logical regression method is more suitable for the practical application of power empty nest user identification.

In the meanwhile, there are still some issue in this research. The recognition accuracy of logistic regression has not reached a high level. In the next stage, it is possible to consider a combination of several classification methods to improve the classification. In addition, how to improve the unbalanced classification effect of power users can also be a further research direction.

Foundation. Basic public welfare research project of Zhejiang Province (LGG20E070003).

References

1. Zhang, D.X., Mao, X., Liu, L.P., Zhang, Y., Liu, K.Y.: Research on development strategy for smart grid big data. In: Proceedings of the CSEE, pp. 2–12 (2015)
2. Li, J.N., Li, W., Li, H.J., He, X.M., Zhang, S.L.: Research on big data acquisition and application of power energy based on big data cloud platform. Electr. Measur. Instrum. 104–109 (2019)
3. Xin, M.M., Zhang, Y.C., Xie, D.: Summary of researches on consumer behavior analysis based on big power data. Electr. Autom. 1–4 + 27 (2019)
4. Tang, J., et al.: Research on the accurate forecasting of monthly electricity demand based on the multi-feature analysis. Power Syst. Prot. Control **45**(16), 145–150 (2017)
5. Li, Q., Qi, L.H., Tian, L., Wang, H., Tian, S.M., Pu, F.P.: An efficient classification method for electricity users based on dimension reduction and clustering. Electric Power Inf. Commun. Technol. 75–79 (2019)
6. Ye, K., Leng, X.W., Xiao, F., Li, X.L., Zhu, L.C.: Intelligent analysis method of power grid monitoring based on big data label technology. Electr. Measur. Instrum. 75–79 (2019)
7. Tan, L., Chen, G., Wang, S.Y., Meng, X.M.: Landslide susceptibility mapping based on logistic regression and support vector machine. J. Eng. Geol. **22**(1), 56–63 (2014)
8. Peng, X.Z., Hu, Z.: Family changes and family policy reconstruction in contemporary China. Soc. Sci. China 113–132, 207 (2015)
9. Wang, H., Yang, Q.X.: Population changes and aging challenges within the seventy years of the founding of new China: a review of literature and policy. J. Macro-Qual. Res. 30–54 (2019)
10. Qiu, Z.J.: A study on the difference of family structure and old-age support among the old in urban and rural areas of China. Hubei Soc. Sci. 46–52 (2018)

11. Cui, H.W., Luan, S.Z.N., Li, Y.F., Cai, Z.J., Cao, Y.: Accurate identification approach to empty-nester mobile-phone users. Math. Model. Appl. 49–62 (2014)
12. Mao, Y., Chen, W.L., Guo, B.L., Chen, Y.X.: A novel logistic regression model based on density estimation. Acta Autom. Sin. **40**(1), 62–72 (2014)
13. Vaidya, A.: Predictive and probabilistic approach using logistic regression: application to prediction of loan approval. In: 2017 8th International Conference on Computing, Communication and Networking Technologies (ICCCNT), Delhi, pp. 1–6 (2017)
14. Li, Z.J.: A small business credit evaluation model based on hierarchical logistic regression. Stat. Decis. 178–182 (2016)
15. Wang, X.J., Bi, S., Xu, Y.K., Sun, Y.J.: Short-term load forecasting based on support vector machines and data mining technology. Electr. Measur. Instrum. 62–67 (2016)
16. Zhang, X.G.: Introduction to statistical learning theory and support vector machines. Acta Autom. Sin. 36–46 (2000)

Study on the Modeling and Online SOC Estimation of the Aluminum Air Battery

Hui Cai[1][⊠], Shuya Cheng[1][⊠], Yuhua Wang[1], Shuxiong Zhang[2], and Weiming Liu[2]

[1] College of Mechanical and Electrical Engineering, China Jiliang University, Hangzhou 310018, China
caihui@cjlu.edu.cn, 1820760593@qq.com, wangyh977@126.com
[2] Peking University - Taizhou Metal Fuel Cell Research Center, Taizhou 317700, China
zsx416@163.com, amo52074@126.com

Abstract. At present, compared with lithium ion batteries, the accuracy of detection, modeling and state of charge (SOC) estimation still need to be improved in aluminum air battery management system. First, the relation curve between open circuit voltage and SOC is calibrated by discharging experiments under different rates. Then, an equivalent circuit parameter identification method by recursive least square method with forgetting factor is presented to identify the model parameters. Finally, with real-time parameter identification of the second order RC battery equivalent circuit model of battery, the real-time variation of open circuit voltage (OCV) value is obtained. A piecewise SOC estimation method combining the online open circuit voltage method and the ampere-hour integral method is proposed, and the SOC estimation error is no bigger than 0.05. The research result shows that this method is not only feasible, but also can effectively judge whether the aluminum air battery becomes invalid or not.

Keywords: Aluminum air battery · Online identification · OCV-SOC curve

1 Introduction

Renewable energy conversion is an effective solution to significantly reduce dependence on fossil fuels, and environmentally friendly electrochemical batteries are an important basic technology for sustainable energy economy [1]. Metal air battery is a kind of electrochemical battery with very promising development prospects, especially aluminum air battery. Aluminum air battery has the strong points of high-power density, abundant raw materials, long life and low cost [2]. At present, the accuracy of detection, modeling and SOC estimation still need to be improved in aluminum air battery supervision system. Therefore, it is particularly important to establish the dynamic model of aluminum air battery management system and estimate the SOC [3]. The research of air battery is of great significance. If we take the time to do in-depth research, it is likely to help us get rid of our dependence on fuel resources. Air battery is an important research direction of energy.

The dynamic model of the battery mainly includes electrochemical model [4], neural network model [5] and so on. The electrochemical model is not suitable for battery

© Springer Nature Singapore Pte Ltd. 2020
M. Fei et al. (Eds.): LSMS 2020/ICSEE 2020 Workshops, CCIS 1303, pp. 108–122, 2020.
https://doi.org/10.1007/978-981-33-6378-6_9

monitoring systems. The neural network model requires much battery data, but aluminum air batteries are difficult to obtain. By contrast, the equivalent circuit model is established according to the relationship of circuit parameters, which can directly simulate the curve of battery external voltage with time elapse and well simulate the nonlinear dynamic characteristics of battery. This method has good effect, good practicability and high stability in the experiment.

The equivalent circuit model is used to analyze the aluminum air battery. According to the different circuit structure, the equivalent circuit model can be subdivided into Rint, Thevenin, PNGV and Massimo Ceraolo model [6]. Rint model is the simplest model, but it can't explain the internal polarization and self-discharge effect of battery well. Thevenin model adds RC parallel ring on the basis of Rint model. However, there is a large error in the dynamic simulation, and the self-discharge factor is not considered in the model. Compared with Thevenin model, PNGV model considers the influence of open circuit voltage and has higher precision. Massimo-ceraolo model is an improvement on Thevenin model, which fully considers the nonlinearity of the model of battery. By increasing the order of RC parallel circuit and considering the polarization reaction, ohm and current accumulation effect, the accuracy of massimo-ceraolo model was further improved. In the model, it is very important to determine the order of shunt capacitor and resistor. Repeated experiments are needed to determine the optimal order, which provides a good foundation for field measurement.

However, it is necessary to determine the appropriate model order to achieve the tradeoff between accuracy and complexity. Reference [7] analyzed the accuracy of the equivalent circuit model of batteries for RC parallel links of the first, second and third orders. The analysis shows that although the first-order RC parallel circuit can describe the nonlinear and dynamic characteristics of the battery, its accuracy is low and cannot meet the research requirement. The second-order RC circuit model is able to track the voltage change of the battery end with high accuracy, which not only requires less response speed and computational complexity, but also can process more quickly. Two more unknowns in the third order RC circuit than in the second order RC circuit will affect the practical implementation of the third order RC model. The second order equivalent model is more suitable for aluminum air cell modeling. By consulting the data, theoretical analysis and repeated experimental analysis, the order of capacitor resistance parallel connection is determined to be 2.

Battery state of charge (SOC) estimation is another important concern of battery management system. There are many state of charge estimation methods. Open circuit voltage method and Ampere-hour integral method are simple. Due to their low estimation accuracy, they are rarely used to estimate battery SOC alone. Although neural network and Kalman filter method have higher precision, they are seldom applied because of high computational complexity and hardware requirements. At present, the research on SOC estimation method in aluminum air battery monitoring system is still at the initial stage. However, due to the unique advantages of aluminum air battery, the research prospect is very good. It is worth further study.

Reference [8] discussed the relationship between SOC and PNGV model. The regress function was adopted for parameter identification of PNGV equivalent circuit models. HPPC (Hybrid Pulse Power Characteristic) test is conducted on aluminum aerodynamic battery under different SOC batteries, and the parameters of PNGV model under different SOC are obtained according to the recorded data. However, the PNGV model is similar to the RC first-order model. The error of model results is large. As in [9], the SOC prediction based on hidden semi-Markov model (HSMM) is adopted as a supplement to the Ampere-hour integral method, so that the accuracy of aluminum air battery in the later period can be guaranteed. However, large amount of data analysis works are needed to refine the decommissioning process to ensure the accuracy of SOC prediction. As in [10], experimental results indicate that, although ACKF can improve the accuracy of the model and reduce the error. This method has some limitations for practical applications. For the above models, there are certain limitations and very desirable advantages. It is necessary to extract its essence and remove its dross to further study it. In the future, air batteries will be applied stably and accurately in the field. For the above models, there are certain limitations and very desirable advantages. It is necessary to extract its essence and remove its dross to further study it. In the future, air batteries will be applied to practice stably and accurately.

Aiming at the problems described in the above references, this paper focuses on the study of aluminum air battery data acquisition, the establishment of equivalent model and SOC estimation. Firstly, a relatively accurate equivalent circuit model of RC air battery is proposed. Then the open circuit voltage of the battery can be obtained by real-time identification of the second-order RC equivalent circuit model. On this basis, the open-circuit voltage (OCV) method, Ampere-time integral method and the piecewise SOC estimation method combining the two methods are proposed and implemented.

2   Research Approach of Equivalent Circuit Modeling and Online SOC Estimation

The structure of an aluminum air cell is shown in Fig. 1(a). In an electrolytic cell, aluminum loses electrons and oxygen gains electrons and loops through the electrolyte. In this paper, ten groups of aluminum air battery composing a pack is the research object with 580 Ah rated capacity, as shown in Fig. 1(b).

The content of this paper consists of two parts, which are equivalent circuit modeling and SOC estimation method of aluminum air battery. Flowchart of research approach is shown in Fig. 2.

The research ideas of equivalent circuit modeling as follows:

Step 1. Discharge data of battery pack are collected (described in Sect. 3.1);
Step 2. The second-order RC Massimo-Ceraolo circuit model is adopted. Its mathematical formula is deduced (Sect. 3.2);
Step 3. The recursive least square method is applied to identify the parameters of the equivalent circuit model online (Sect. 3.3).

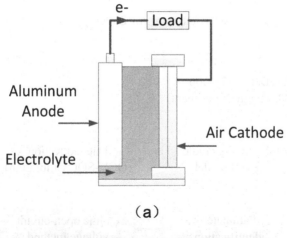

(a)

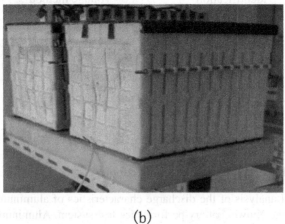

(b)

Fig. 1. (a) Structure of an aluminum air cell; (b) Aluminum air battery pack.

The research ideas of SOC estimation as follows:

Step 1. The curves of OCV and SOC are obtained through calibration experiments (Sect. 4.1);
Step 2. With the real time OCV value, an online open circuit voltage method is proposed to estimate SOC (Sect. 4.2);
Step 3. A piecewise SOC estimation method combining online open circuit voltage method and Ampere hour integral method is proposed (Sect. 4.3).

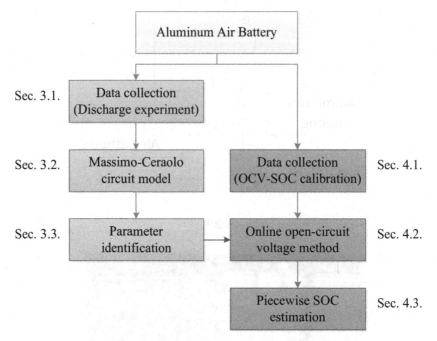

Fig. 2. The flowchart of research approach.

3 Equivalent Circuit Modeling

3.1 Discharge Experiment and Data Collection

The experimental analysis of the discharge characteristics of aluminum air battery is carried out with the Xinwei battery performance test system. Aluminum air battery is discharged under the condition of good ventilation. The experimental result is shown in Fig. 3, and the voltage curve obtained is consistent with the discharge results of aluminum-air batteries.

The bottom Figure of Fig. 3 shows the current variation curve of the battery pack under the condition of good ventilation. Firstly, the discharge current increases to 15 A at the rate of 1 A/Min, and then increases to 50 A at the rate of 5 A/Min. Finally, the discharge current remains at constant 50 A until the end of the discharge. The upper Figure of Fig. 3 shows the curve of output terminal voltage of aluminum air battery under simulated working condition during discharging. The initial terminal voltage is high, then stable at about 11 V, and finally the voltage drops to 2.9 V, and the battery pack stops working. The voltage plateau phase (the voltage varies from 11.8 V to 10.6 V, as shown with dotted line) lasts a long time, about 10 h. The sharp drop of voltage in the last stage (upcoming failure phase) is mainly caused by the consumption of aluminum plate, the accumulation of sediments and the drop of electrolyte concentration. Through discharge experiments, it can be verified that the aluminum air battery can discharge for a long time with higher specific energy. It can be seen from the figure that the stable process

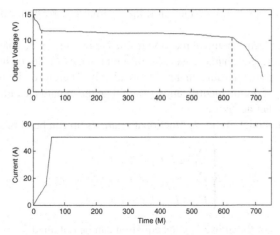

Fig. 3. Terminal voltage and discharge current

of discharge can reach 600 min, that is, 10 h. If the batteries are properly and effectively connected in series and parallel, a longer stable application time can be obtained.

3.2 Second-Order RC Massimo-Ceraolo Circuit Model

After comprehensive analysis of the accuracy and computational complexity of each model, considering data processing capacity of the battery energy management system, this paper adopts the second-order RC Massimo-Ceraolo circuit model, deduces its mathematical formula and conducts modeling and simulation.

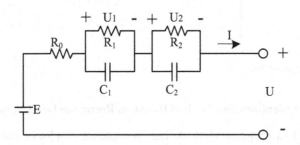

Fig. 4. Second-order RC Massimo-Ceraolo circuit model

Figure 4 shows second order RC model of aluminum air battery, E as the ideal voltage source (Open-circuit voltage is equal to E), resistance R_0 for ohm internal resistance, R_1 for polarization resistance, C_1 for polarization capacitance (R_1 and C_1 parallel simulation inside the battery for the creation and depletion of polarization), U as the terminal voltage. When the battery is in a stable state, the measured terminal voltage is similar to the open circuit voltage of the battery (Battery open-circuit voltage measurement requires the battery to be in nonworking state until the internal polarization effect completely disappears).

Due to violent electrochemical reaction inside battery during discharge process, the battery exhibits different characteristics at different stages of the battery polarization reaction, and its parameter value changes dramatically. Therefore, the proposed model adopts the method of online identification to simulate the dynamic characteristics of the air battery with higher accuracy.

Because of the model in Fig. 4, functional relationship can be obtained as follows:

$$\begin{cases} \dot{U}_1 = I/C_1 - U_1/C_1R_1 \\ \dot{U}_2 = I/C_2 - U_2/C_2R_2 \\ U = E - U_1 - U_2 - IR_0 \end{cases} \tag{1}$$

After discretization of the above (1), the equation can be obtained:

$$\begin{bmatrix} U_1, k \\ U_2, k \end{bmatrix} = \begin{bmatrix} a_1 & 0 \\ 0 & a_2 \end{bmatrix} \begin{bmatrix} U_1, k-1 \\ U_2, k-1 \end{bmatrix} + \begin{bmatrix} b_1 \\ b_2 \end{bmatrix} I_{k-1} \tag{2}$$

$$U(k) = d - U_1(k) - U_2(k) - cI(k) \tag{3}$$

where, a1, a2, b1, b2, c and d are coefficients of above equation of state, and their values are as follows:

$$\begin{cases} a_1 = \exp.(-\Delta t/R_1C_1) \\ a_2 = \exp.(-\Delta t/R_2C_2) \\ b_1 = R_1(1 - \exp.(-\Delta t/R_1C_1)) \\ b_2 = R_2(1 - \exp.(-\Delta t/R_2C_2)) \\ c = R_0 \\ d = E \end{cases} \tag{4}$$

In (4), Δt is sampling time.

3.3   Parameter Identification Method Based on Recursive Least Square Method

The Recursive Least Square Method. Due to existence of a lot of unknown parameters in equivalent circuit model of aluminum air battery, these parameters cannot be directly measured, so it is necessary to identify these parameters. The recursive least square method with forgetting factor was applied to identify the parameters of RC circuit of second order aluminum air battery.

By further simplifying (2) and (3), the discretized difference equation of system be obtained:

$$U(k) = h_0 + h_1 U(k-1) + h_2 U(k-2) + h_3 I(k) + h_4 I(k-1) + h_5 I(k-2) \tag{5}$$

$$\begin{cases} h_0 = (1 - (a_1 + a_2) + a_1 a_2)d \\ h_1 = a_1 + a_2 \\ h_2 = -a_1 a_2 \\ h_3 = c \\ h_4 = b_1 + b_2 - (a_1 + a_2)c \\ h_5 = a_1 a_2 c - b_1 a_2 - b_2 a_1 \end{cases} \tag{6}$$

In (5), I (k) is model input, U (k) is model output, let:

$$\begin{cases} \varphi(k) = [1, U(k-1), U(k-2), I(k), I(k-1), I(k-2)]^T \\ \theta = [h_0, h_1, h_2, h_3, h_4, h_5]^T \end{cases} \tag{7}$$

In the process of system identification, a group of new parameter estimation values are obtained, and the old parameter estimation values are modified according to the recursive formula. Finally, the new parameter estimation values can be obtained, so as to complete the real-time parameter estimation. The value range of forgetting factor λ is between 0.950 and 0.999. The specific recurrence formula is as follows:

$$\theta(k) = \theta(k-1) + K(k)[U(k) - \varphi^T(k)\theta(k-1)] \tag{8}$$

$$K(k) = \frac{P(k-1)\varphi(k)}{\lambda + \varphi^T(k)P(k-1)\varphi(k)} \tag{9}$$

$$P(k) = P(k-1)[1 - K(k)\varphi^T(k)]/\lambda \tag{10}$$

The process of online identification is as follows:

a) Import battery measurement data.
b) Initialing the coefficients used in this method:

$$\theta(0) = [0.4, 0.4, 0.4, 0.4, 0.4, 0.4];$$
$$K(0) = [-0.32, 0.5, -0.02, -0.01, 0, -0.01];$$
$$P(0) = 10I;$$
$$\lambda = 0.992$$

c) Run the recursive process of parameter identification algorithm, get the values of the 5 difference equation parameters h0–h5 at time k through the recursive formula.
d) Through the conversion (4) and (6), the parameters with time can be finally obtained.

Then model of aluminum air battery can be established with these identified model parameters.

Results of Parameter Identification. The original experimental data used for identification are the open circuit voltage and discharge current data mentioned in Sect. 3.1, and the program is written according to the above method for parameter identification. Figure 5 are the results of circuit parameters identified based on the Massimo-Ceraolo circuit model.

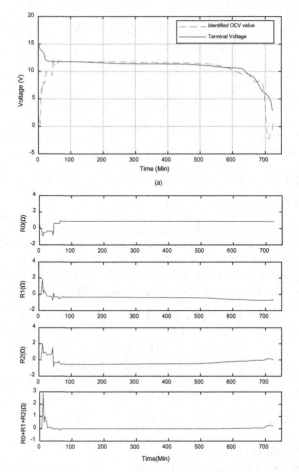

Fig. 5. (a) Identified OCV and terminal voltage; (b) The identification result of R0, R1, R2 and total resistance

The broken line in Fig. 5(a) is the identified OCV value and the full line is the terminal voltage drawn again for reference. According to the circuit principle, the variation trend of OCV curve should be consistent with that of terminal voltage, meanwhile the value of OCV should be slightly larger than the terminal voltage. At the early stage of identification, due to self-training of the algorithm parameters, the initial identified OCV value differs greatly from the terminal voltage. In the plateau phase, OCV value is indeed slightly larger than value of the terminal voltage. So the result of OCV identified is reasonable. However, at the end of identification, due to the gradual decline of the correction ability, the fitting ability will be slightly worse. Therefore, in the fitting process, from about 50 min to about 600 min, the fitting effect is good, and the OCV value can be accurately estimated.

Figure 5(b) and Fig. 6 are the identified resistance and capacitance parameters of the equivalent circuit. During the voltage plateau period, the circuit parameters basically keep stable. Among them, the three resistors R0, R1 and R2 are also stable in the voltage stability stage. The resistance R1 and R2 fluctuate slightly, but it does not affect the whole model. The stability of the whole model is determined by the stability of the total resistance. The sum of the three resistances R0, R1 and R2, can be regarded as the total resistance. The total resistance is about 0.002 Ω during the plateau phase and can vary to 0.27 Ω at the end of upcoming failure phase, but the increment is relatively small, as shown in Fig. 5(b). This phenomenon is somehow consistent with the fact that the internal resistance of aluminum-air battery increases when it is about to fail. The capacitor C1 is about −4.1f in the stable stage of the platform, and only slightly increased in the coming fault stage, which has negligible impact on the whole model. Capacitor C2 is about 3.45F in the plateau stage, but it increases rapidly in the upcoming failure stage, such as 6.2F, 81.5F and 228.8F at 600, 650 and 680 min respectively. Therefore, C2 is suitable as an indicator to judge the imminent failure of aluminum air battery.

According to the above analysis, it is verified the method can identify parameters of the second-order RC model. With OCV value observed, the following SOC estimation can be feasible.

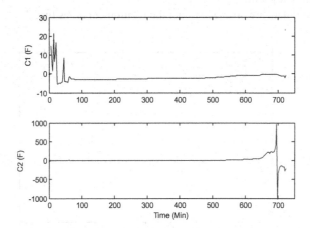

Fig. 6. The identification result of C0 and C1

4   SOC Estimation Based on Second-Order RC Massimo-Ceraolo Circuit Model

4.1   Data Collection and OCV-SOC Calibration

Before the model parameter identification of the aluminum air battery pack, it is necessary to obtain the relationship curve between SOC and OCV. For batteries, there is a fixed monotone nonlinear function relation between OCV and SOC.

Through the analysis of discharge of the battery pack (in Sect. 3.1), it is found that terminal voltage of the aluminum air battery pack tends to be stable when it is one more hour after the end of discharge. And at this point the polarization reaction inside the battery can be considered to have basically disappeared. Then the corresponding OCV value of the battery can be obtained after one hour of static treatment, and OCV can be considered equal to the terminal voltage at this moment. Intermittent discharge method was chosen for the OCV-SOC calibration experiment. When the experimental environment is 20 ± 5 °C, constant current discharge was carried out on the battery at discharge rate of 0.1C–0.4C respectively (0.1C represents 10% SOC), and the SOC and corresponding OCV values of the aluminum air battery in each discharge period were recorded in real time. The data obtained from the experiment are sorted out in Matlab to obtain the OCV-SOC curve of the battery. Specific steps are as follows:

a) In the case of ensuring the full power of battery pack, constant discharge is carried out at above discharge rate;
b) Each 5% reduction in the battery SOC needs to be left static settlement for one hour;
c) Repeat the above two steps until the SOC is 5%;
d) Continue to discharge until it reaches the cut-off voltage of 2.9 V (corresponding to 0.1C discharge rate), then after one hour static settlement, record the terminal voltage at the moment;
e) Experimental data are arranged to get the relation curve, as Fig. 7.

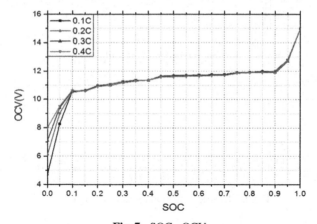

Fig. 7. SOC - OCV

Repeated experiments were carried out on aluminum air batteries at different discharge rates. The experimental results show that the OCV-SOC curves only have discrepancy in the last phase (upcoming failure phase) with rapidly decreasing voltage, but have good repeatability in other phase. When the voltage in the upcoming failure phase is too small for the battery pack to output enough power, this battery pack should be shut down and replaced with a new one. So it's unnecessary to think about the discrepancy in the last stage. In a word, the ocv-soc curve is relatively stable in the stable process of battery discharge.

4.2 Online Open-Circuit Voltage Method

Before discussing the open-circuit voltage method, it is necessary to briefly introduce the Ampere-hour integral method. Principle of this method is that it does not take into account the external structure and chemical reaction of the battery, only the residual electric quantity obtained by recording the current flowing through the battery for a long time and integrating the current. The calculation formula is as follows:

$$SOC = SOC_0 - \int_0^t Idt \bigg/ Q_{rated} \tag{11}$$

where: SOC_0 is initial power of the air battery; Q_{rated} is rated capacity of air battery; I is the discharge current of air battery.

Accuracy of Ampere-hour integral method is related to the initial capacity of the battery and the accuracy of current detection. The cumulative error will be generated with the increase of discharge time. There will be a serious deviation between the estimated and the actual value of SOC in later stage because of inaccurate Q_{rated} value.

Open-circuit voltage method is another classical one to get the SOC. As can be seen from Sect. 4.1, the measurement of battery OCV requires the battery to be left in the nonworking state for a long time until battery reaches the equilibrium state. At this time, the terminal voltage is similar to the value of open circuit voltage. Because of the unique relationship between battery OCV and SOC, the battery SOC can be obtained. The disadvantage is that the air battery needs to be out of use for a long time to reach a balanced and stable state, and the SOC of the battery needs to be estimated in real time during the measurement process. This defect will bring great difficulties to the SOC measurement. This method can achieve a certain accuracy at the beginning and final stage of discharge, so the open-circuit voltage method is generally used to calculate initial value of SOC [11].

Definitely, OCV cannot be directly measured in the practical operation of aluminum air batteries. Therefore, the traditional open-circuit voltage method isn't suitable for real-time SOC estimation. Fortunately, by adopting the above equivalent circuit model and parameter identification method, OCV can be observed online. The SOC value can be acquired with the OCV-SOC curve online too.

SOC estimation waveform obtained by this method is shown in Fig. 8(a). The full line is the result of the open-circuit voltage method. By contrast, the broken line is the result of the Ampere-hour integral method with the discharge current data, which can be considered to be the actual value because it is obtained in a laboratory environment. As in Fig. 8(a), it is obvious that the SOC obtained in the beginning phase is inaccurate due to the identification transition process. Because the convergence of parameter identification takes time, there is a big error in the transition process. But it doesn't matter because users are more concerned about the health of batteries in the long term. And the error of the two from the beginning of the maximum error drops sharply, and quickly tends to stabilize a small error. After about 180 min, the error of the open-circuit voltage method is no more than 0.12, so the result is acceptable. In the upcoming failure stage, the relatively small SOC value of open circuit voltage method is conservative, so it can effectively judge whether the aluminum air battery fails.

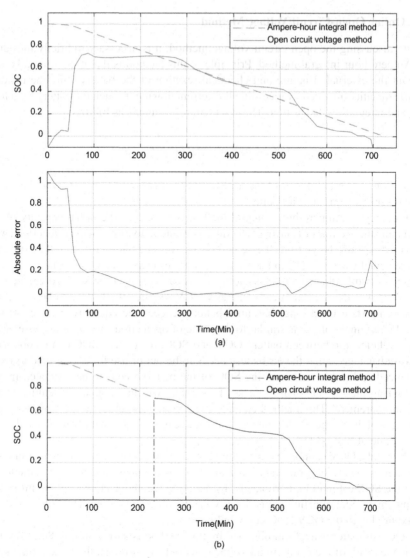

Fig. 8. (a) SOC estimation result of two mentioned method; (b) The curve of piecewise SOC estimation

Compared with existing reference on li-battery SOC estimation, it seems that accuracy of our method is smaller than the EKF method, but slightly larger than the MAEKF method. In addition, this paper does not rely on kalman filter, and mainly studies online parameter identification and SOC estimation. Compared with existing reference on SOC estimation for lithium ion batteries, online parameter identification can avoid complicated process and has better accuracy, which provides preparation for its practical application.

4.3 Piecewise SOC Estimation

Based on the analysis in above section, both Ampere-hour integral method and open-circuit voltage method commonly used in aluminum air batteries have some disadvantages when used alone. Therefore, a comprehensive method is designed in this paper. Different methods are used for different stages in the discharge process of aluminum air battery. As follows:

a) Firstly, in the initial stage of discharge, the battery is in a long-term nonworking state, the traditional open circuit voltage method (shown in Fig. 3) can be used to fix the beginning SOC value (SOC_0) of the battery.

b) After SOC_0 is known, the Ampere-hour integral method is adopted to evaluate the SOC of battery, and current terminal voltage value is tested to fix whether air battery enters next discharge stage;

c) Ampere-hour integral method will bring cumulative error as the progress of discharge process. Therefore, in the last stage, it is more suitable to estimate SOC with the proposed open circuit voltage method.

In the model, SOC_0 is first obtained through open circuit voltage method. Then, according to value of terminal voltage, estimation process is divided into two parts: the Ampere-hour integral method and the proposed open circuit voltage method. Final result is in Fig. 8(b). The green dotted line represents the ampere hour integration method, and the blue implementation represents the open circuit voltage method. The vertical dashed line is the boundary of the two methods, which is determined by the first intersection between the curves of two methods. It can be clearly seen that this method can effectively solve the large unstable error in the initial stage of identification.

Note that Fig. 8(b) is just the outcome according to discharge test results in this paper. And actual Ampere-hour integral curve varies with the actual conditions. It is necessary to measure repeatedly in practice to determine the effectiveness of the method.

5 Conclusion and Future Work

In paper, a more accurate second-order RC Massimo-Ceraolo equivalent circuit model is established after electrical characteristics of aluminum air battery were tested. The recursive least square method with forgetting factor is adopted to realize online identification of the model parameters. Experimental results show that this method is a good online estimation method with good accuracy. Moreover, the method is simple and low complexity, and can be well reproduced in any practical application. The results show that the second-order RC equivalent circuit model can properly simulate electrical characteristics of aluminum air battery and achieve open circuit voltage in real time. And in the real-time process can be based on the current data for effective correction, making the results more and more accurate, the error is smaller and smaller. Based on it, a piecewise SOC estimation method by combining open circuit voltage method based on second-order RC equivalent circuit model with Ampere-hour integral method is proposed. Simulation results show that the SOC estimation method overcomes the shortcomings of Ampere-hour integral method and traditional open-circuit voltage method. The piecewise estimation method solves the large error in the initial stage of identification, and uses capacitance data to terminate the model before the final fault starts. The research of air battery is of great significance.

If we take the time to do in-depth research, it is likely to help us get rid of our dependence on fuel resources. Air battery energy is an important research direction. Due to the unique advantages of aluminum air battery, its research prospect is very broad. It is worth studying further. So far, the aluminum air models have some limitations and very desirable advantages. In order to further study, it is necessary to extract its essence and remove its dross. In the future, air battery will be applied stably and accurately in practice. There are still some future prospects worth further study in the research. For example, the effects of monomer inconsistencies and grouping patterns on cell modeling are studied. Consider the temperature running model. The initial identification coefficients of different types of aluminum-air batteries were studied.

Acknowledgments. This research is supported by the Zhejiang Provincial public welfare Foundation (LGG19E070004), China.

References

1. Elia, G.A., Marquardt, K., Hoeppner, K., et al.: An overview and future perspectives of aluminum batteries. Adv. Mater. **28**, 7564–7579 (2018)
2. Gelman, D., Shvartsev, B., Wallwater, I., et al.: An aluminum - ionic liquid interface sustaining a durable Al-air battery. J. Power Sources **364**, 110–120 (2017)
3. Hu, X., Feng, F., Liu, K., et al.: State estimation for advanced battery management: key challenges and future trends. Renew. Sustain. Energy Rev. **114**, 1–13 (2019)
4. Li, H., Zhang, W., Yang, X., et al.: State of charge estimation for lithium-ion battery using an electrochemical model based on electrical double layer effect. Electrochim. Acta **326**, 1–12 (2019)
5. Cheng, C., Rui, X., Ruixin, Y., et al.: State-of-charge estimation of lithium-ion battery using an improved neural network model and extended Kalman filter. J. Clean. Prod. **234**, 1153–1164 (2019)
6. Ceraolo, M.: New dynamical models of lead-acid batteries. IEEE Trans. Power Syst. **15**, 1184–1190 (2000)
7. Qiaoqiao, X.: Research and practice of residual capacity estimation algorithm for lithium ion power batteries. Chongqing University, Chongqing (2013)
8. Yi, M.: Research on motor drive power system based on aluminum aerodynamic battery. University of Electronic Science and Technology (2016)
9. Yuping, Z., Dong, C., Yang, L., et al.: Estimation of SOC of aluminum empty batteries based on HSMM. J. Univ. Electron. Sci. Technol. **46**, 380–385 (2017)
10. Xia, B., Wang, H., Tian, Y., et al.: State of charge estimation of lithium-ion batteries using an adaptive cubature Kalman filter. Energies **8**, 5916–5936 (2015)
11. Yingxu, J.: Review of SOC estimation methods for batteries. Electr. Measur. Instrum. **51**, 18–22 (2014)

Modeling of Main Steam Temperature Using an Improved Fuzzy Particle Swarm Optimization Algorithm

Hanyu Mi, Chen Peng[✉], and Chuanliang Cheng

School of Mechatronic Engineering and Automation, Shanghai University,
Shanghai 200444, China
c.peng@shu.edu.cn

Abstract. Reliable control of the main steam temperature is a necessary condition to achieve the safe and economic performance of the thermal power plants. For the purpose of improving accuracy of control system, an improved fuzzy particle swarm optimization (FPSO) algorithm is proposed for the modeling of the main steam temperature in the ultra-supercritical (USC) units in this paper. Different from the traditional methods, the nonlinear problem is effectively solved by a fuzzy k-means network which contains local models and fuzzy rules. In the network, an improved particle swarm optimization (PSO) that has faster convergence speed is introduced to build the local models. And the nonlinear dynamic process of the main steam temperature is elaborately approximated by the fuzzy fusion of the local linear models. Simulation results show that the improved FPSO algorithm proposed in this paper can achieve better performance with higher convergence speed and identification accuracy, which is conducive to the subsequent control of boiler combustion process.

Keywords: Main steam temperature · Particle swarm optimization · Ultra-supercritical unit · Fuzzy k-means network · Fuzzy fusion

1 Introduction

The combustion of the coal-fired power unit is a process of energy conversion. During the process, the water is heated into high temperature steam through the coal combustion, where chemical energy is converted into thermal energy. And then the high temperature steam enter into the turbine to drive the generator, in which thermal energy is transferred into mechanical energy. Through the generator operation, the mechanical energy is converted into electrical energy in the end. Obviously, high temperature steam plays a key role in this process. Consequently, the effective control of steam temperature has become an essential part of the economical and safe operation of thermal power plants. Most of the control methods introduced in the previous literature are based on accurate models [1,2]. So modeling has become a hot topic in recent years. Therefore, it

© Springer Nature Singapore Pte Ltd. 2020
M. Fei et al. (Eds.): LSMS 2020/ICSEE 2020 Workshops, CCIS 1303, pp. 123–136, 2020.
https://doi.org/10.1007/978-981-33-6378-6_10

is of great significance to establish a model of the main steam temperature not only for the control and optimization of the combustion, but also in line with the current energy transformation and the requirements of energy conservation and environment sustainable development.

At present, ultra-supercritical unit (USC) is an advanced and efficient coal-fired power generation technology. Ultra-supercritical refers to higher temperature and higher pressure compared with supercritical units. In the furnace of USC, the steam temperature is not lower than 593 °C or the steam pressure is not lower than 31 MPa. In other words, ultra-supercritical units have higher thermal efficiency and lower energy consumption than supercritical units. Therefore, it is of great importance to establish a suitable model for the boiler of USC.

In recent years, many predecessors have carried out a series of researches on boiler combustion process modeling. The least square method is a traditional model identification method. The recursive least squares (RLS) method is introduced to identify models in boiler and obtained relatively good results [3]. The correlation analysis algorithm is improved and applied to the recognition of nonlinear systems [4]. Based on the maximum likelihood method, the model parameters of coal-fired power plants can be identified effectively [5,6]. The stochastic approximation method and the derivative method based on its basic idea can also identify the model parameters of nonlinear objects such as boiler [7,8]. Neural network is also a common method in boiler object modeling. Because of its strong adaptability to nonlinear problems, it has been widely used in boiler modeling. Sun et al. conducted combustion experiments according to the combustion mechanism and established the NOx emission model based on the test results [9].

Due to its high precision and easy realization, intelligent algorithm is more and more widely used in model identification. PSO is a typical swarm intelligence algorithm, which has been widely used in many fields such as parameter identification and system control due to its fast convergence and easy implementation. On the basis of elementary PSO, Huang et al. incorporated cuckoo searching (CS) algorithm to solve the problem that PSO is prone to local optimization. A new algorithm combining PSO with cuckoo algorithm (cuckoo) is proposed to improve the identification accuracy [10]. Liu et al. introduced a kind of electromagnetic mechanism algorithm into PSO and proposed a kind of electromagnetic mechanism particle swarm optimization (EMPSO) algorithm to identify the thermal process of circulating fluidized bed boiler. The performance of particle swarm is improved and the recognition effect is better [11].

In this paper, an improved fuzzy particle swarm optimization (FPSO) algorithm is used for the parameter identification of main steam temperature of ultra-supercritical units. PSO is chosen as it possesses the advantages of less adjustment parameters, higher identification accuracy and easier realization by computer. At the same time, aiming at the shortcoming of PSO which is easy to fall into local optimality, a uniform method is used to improve the initialization procedure. According to the strong nonlinear characteristics of the combustion process of power station boiler, the dynamic process of boiler combustion is divided into several linearized regions by clustering method. Then model of each local region is modeled separately. Then the fuzzy idea is introduced to the fusion of the linearized region model, so as to obtain the model of the whole dynamic process of the main steam temperature. The main contributions of this paper are as follows:

- A homogenization method is proposed to improve the initialization steps of traditional PSO algorithm, so that the distribution of particles becomes more uniform. Thus, the optimization process of particle searching is accelerated and the convergence speed of the algorithm is improved.
- The k-means fuzzy network is introduced to the improved algorithm. The linearization and fuzzy fusion of local model enhance the adaptability of the algorithm to nonlinear system. Thus, the limitation of traditional algorithm in nonlinear system identification is improved.

The structure for the remainder of this paper is as follows: the mathematical model structure of the system to be identified and the model description of the thermal process are introduced in Sect. 2. The design and improvement of PSO are described in Sect. 3. The combination of fuzzy theory in algorithm is given in Sect. 4. A simulation example is presented in Sect. 5. And the conclusion is given in Sect. 6.

2 Model Design

2.1 System Identification

The method of establishing mathematical model by determining the relationship between system input data and output data is the definition of system identification. Because the operating mechanism of the system is not considered in this simple identification process, the established model is also known as the black box model [12]. In the process of system identification, the mathematical model structure which can best reflect the dynamic process of the system is analyzed according to the collected input and output data. After the mathematical model structure is determined, the parameters of the model can be identified by the specific optimization method.

In the Fig. 1, θ is the actual value of system parameters, $\hat{\theta}$ is the estimated value of model parameters, $u(k)$ is the input data at time k, $y(k)$ is the output data at time k, $\hat{y}(k)$ is the estimated value of model at time k, $e(k)$ is the difference between the estimated output value of model and the actual output data of the system at time k, and $v(k)$ is the noise disturbance.

2.2 Model Description

In the main steam temperature system, the steam is heated by the water in the drum. As is shown in the figure of the generation of main steam temperature steam, the steam temperature, which passing through the primary superheater,

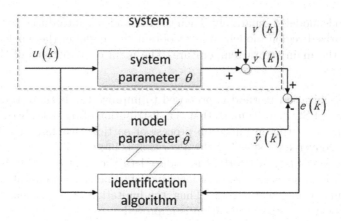

Fig. 1. Basic principle of model identification

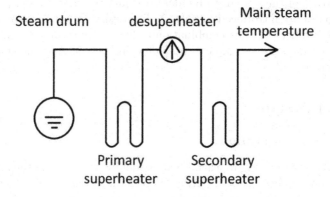

Fig. 2. Generation of the main steam temperature

the desuperheater and the secondary superheater in turn, is the main steam temperature [13, 14].

The main steam temperature is mainly affected by water:fuel ratio, superheater spray flow rate and other factors. The water:fuel ratio can affect the main steam temperature by controlling the intermediate point temperature [15]. Therefore, in the system model identification, intermediate point temperature, primary superheater spray flow and secondary superheater spray flow are selected as the input variables, and the main steam temperature is used as the output variable. The structure of the multiple input single output(MISO) model is as below:

$$y(k) = f\left(\sum_{i=1}^{n} a_i y(k-1) + \sum_{j=1}^{m} b_m u(k-j+1) + \zeta(k)\right) \tag{1}$$

where f is the nonlinear fusion function based on fuzzy idea, $y(k)$ is the k-th output value of the system, $u(k)$ is the k-th input value of the system, n is the output order of the system, and n is 1 in this paper, m is the input order of the system, and $\zeta(k)$ is the noise of the system. In the Fig. 2, K is the number of linearized regions divided.

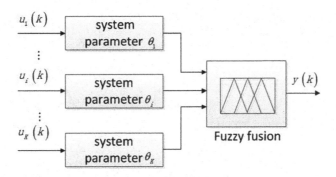

Fig. 3. System framework

3   Fuzzy k-Means Network

Because of the strong nonlinearity of the dynamic process of boiler combustion, it is difficult to identify the model parameters. Therefore, the fuzzy theory is introduced to solve this problem [16]. The structure of the fuzzy k-means network is shown in the Fig. 3.

The first layer is the input layer, where x_i represents the i-th input language variable. In the modeling of ultra-supercritical units, the operation area is divided into several regions according to the main team temperature, which are: 'low', 'medium low', 'medium', 'medium high', and 'high', respectively.

The second layer is the fuzzy layer. Each set of nodes represents a local region. The endpoints of each region are determined by k-means clustering algorithm. And the trigonometric function is selected as the fuzzy membership function as is shown in the Fig. 4.

K-means clustering algorithm is a representative kind of clustering algorithm based on distance [17]. The distance plays a part of the evaluation index of similarity. In other words, the closer the two objects are, the more similar they are. The basic steps of k-means clustering algorithm are as follows:

- Determine the K value, which is the number of clusters of the data.
- K objects were chosen in random from the data set as the initial clustering center.
- For each point in the data set, the distance between it and each center of mass is calculated, and it is divided into the cluster belonging to the nearest cluster center.

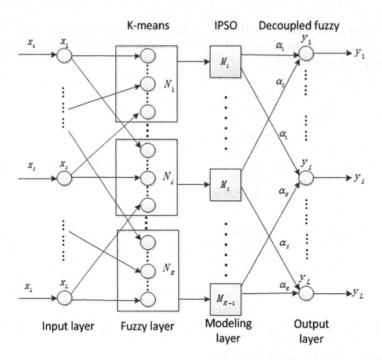

Fig. 4. System network structure

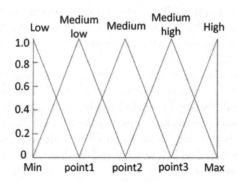

Fig. 5. Fuzzy membership function

- There are K clusters after all the data being divided, and then the center of mass of each cluster is recalculated.
- If the distance between the newly calculated center of mass and the original center of mass is less than the threshold, the clustering can be considered to have reached the desired result, and the algorithm is terminated.
- If not, repeat steps 3–5 until the centers of the clusters remain unchanged or the iterations number is reached (Fig. 5).

The third layer is the modeling layer. The modeling method will be introduced in detail in the next section. The nodes of the third layer represent the fuzzy operation region submodel, which is inferred according to the following fuzzy rules:

R_i: IF the main team temperature belongs to condition i, THEN the local submodel i:

$$\hat{Y}_i(k) = \sum_{j=1}^{n} A_{ij} Y(k-j) + \sum_{j=1}^{m} B_{ij} U(k-j+1) + V(k) \tag{2}$$

where $\hat{Y}_i(k)$ is the output of the i-th local model, $Y(k-j)$ is the input of the system process, $U(k-j+1)$ is the output of the system process, n is the output order of the system, m is the input order of the system, A_{ij} and B_{ij} are the local model parameters of the main steam temperature in the i-th linearized region.

The fourth layer is the output layer, that is, the decoupled fuzzy layer. The output inferred by the fuzzy rules of this layer can be described as the following form:

$$\hat{Y}(k) = \frac{\sum_{i=1}^{K} \alpha_i \hat{Y}_i}{\sum_{i=1}^{K} \alpha_i} = \sum_{i=1}^{K} \bar{\alpha}_i \hat{Y}_i \tag{3}$$

where K is the number of local regions, and α_i is the membership value of each local region.

4   Particle Swarm Optimization (PSO) Design

In this section, an improved particle swarm optimization(IPSO) design is proposed to model the dynamic process of the main steam temperature in each local region [18]. In 1995, PSO was proposed by J. Kennedy and R. C. Eberhart in the first place. The concept comes from the study of the predatory behavior of birds [19]. The main contents of the traditional particle swarm optimization include fitness function, initialization population, position and speed update. The core of PSO algorithm is its speed and position updating formula, which are described as:

$$V_i^{k+1} = \omega \times V_i^k + c_1 \times r_1 \times (P_i^k - X_i^k) + c_2 \times r_2 \times (P_g^k - X_i^k) \tag{4}$$

$$X_i^{k+1} = X_i^k + V_i^{k+1} \tag{5}$$

4.1   Fitness Function

Fitness function is the key connecting factor between algorithm and application case, which is used to measure the performance of an individual. Under the same input excitation, the sum of the errors between the actual output of the system and the model output is the fitness function. The smaller the fitness function is, the smaller the error is. The fitness function is described as:

$$J = \sum_{i=1}^{L} ((y(i) - \hat{y}(i))^2 \tag{6}$$

where $y(i)$ is the real output value of the system, $\hat{y}(i)$ is the estimated output value of the model, and L is the length of the input and output data of the system.

4.2 Population Initialization

Each particle in PSO represents the system model parameter of the search. The dimension of the target search space represents the number of parameters. The number of particles in the population means that there are m particles representing the solution of potential problems. In the linearized region, the output is described as:

$$y_i(k) = a_1 y(k-1) + b_{11} u_1(k) + b_{12} u_2(k) + b_{13} u_3(k) + b_{21} u_1(k) + b_{22} u_2(k) + b_{23} u_3(k) \tag{7}$$

According to the above equation, the particle is composed of 7 parameters to be identified, described as:

$$\theta_i = [a_{i1}, b_{i11}, b_{i12}, b_{i13}, b_{i21}, b_{i22}, b_{i23}] \tag{8}$$

The initial value of each particle's position parameter takes a random number between the position boundaries.

Since the elementary PSO is easy to fall into local optimality and is slow to the convergence, its application in model parameter identification is limited. Therefore, adding uniformization to the initialization step of elementary PSO algorithm is proposed in this paper, so that the particles can be distributed as far as possible throughout the search space. Thus, the probability of the algorithm falling into local optimum is reduced, and the convergence rate is improved. The operation of the uniform initialization step is described as follows:

$$X_{i.initial} = X_{min} + \frac{X_{max} - X_{min}}{m} \times (i-1) + \frac{X_{max} - X_{min}}{m} \times rand() \tag{9}$$

where $X_{i.initial}$ is the initialization position of the i-th particle; X_{min} and X_{max} is the position boundary of the particle; $rand()$ is a random number between 0 and 1.

4.3 Algorithm Flow

Therefore, on the basis of the primary particle swarm optimization(PSO) algorithm, the improved fuzzy particle swarm optimization(FPSO) algorithm proposed in this paper can be obtained by combining fuzzy theory and uniform initialization steps. The whole process implementation is shown as follow:

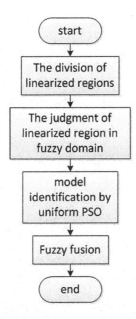

Fig. 6. Algorithm flow of improved FPSO

5   An Example

In this section, the proposed FGA is employed to model the dynamic process of the main stream temperature of a 1000MW USC unit. The experimental simulation platform in this section is Matlab software. The proposed algorithm is used to identify the main steam temperature model parameters of a 1000MW ultra-supercritical unit and identify its dynamic process. In this example, 1000 sets of data were used for training, and 600 sets of data were used to test the model. In the IPSO, the spatial dimension is 7, the population number is 50, the maximum number of iterations is 100, the inertia weight is 0.8, and the learning factor is 0.5.

5.1   Uniform PSO Simulation

Taking the NS region as an example, the simple particle swarm optimization algorithm and the uniformly initialized particle swarm optimization algorithm are respectively used to search for optimization. The convergence process is shown in Fig. 6.

It can be seen from the Fig. 7 that the uniform PSO converges faster than the elementary PSO, that is because the uniform distribution of the particle making the searching process more reasonable. Through the results, it can be concluded that uniform initialization can improve the convergence rate of PSO.

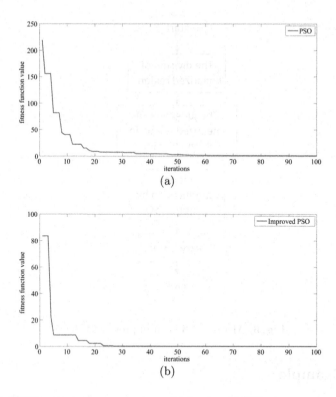

Fig. 7. (a) PSO convergence procedure; (b) Improved PSO convergence procedure

5.2   Training Data Simulation

After 1000 sets of data training, the fitting results of the dynamic process can be obtained by the elementary PSO and the improved PSO respectively.

It can be seen from the Fig. 8, Fig. 9 and Table 1 that the improved FPSO algorithm is more accurate than the elementary PSO algorithm for model identification of the main steam temperature dynamic process in ultra-supercritical unit. This is due to the whole nonlinear region is linearized locally by proposed improved FPSO, which reduces the nonlinearity to a minimum.

Table 1. Comparison of training identification error between PSO and improved PSO.

	MAE	MSE
PSO	0.4343	0.1573
Improved FPSO	0.4096	0.1453

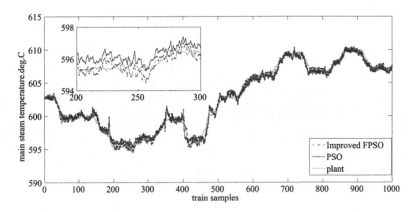

Fig. 8. Comparison of the training identification results between PSO and improved FPSO

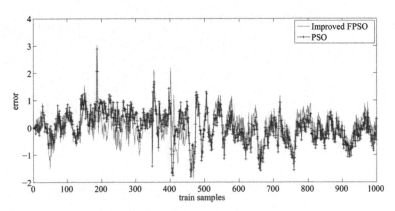

Fig. 9. Comparison of the training identification errors between PSO and improved FPSO

5.3   Test Data Simulation

A certain amount of test data is needed to confirm the result of system model identification. In this paper, 600 groups of data were used to test the identification results.

Table 2. Comparison of test identification error between PSO and improved PSO.

	MAE	MSE
PSO	0.5156	0.2124
Improved FPSO	0.4524	0.1689

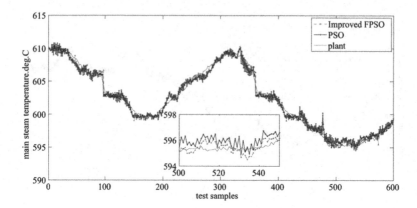

Fig. 10. Comparison of the training identification results between PSO and improved FPSO

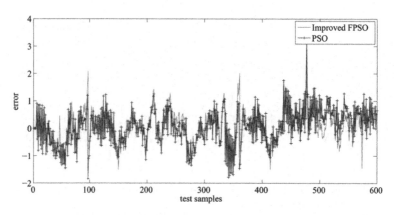

Fig. 11. Comparison of the test identification errors between PSO and improved FPSO

It can be seen from Fig. 10, Fig. 11 and Table 2, the test data results are consistent with the training data results. Therefore, it can be concluded that the improved PSO proposed in this paper can identify the model of the temperature of main steam in ultra-supercritical units with effect, and the identification accuracy of the improved PSO is higher than that of the elementary PSO.

The simulation results show that the nonlinear characteristics of the main steam temperature should be fully considered in the model identification. In addition to selecting the appropriate model identification method, it is especially important to add the nonlinear processing method based on the existing model identification method. In this paper, on the basis of elementary particle swarm optimization (PSO), linear partitioning and fuzzy fusion steps are added to improve PSO algorithm. The simulation results also prove the superiority of the improved particle swarm optimization algorithm.

6   Conclusion

In this paper, an improved fuzzy particle swarm optimization(FPSO) algorithm is proposed to model the main steam temperature of a 1000MW ultra supercritical power plant. Uniform initialization step and fuzzy ideas have been incorporated into the improved FPSO proposed compared with the elementary PSO. The FPSO is quite suitable for nonlinear processes, with the whole operating region being partitioned into several local linear operating region. The simulation results have showed that the improved FPSO algorithm can identify the main steam temperature model effectively and improve the convergence speed and identification accuracy. It has laid a good foundation for the subsequent control system of the main steam temperature in ultra-supercritical units.

References

1. Sun, L.: Multi-objective optimization for advanced superheater steam temperature control in a 300 MW power plant. Appl. Energy **208**(15), 592–606 (2017). https://doi.org/10.1016/j.apenergy.2017.09.095
2. Qian, H., Feng, Y., Zheng, Z.: Design of adaptive predictive controller for superheated steam temperature control in thermal power plant. In: Li, K., Xue, Y., Cui, S., Niu, Q., Yang, Z., Luk, P. (eds.) LSMS/ICSEE -2017. CCIS, vol. 763, pp. 409–419. Springer, Singapore (2017). https://doi.org/10.1007/978-981-10-6364-0_41
3. Marjanovic, A.: Control of thermal power plant combustion distribution using extremum seeking. IEEE Trans. Control Syst. Technol. **25**(5), 1670–1682 (2016)
4. Lang, Z.Q., Futterer, M., Billings, S.A.: The identification of a class of nonlinear systems using a correlation analysis approach. In: 16th IFAC World Congress 2005, IFAC, vol. 38, pp. 208–212. (2005). https://doi.org/10.3182/20050703-6-cz-1902.00035
5. Haryanto, A., Turnip, A., Hong, K.S.: Parameter identification of a superheater boiler system based on Wiener-Hammerstein model using maximum likelihood method. In: 7th Asian Control Conference 2009, ASCC, vol. 2009, pp. 1346–1351 (2009)
6. Chandrasekharan, S.: Parametric identification of integrated model of a coal-fired boiler in a thermal power plant. J. Pow. Energy **2**(5), 99–110 (2019). https://doi.org/10.1177/0957650919870383
7. Zhao, W.X.: Recursive identification for nonlinear ARX systems based on stochastic approximation algorithm. IEEE Trans. Autom. Control **55**(6), 1287–1299 (2010). https://doi.org/10.1109/TAC.2010.2042236
8. Greblicki, W.: Hammerstein system identification with stochastic approximation. Int. J. Model. Simul. **24**(3), 131–138 (2004). https://doi.org/10.1080/02286203.2004.11442297
9. Sun, W.F., Lv, J.S.: Application of fuzzy artificial neural networks model for modeling NOx emissions in a power station boiler on generation mechanism. In: 2015 8th International Symposium on Computational Intelligence and Design 2015, ISCID, vol. 2, pp. 319–322. (2015). https://doi.org/10.1109/ISCID.2015.77
10. Huang C., Zhang T., Dang X: Model identification of typical thermal process in thermal power plant based on PSO-CS fusion algorithm. In: Chinese Control And Decision Conference 2018, CCDC, vol. 2018, pp. 3847–3852, (2018). https://doi.org/10.1109/CCDC.2018.8407791

11. Liu, C., Sun, X.: Electromagnetism-like mechanism particle swarm optimization and application in thermal process model identification. In: Chinese Control and Decision Conference 2015, CCDC, vol. 2010, pp. 2966–2970, (2016). https://doi.org/10.1109/CCDC.2010.5498676

12. Hou, Z.: Application of neural network for boiler combustion system identification. J. Harbin Inst. Technol. **36**(2), 231–233 (2004)

13. Kong, X.B.: Distributed supervisory predictive control of main steam temperature for ultra-supercritical unit. J. Shanghai Jiao Tong Univ. **51**(10), 1252–1259 (2017). https://doi.org/10.16183/j.cnki.jsjtu.2017.10.015

14. Zhang, J.: Main steam temperature control of thermal power boiler based on dynamic matrix control. J. Xi'an Jiaotong Univ. **52**(10), 96–102 (2019). https://doi.org/10.7652/xjtuxb201910013

15. Hu, W.: Multimodel parameters identification for main steam temperature of ultra-supercritical units using an improved genetic algorithm. J. Energy Eng. **139**(4), 290–298 (2013)

16. Liu, X., Tu, X., Hou, G.: The dynamic neural network model of a ultra supercritical steam boiler unit. In: Proceedings of the 2011 American Control Conference 2011, IEEE, vol. 2011, pp. 2474–2479. (2011). https://doi.org/10.1109/acc.2011.5991224

17. Yang, Y.: Consensus clustering based on constrained self-organizing map and improved Cop-Kmeans ensemble in intelligent decision support systems. Knowl. Based Syst. **32**, 101–115 (2012)

18. Vinatoru, M., Maican, C., Canureci, G.: Heat exchange model for a power station boiler. In: Proceedings of the 2011 American Control Conference, IEEE, vol. 2012, pp. 26–31, (2012). https://doi.org/10.1109/aqtr.2012.6237670

19. Kennedy, J., Eberhart, R.: Particle swarm optimization. In: Proceedings of ICNN 1995-International Conference on Neural Networks 1995, IEEE, vol. 4, pp. 1942–1948. (1995). https://doi.org/10.1109/ICNN.1995.488968

Research on Testing Method of Low Voltage IGBT Module Parameter

Pengcheng Niu[1,2], Wengen Gao[2], and Jie Xu[2(✉)]

[1] Department of Electrical and Computer Engineering, Faculty of Engineering,
Western University, London, ON, Canada
[2] Key Laboratory of Advanced Perception and Intelligent Control of High-End Equipment,
Ministry of Education, Anhui Polytechnic University, Wuhu, China
xujie_1288@126.com

Abstract. For the safe and rapid testing of multi-module power devices in complex systems, low-voltage testing is performed through a test experimental platform for insulated gate bipolar transistors, an inverter composed of IGBT modules, and the low-voltage experiment results confirm whether the power device works normally with rated conditions. A new method for rapidly detecting the IGBT modules is presented in this paper by monitoring inverter output, which uses low voltage input test. This paper proposes the overall design scheme, key function design and circuit principle analysis of this tester. This paper analyzes the test conditions for power devices during low-voltage testing, and provides a reference for low-voltage testing. The realization of this method has practical application value prospects for developing new products of power devices.

Keywords: IGBT module · Voltage inverter drive · Rapid test platform

1 Introduction

With the widespread use of power devices, the performance of power devices is becoming more and more important. At present, power devices are used more and more widely, involving new types of micro grids, smart grids, energy internet, and smart grids. High current and high voltage continue to develop. The parameters including power, frequency, voltage, and current are changing. It is important for device safety. It is also more important to propose a fast, efficient, and simple high-power IGBT test method for core devices.

According to the improvement of the required voltage and capacity level, IGBT devices with higher power levels, higher performance and higher reliability continue to appear [1–3]. The static characteristics and dynamics of IGBT modules are large-capacity power electronic equipment. The important test performance of the core components directly affects the electrical performance of the system [4–7]. Further, the reliability of the IGBT module has become a research hotspot in the field of power electronics [8–12]. Extensive and in-depth research on IGBT switching characteristics has been carried out [13, 14].

© Springer Nature Singapore Pte Ltd. 2020
M. Fei et al. (Eds.): LSMS 2020/ICSEE 2020 Workshops, CCIS 1303, pp. 137–156, 2020.
https://doi.org/10.1007/978-981-33-6378-6_11

This paper analyzes the dynamic characteristics of IGBT with low voltage input, and further conducts the inverter characteristics experiment through the inverter experiment. The current research is mainly in the field of rated voltage, and there is no reference to a test system that is under low voltage and normal operation, which is also a new study proposed by special environments [15–19]. This paper introduces the design scheme, key function design and principle analysis of low-voltage power devices in detail, which has practical application value and broad market prospects [20–22].

During the use of IGBT, many problems have been discovered that affect the widespread use of IGBTs. Among them are mainly the reasonable selection and correct design of IGBT drive circuits. The distribution parameters, operating temperature, and external signal sampling of the device will affect the IGBT test results.

2 Low-Voltage Input Performance Analysis

At present, it is mainly focused on the IGBT test of medium and high voltage, and it rarely involves the switching characteristic test of low voltage power IGBT modules. Currently, it operates safely with the device, and protects the low voltage test. For the low voltage work dead zone, the impact of the low voltage electrical input test on the high voltage test, test confirmation of device performance, IGBT off state, on state, off state.

During the power switching process, depending on the characteristics, grades, and structure of the device, the switching time may vary by dozens. At present, there are mainly characteristics analysis such as short switching process time, large current and voltage change rate [23–25]. At present, most research collections focus on high voltage and ultrahigh voltage. The research on low voltage is not carried out systematically. The impact of low voltage testing of devices on the system is also very important. It is mainly reflected in the test of whether the power device IGBT is operating normally and the low. The working effective domain value of the voltage input. This paper analyzes the measured waveforms of the experimental system [26–29].

$$V_{rating} = \sqrt{2}\varphi V_{acLC} + V_S + V_{reg} + a \tag{1}$$

V_{rating} denotes the rated voltage, V_{acLC} is AC low voltage input threshold, V_S represents voltage overshoot, V_{reg} is Voltage increase due to regeneration effect, φ represents the power supply constraint coefficient, a is Margin. The maximum peak voltage must not exceed 80% of the rated voltage, DC voltage does not exceed 50%–60% of rated voltage.

By solving Eq. (1), we can obtain,

1) $\varphi = 0$, where,

$$V_{rating} = V_S + V_{reg} + a$$

2) $\varphi = 1$, then,

$$V_{rating} = \sqrt{2}V_{acLC} + V_S + V_{reg} + a$$

3) $\varphi = 0.15$

$$V_{rating} = 0.21V_{acLC} + V_S + V_{reg} + a$$

It has the lowest value for effective operation, which is the effectiveness performance value, the voltage test is based on the normal operation of the device. The rectified DC value is 15 V, and the trigger angle is 90°.

This is the limit state, we can get a constant value.

$$I_{peak} = P_{inv} \times \sigma + (\varphi V_{acLC})/\sqrt{3} \times \sqrt{2} \times \xi \tag{2}$$

I_{peak} is the peak current of the inverter, P_{inv} is the capacity, σ is the overload factor, V_{acLC} is the effective value of AC input threshold, ξ is ripple coefficient. Maximum repeatable current peak should not exceed 70–80% of rated current.

The characteristic of the field φV_{acLC} value is the constant value, which is the limit value of Eq. (2).

By solving Eq. (2), we can obtain,

1) $\varphi = 0$, where,

$$I_{peak} = P_{inv} \times \sigma$$

2) $\varphi = 1$, Then,

$$I_{peak} = P_{inv} \times \sigma + V_{acLC} \Big/ \sqrt{3} \times \sqrt{2} \times \xi$$

3) $\varphi = 0.15$, we can obtain,

$$I_{peak} = P_{inv} \times \sigma + 0.15V_{acLC} \Big/ \sqrt{3} \times \sqrt{2} \times \xi$$

From the above equations, it can be seen that through the analysis of the input voltage value, the system output has a numerical field, which explains the effect of the value on the system output, and further shows that the equation has a constant characteristic. The system has normal output characteristics (Fig. 1).

3 IGBT Control and Implementation Circuit

IGBT is a voltage-type device. The turning on and off of the device is controlled by changing the gate insulated gate voltage. Different voltage and current levels have different gate equivalent input charge values and equivalent input capacitance values, which require different chip driving capabilities. This paper uses the characteristics of low voltage input.

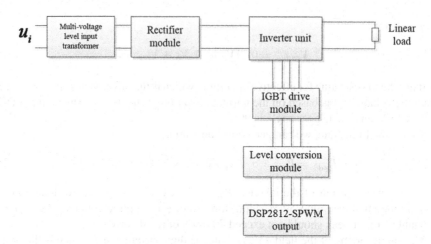

Fig. 1. System control block diagram.

This paper uses adjustable voltage input, the test voltage is 24 V, 48 V, 110 V, 220 V. Sampling of voltage and frequency signals uses voltage sensors. The input voltage goes through the rectifier module to achieve AC to DC, and then connects to the load by the inverter unit. Here is an adjustable load. The single-phase inverter unit inputs the SPWM generated by the DSP to the level conversion module for the level conversion stable.

An IGBT drive signal level conversion device, as shown in Fig. 2, includes: a peripheral circuit, an IGBT drive module and a PWM output module, the IGBT drive module which is connected to the peripheral circuit and a PWM output module, and the level conversion device. A level conversion module is further included, and the IGBT drive module is connected to the PWM output module via the level conversion module. The level conversion module is used to output the high level output controlled by the PWM.

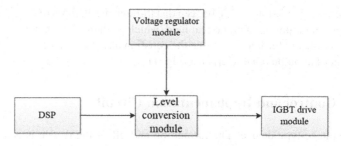

Fig. 2. Schematic diagram of signal flow of the conversion module.

The output module of 3.3 V signal is converted into a high-level PWM signal of 5 V, which ensures the stable working core DSP, and inputs the IGBT drive module.

The level conversion module converts the PWM signal with a high level of 3.3 V output by the PWM output module into a PWM signal with a high level of 5 V to make up for the voltage difference between the PWM output module and the IGBT drive module, which enables the IGBT to drive. The combination of the module and the PWM output module can work better, reduce the failure rate, and drive the peripheral circuit to work normally. This paper uses level conversion to achieve a stable PWM input signal for the system.

The device junction temperature Tj can be adjusted from 25 °C to 175 °C. The test platform is controlled by the DSP, and the data of the test platform is displayed by the oscilloscope to measure the indicators of each project.

In the sampling system block diagram, the signal passes through the sample-and-hold module, zero-crossing protection, A/D conversion module, phase-locked loop part, which enter the addition circuit module, then enters the core DSP controller after in Fig. 3.

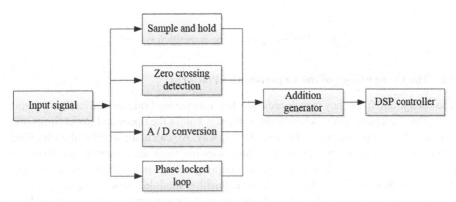

Fig. 3. Sampling system block diagram.

In order to connect the collected input signal to the DSP controller, the sampled signal in this article enters into the comparison circuit and compares the magnitude of the two voltages. Among them, the high or low level of the output voltage is used to indicate the magnitude relationship between the two input voltages. When the input voltage is higher than the "−" input, the voltage comparator output is high, when the "+" input is lower than the "−" input, the voltage comparator output is low, it can work in linear workspace and nonlinear workspace.

As shown in Fig. 4, the sampling control part of the system includes the following parts, the input voltage for frequency and voltage sampling, and then connected to the control part, using external voltage power supply. Here is a single-phase signal acquisition, with modular function, which can achieve three-phase signal acquisition, line voltage, phase voltage acquisition, signal access to the controller simultaneously, with integrated and modular characteristics. This system has compatible characteristics. Through the sampling module integration method, three identical single-phase modules are used to detect one phase respectively, which facilitates timely confirmation of the faulty module in the event of a fault and facilitates the control of the corresponding single-phase acquisition module to re-test the data.

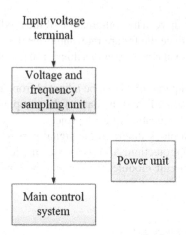

Fig. 4. Block diagram of the acquisition system.

3.1 The Composition of the Experimental Platform

The voltage and frequency signals are collected from the load side and fed back to the grid side. This paper calculates whether the voltage is within the upper and lower limits and whether the compensation is effective. In this way, we calculate whether the sampling voltage is effective. If the sampling voltage does not meet the requirements, enter the grid-side calculation cycle again until the sampling voltage meets the requirements and then the sample in Fig. 5. The provided acquisition module version detection system provides accurate detection of the acquisition module version, simple operation, high efficiency, and low labor costs.

In the current research, the electrical parameters of power devices are mainly voltage overshoot, switching loss and on-state loss, turn-on time and turn-off time. Currently, offline, quasi-online, and online testing systems are used to quickly detect the system [27–29].

The first is an offline test system. At present, many power device suppliers use a dedicated device experimental test platform. Generally, an off-line test system is often used to test the power device, and part of the power of the power device is tested.

The second is a quasi-online test system. It adopts a real-device, analog, or a combination of special device experimental test platform, through the dual pulse control signal, and through the field test recording instrument into the working state record, the full function of power devices can be tested.

The third is an online test system: a true full-featured test power device, which mainly involves testing the transient characteristics of the power device such as the on state and the off state. It has the power device operating parameters under a certain period of continuous operation. Dedicated experimental test platform, synchronizing the record with real-time full-process test data.

The fourth is a fast test platform. The actual real platform test platform is used to continuously detect and record the power devices through the control and test platform to get the power devices on and off. The test platform with basic functions can have a functional experimental platform, which is easy to implement and operate.

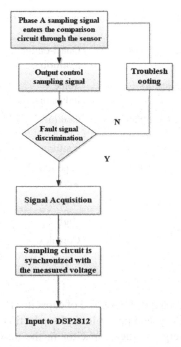

Fig. 5. Signal acquisition system flowchart.

From Table 1, it is clear that the different types of test platforms for power devices are on the experimental platform. According to different test-oriented functional indicators, different experimental platforms are selected. This is the most important feature of IGBT power device experimental platforms. For different application environments, different experimental platforms are used. For some functional indicators, it is advisable to use offline experimental platforms, but for real working environments, quasi-online platforms are more suitable, and comprehensive. For the power device experiment platform, a fast test experiment platform is appropriate. This paper uses a rapid test experimental platform with some functions of power devices.

The sampling circuit uses VSM025A Hall voltage sensor for detecting DC, AC or pulse voltage of 10–500 V. The control chip used in this main control board is the TMS320F2812 chip, with a processing frequency of 150 MHz, and an on-chip ROM of 256bit and 16 bits.

Signal acquisition is of great significance for feedback control and data statistics of power systems. How to effectively collect a variety of different voltage signals is an important step to solve many problems.

The hardware test platform has 5 parts, as shown in Fig. 6, which are power module, rectifier module, inverter module, sampling module and control module. Setting the time axis, channel sensitivity, and trigger method of the oscilloscope are necessary for the operation, and saving the waveform data recorded by the oscilloscope to the computer can help researchers analyze the output. The energy storage capacitor on the DC side plays an important role in providing a stable voltage to the load. It mainly plays the role

Table 1. Reduction in maximum temperature of different locations.

Content	Offline experiment platform	Quasi-online platform	Online test platform	Rapid test platform
Test indicators	Some indicators	Some indicators	All indicators	Main indicators
Working device	Dedicated test platform	Control signal/test platform	Actual working device	Actual working device
Testing time	Non-working time	A period of work	A period of work	A period of work
Working condition	Non-simulation/non-real	Analog/real	Real	Real
Experimental cost	Medium	High	High	Low
Degree of system implementation	Simple	Complex	Complex	Simple

of non-functional energy exchange. The choice of capacitor is related to t the magnitude of current, switching frequency, and power limit. Its selection can refer to the selection method of the DC side energy storage capacitor in the maximum power tracker. In addition, the ripples of the output current and the dynamic characteristics of the inverter affect the inductance parameters of the low-pass filter. This experimental platform tests IGBT power devices, inverter control, and signal acquisition.

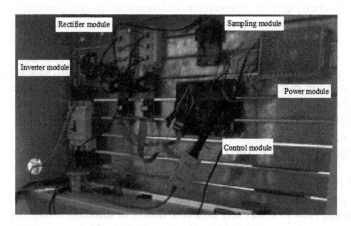

Fig. 6. System experiment platform.

3.2 Characteristics of the Experimental Platform

In order to quickly test the low voltage input and test the IGBT module, it provides a basic experimental platform. It has the experimental characteristics of brief operation, high testing efficiency. In order to further study the device in complex system environment, it has better experimental characteristics. Among them, two 2-bridge IGBTs constitute a single-phase inverter module. During the turn-on delay period, the power supply voltage Vcc is 600 V, the gate emitter voltage VGE is 15 V, and Ic is 100 A, turn-on equivalent resistance RGon is 6.8 Ω, and the maximum on-time is 260 ns.

4 Analysis of Results

The experimental analysis by setting the voltage sampling unit is shown in Fig. 7. With high accuracy measurement, the sampling frequency can change with the frequency of the sampling signal, keep the number of samples in one cycle.

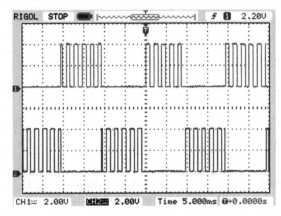

Fig. 7. The output of sampling pulse and reversed phase sampling pulse.

The two complementary synchronous sampling control pulse signals are generated by the phase-locked loop in the sampling circuit, which can achieve effective sampling with the measured signal. Rectified by Rectifier circuit, the sampled load voltage has been rectified as follows in Fig. 8.

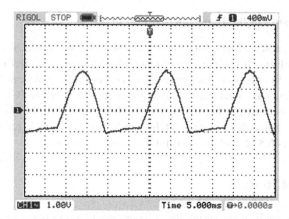

Fig. 8. The wave of rectified load voltage.

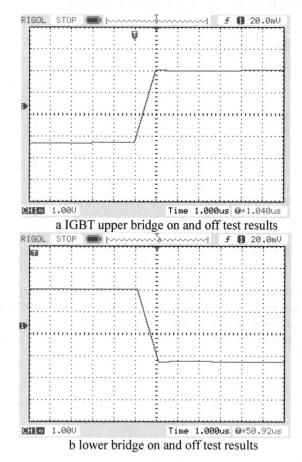

a IGBT upper bridge on and off test results

b lower bridge on and off test results

Fig. 9. PWM input waveform, 3.3 V, 10 kHz.

The output of the rectified waveform is realized here, and a positive waveform can be obtained.

In order to maximize the efficiency of IGBT, the inhibition of miller platform is one of the most important criteria to improve efficiency. By testing different frequency of PWM input, the correct frequency of PWM pulses can be found.

In the equipment, the input voltage of IGBT varies from 15 to 25 V, because the system is designed for proving the eligibility of virtual comparison rather than in real conditions.

In this experiment, the upper and lower bridges of the IGBT module were tested for continuity. The PWM input provides 3.3 V, 10 kHz, as shown in Fig. 9, Fig. 10, and Fig. 11, the analysis shows that the rise time 4 us and the fall time 4 us.

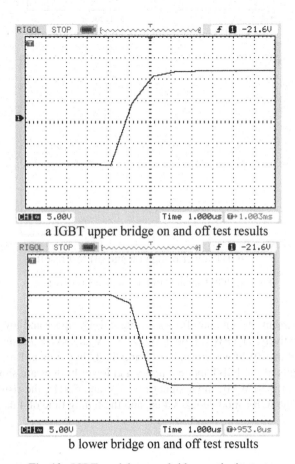

a IGBT upper bridge on and off test results

b lower bridge on and off test results

Fig. 10. IGBT module upper bridge continuity test.

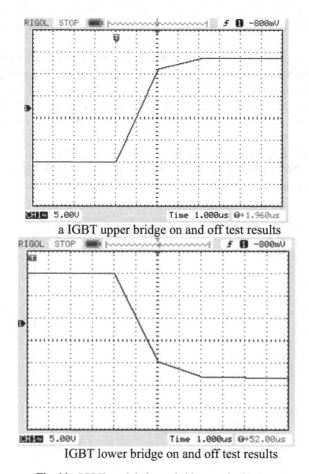

a IGBT upper bridge on and off test results

IGBT lower bridge on and off test results

Fig. 11. IGBT module lower bridge continuity test.

In this experiment, the on and off tests of the IGBT's upper and lower bridges are conducted respectively. The PWM input provides 3.3 V and 5 kHz. As shown in Fig. 12, the rising time of the PWM waveform is 1 us and the falling time is 1 us.

The test results of the IGBT module upper and lower bridges are shown in Fig. 13 and Fig. 14. After the PWML input signal is 3.3 V and 5 kHz, the rising time of the upper bridge of the IGBT module takes 4 us, and the falling time of 4 us.

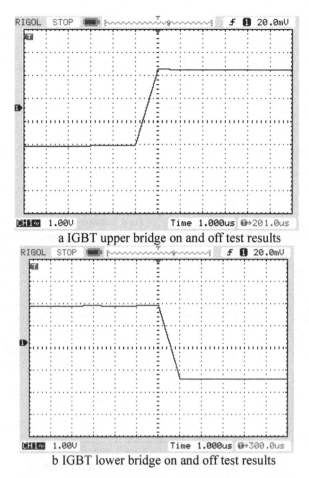

a IGBT upper bridge on and off test results

b IGBT lower bridge on and off test results

Fig. 12. PWM input 3.3 V, 5 kHz.

IGBT Module 1 uses 4 us for the rising edge time of the lower bridge and 4 us for the falling edge time, and the on and off times of its upper and lower bridges are within 4 us.

The above analysis shows that in order to ensure that the upper and lower bridges of the IGBT module are not connected at the same time, and to improve the performance of the Miller platform, the system dead time parameter is set to 8 us, but the on and off times of the module are not changed before and after the frequency increase.

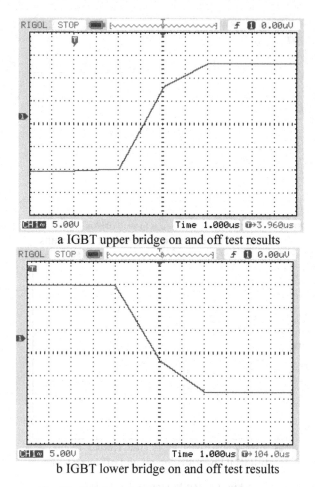

a IGBT upper bridge on and off test results

b IGBT lower bridge on and off test results

Fig. 13. IGBT module upper bridge continuity test.

Two IGBT drives, two PWM drives, and complementary, the amplitude reaches 20 V, the frequency is 6.4 K, the PWM drive is complementary, and the local is placed large in Fig. 15 and Fig. 16. There is a dead zone within this 4.8 us, which helps reduce and improve the reliability of the drive.

Here, the IGBT drive failed, complementarity was not achieved, and complementary PWM drive occurred. The low-voltage implementation needs to be tested. First, test the PWM and transfer to the IGBT drive. The voltage amplitude at this time is 2.60 V.

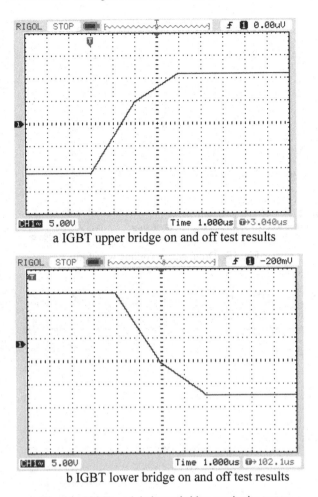

a IGBT upper bridge on and off test results

b IGBT lower bridge on and off test results

Fig. 14. IGBT module lower bridge continuity test.

After the adjustment, the IGBT drive waveform, low voltage drive and other tests have been passed. Setting the input AC voltage with 48 V, output voltage waveform, output 4.6 V, 1.6 kHz. The output has sine wave characteristics, pulse wave characteristics, and stable waveforms. When low-voltage input is used, the inverter output is stable, achieving low-voltage input and inverter-stable output characteristics.

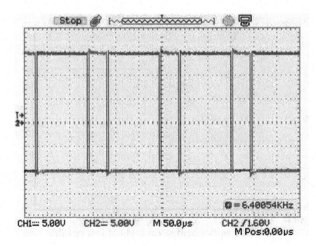

Fig. 15. Driving part working waveform.

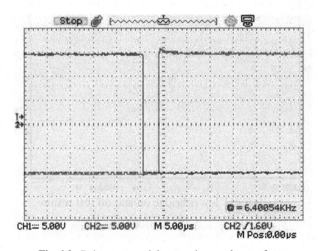

Fig. 16. Drive part partial expansion work waveform.

The experimental results of the inverter part are shown in Fig. 17 and Fig. 18, 4.62 V, 1.6 kHz, and a partial graph, which can output a pulse shape, and the pulse amplitude gradually changes to both sides, the middle ratio is larger, and the ratio at both ends gradually decreases.

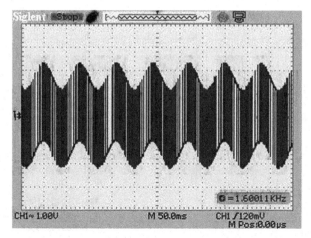

Fig. 17. Low voltage input inverter output waveform.

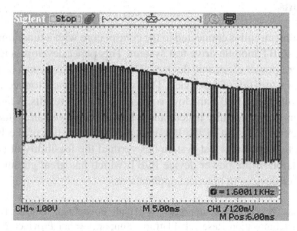

Fig. 18. Low-voltage input inverter output waveform partial expansion diagram.

Setting a low voltage input of 110 V, the experimental results show that the inverter output waveform has a pulse characteristic, generating a constant pulse width ratio. The waveform in Fig. 19 has the same pulse amplitude. Similar duty cycle, the experimental output value is 345 V, and the experimental results have stable and balanced pulse width performance.

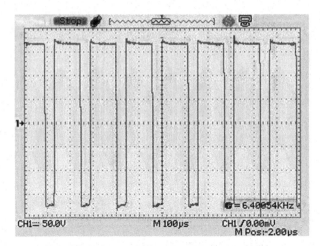

Fig. 19. Low voltage inverter output waveform.

5 Conclusions

Each functional unit circuit is connected to the control core with a functional interface to form a complete IGBT module drive test system. This paper develops a test system for switching losses of power electronic devices. The rapid test device can be used to study the switching characteristics of the device. The system is a user-customized platform that provides a low-voltage test IGBT and low-voltage inverter test results. A fast low-voltage test IGBT is been experimentally tested, and a low-voltage test can be used to quickly test the IGBT device. The test of the low-voltage characteristics of the IGBT module switching characteristics rapid test system has verified the feasibility of this method in the normal operation of the power device IGBT, which has great guiding significance for the selection of devices in practical applications.

Acknowledgments. This work is supported by Anhui Polytechnic University of Engineering Energy Internet Innovation Research Platform Support, Anhui Province Natural Science Foundation (1908085MF215), Pre-research of National Natural Science Fund Project of Anhui Polytechnic University (Xjky02201905).

References

1. Bai, H., Liu, C., Rathore, A.K.: An FPGA-based IGBT behavioral model with high transient resolution for real-time simulation of power electronic circuits. IEEE Trans. Ind. Electron. **66**(8), 6581–6591 (2019)
2. Al'Akayshee, Q., Reynolds, S., Golland, A., et al.: Advance high power semiconductors devices in drives and power conversion. In: 6th IET International Conference on Power Electronics, Machines and Drives, Briston, UK, pp. 1–4 (2012)
3. Scharnholz, S., Schneider, R., Spahn, E., et al.: Investigation of IGBT-devices for pulsed power applications. In: 14th IEEE International Pulsed Power Conference, Dallas, TX, pp. 349–352 (2003)

4. Eicher, S., Rahimo, M., Tsyplakov, E., et al.: 4.5 kV press pack IGBT designed for ruggedness and reliability. In: 39th IAS Annual Meeting Industry Applications Conference, Seattle, WA, pp. 1534–1539 (2004)
5. Zeng, Z., Zheng, W., Zhao, R., et al.: Modeling, modulation, and control of the three-phase four-switch PWM rectifier under balanced voltage. IEEE Trans. Power Electron. **31**(7), 4892–4905 (2015)
6. Wang, Z., Shi, X., Tolbert, L.M., et al.: A di/dt feedback based active gate driver for smart switching and fast overcurrent protection of IGBT modules. IEEE Trans. Power Electron. **29**(7), 3720–3732 (2014)
7. Luo, H., Li, W., He, X.: Online high-power P-i-N diode chip temperature extraction and prediction method with maximum recovery current di/dt. IEEE Trans. Power Electron. **30**(5), 2395–2404 (2015)
8. Bryant, A., Yang, S., Mawby, P., et al.: Investigation into IGBT dv/dt during turn-off and its temperature dependence. IEEE Trans. Power Electron. **26**(10), 3019–3031 (2011)
9. Huang, A.Q., Crow, M.L., Heydt, G.T., et al.: The future renewable electric energy delivery and management (FREEDM) system: the energy internet. Proc. IEEE **99**(1), 138–144 (2011)
10. Yang, S., Bryant, A., Mawby, P., et al.: An industry-based survey of reliability in power electronic converters. IEEE Trans. Ind. Appl. **47**(3), 1441–1451 (2011)
11. Barlini, D., Ciappa, M., Castellazzi, A., et al.: New technique for the measurement of the static and of the transient junction temperature in IGBT devices under operating conditions. Microelectron. Reliab. **46**(9), 1772–1777 (2006)
12. Avenas, Y., Dupont, L., Khatir, Z.: Temperature measurement of power semiconductor devices by thermo-sensitive electrical parameters-a review. IEEE Trans. Power Electron. **27**(6), 3081–3092 (2012)
13. Sheng, K., Williams, B.W., Finney, S.J.: A review of IGBT models. IEEE Trans. Power Electron. **15**(6), 1250–1266 (2000)
14. Lu, L., Bryant, A., Hudgins, J.L., et al.: Physics-based model of planar-gate IGBT including MOS side two-dimensional effects. IEEE Trans. Ind. Appl. **46**(6), 2556–2567 (2010)
15. Castellazzi, A., Batista, E., Ciappa, M., et al.: Full electro-thermal model of a 6. 5 kV field-stop IGBT module. In: Proceedings of the 39th IEEE Power Electronics Specialist Conference, Rhodes, Greece, pp. 392–397 (2008)
16. Cotorogea, M.: Physics-based spice-model for IGBTs with transparent emitter. IEEE Trans. Power Electron. **24**(12), 2821–2832 (2009)
17. Sheng, K., Williams, B.W., Finney, S.J.: A review of IGBT models. Power Electron. **15**(6), 1250–1266 (2000)
18. Marinov, A., Valchev, V.: Power loss reduction in electronic inverters through IGBT-MOSFET combination. Procedia Earth Planet. Sci. **1**(1), 1539–1543 (2009)
19. Mandeya, R., Chen, C., Pickert, V., et al.: Gate-emitter pre-threshold voltage as a health sensitive parameter for IGBT chip failure monitoring in high voltage multichip IGBT power modules. IEEE Trans. Power Electron. **34**(9), 9158–9169 (2018)
20. Rosas-Caro, J., Mancilla-David, F., Ramirez-Arredondo, J., Bakir, A.M.: Two-switch three-phase ac-link dynamic voltage restorer. Power Electron. IET **5**(9), 1754–1763 (2012)
21. Huang, J., Huang, H., Chen, X.B.: Simulation study of a low ON-state voltage superjunction IGBT with self-biased PMOS. IEEE Trans. Electron. Devices **66**(7), 3242–3246 (2019)
22. Tang, Z., Yu, X., Jin, X.: Compensation strategy and simulation of dynamic voltage restorer. Power Syst. Autom. **32**(4), 63–66 (2004)
23. Yang, Y., Ruan, Y., Tang, Y.Y., et al.: Three-phase grid-connected inverters based on PLL and virtual grid flux. Trans. China Electrotech. Soc. **25**(4), 109–114 (2010)
24. Busca, C.: Modeling lifetime of high power IGBTS in wind power applications anoverview. In: IEEE International Symposium on Industrial Electronics Records, pp. 1408–1413 (2011)

25. Bazzi, A.M., Krein, P.T., Kimball, J.W., et al.: IGBT and diode loss estimation under hysteresis switching. IEEE Trans. Power Electron. **27**(3), 1044–1048 (2012)
26. Mitsubishi Electric IGBT Modules Application Note. http://www.mitsubihsielectric.com/cn
27. Infineon IGBT datasheet (2014). http://www.Infineon.com/cms/en/product/index.html
28. Sheng, K., Williams, B.W., Finney, S.J.: A review of IGBT models. IEEE Trans. Power Electron. **15**(6), 1250–1266 (2000)
29. Chen, G., Han, D., Mei, Y.H.: Transient thermal performance of IGBT power modules attached by low-temperature sintered nanosilver. IEEE Trans. Device Mater. Reliab. **12**(1), 124–132 (2011)

Robust Stability of Single Phase PWM Rectifier Based on H_∞ Loop Shaping

Motaz Musa Ibrahim, Lei Ma$^{(\boxtimes)}$, Haoran Liu, and Na Qin

Southwest Jiaotong University, Chengdu 610031, China
mootaz99@hotmail.com, malei@swjtu.edu.cn

Abstract. Control of grid-side rectifiers for locomotive applications must cope with parametric fluctuations, dramatic change of operation condition such as grid voltage, as well as malfunctions and faults of switching devices and sensors. With the introduction of vehicular electronic transformers, fragile power-electronics is directly connected to the grid, even though with a filtering coil. This makes the afore mentioned problems much more serious. Thus the robustness and fault tolerance become more and more important for this kind of devices. Robust H_∞ loop shaping controller is designed for current loop control of single phase PWM rectifier. The main objective is to study the performance and robustness against parametric uncertainties based on ν gap metric.

Keywords: PWM rectifier · Loop shaping · Robustness · Gap metric

1 Introduction

Single phase PWM rectifiers are widely used in traction systems. High speed train works in harsh environment faults, disturbances and parametric uncertainties occur frequently during the train operation. It is necessary to address robust control for the rectifiers to enhance the performance and robustness of the traction system to keep safe and stable operation of the train in such circumstances.

H_∞ control proved high reliability in traction systems and many other applications. H_∞ controller designed by [1] justified that disturbance rejection can be achieved. H_∞ control is proposed by [2] to suppress low frequency oscillation in traction system. An attitude controller for unmanned helicopter using H_∞ loop shaping method is proposed by [3], pre-compensator and post-compensator have been chosen to improve the singular values of the nominal system to get a desired loop shape, the controller achieved good performance. A control strategy based on H_∞ loop shaping is designed by [4], robust stability and robust performance is guaranteed, also in the design procedure pre-compensator and post-compensator are selected to make the singular values of the nominal controlled system become the desired open loop gain.

This work is supported by the NSFC under grant 61733015 and U1934204.

© Springer Nature Singapore Pte Ltd. 2020
M. Fei et al. (Eds.): LSMS 2020/ICSEE 2020 Workshops, CCIS 1303, pp. 157–168, 2020.
https://doi.org/10.1007/978-981-33-6378-6_12

Gap metric is introduced in 1981 by [5] to control literature to study the robustness of feedback systems, which is used to show that admissible uncertainties are constrained in the gap [6]. A new metric proposed by [7], known as ν gap metric has frequency response explanation and not greater than gap metric.

The motivation of using ν gap metric in this work is to answer the question: given a controller subjected to parametric uncertainties within a certain range will the system be stable? and what is the worst case uncertainty before the system loss stability? In this work grid side inductance variation is taken as examples for investigation. This variation in certain range leads to set of perturbed plants. H_∞ loop shaping controller stabilizing capability is investigated based on ν gap metric, the ability of the controller to stabilize the system depends on its stability margin. This must be greater than the ν gap metric between the nominal plant and perturbed plant, because the controller is originally designed for the nominal plant if it is required to stabilize the perturbed plant the condition must be satisfied.

The paper is organized as follows. Section 2 gives a mathematical model of single phase PWM rectifier in synchronously rotating coordinate. Section 3 about the controller design Sect. 4 shows the simulation results and discussion and finally the conclusion.

2   Mathematical Model

The basic circuit of single phase PWM rectifier is described by Fig. 1, e_s is the AC side input voltage L, R represent the inductance and resistance of AC side, C is the capacitance of DC side and R_L is the load resistance.

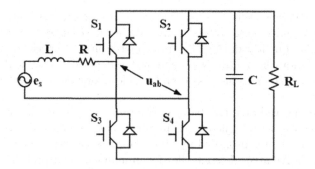

Fig. 1. Architecture of single phase PWM rectifier

According to Fig. 1 Kirchoff's voltage law can be written as:

$$L\frac{di_s}{dt} = e_s - Ri_s - u_{ab} \tag{1}$$

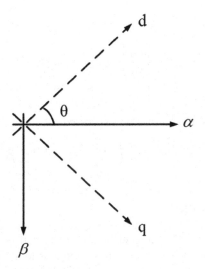

Fig. 2. $\alpha\beta$ and dq reference Frame

The transformation from stationary coordinate to synchronously rotating coordinate as shown in Fig. 2 can be expressed as [8]

$$\begin{bmatrix} d \\ q \end{bmatrix} = \begin{bmatrix} \cos(\theta) & \sin(\theta) \\ -\sin(\theta) & \cos(\theta) \end{bmatrix} \begin{bmatrix} \alpha \\ \beta \end{bmatrix} \tag{2}$$

$$\begin{bmatrix} \alpha \\ \beta \end{bmatrix} = \begin{bmatrix} \cos(\theta) & -\sin(\theta) \\ \sin(\theta) & \cos(\theta) \end{bmatrix} \begin{bmatrix} d \\ q \end{bmatrix} \tag{3}$$

Equation (1) can be written in dq component as:

$$\begin{aligned} e_{sd} &= L\frac{di_{sd}}{dt} + Ri_{sd} - \omega Li_{sq} + u_{abd} \\ e_{sq} &= L\frac{di_{sq}}{dt} + Ri_{sq} + \omega Li_{sd} + u_{abq} \end{aligned} \tag{4}$$

The previous equation can be written as:

$$\begin{aligned} L\frac{di_{sd}}{dt} &= v_d - Ri_{sd} + \omega Li_{sq} \\ L\frac{di_{sq}}{dt} &= v_q - Ri_{sq} + \omega Li_{sd} \end{aligned} \tag{5}$$

The state space equation can be expressed as:

$$\dot{x} = \begin{bmatrix} -\frac{R}{L} & \omega \\ -\omega & -\frac{R}{L} \end{bmatrix} x + \begin{bmatrix} \frac{1}{L} & 0 \\ 0 & \frac{1}{L} \end{bmatrix} u$$

$$y = \begin{bmatrix} 1 & 0 \\ 0 & 1 \end{bmatrix} x$$

where:

$$x = \begin{bmatrix} i_{sd} \\ i_{sq} \end{bmatrix}$$
$$u = \begin{bmatrix} v_d \\ v_q \end{bmatrix} = \begin{bmatrix} e_s - u_{abd} \\ -u_{abq} \end{bmatrix}$$

Then the state space matrices can be expressed as:

$$A = \begin{bmatrix} -\frac{R}{L} & \omega \\ -\omega & -\frac{R}{L} \end{bmatrix}, B = \begin{bmatrix} \frac{1}{L} & 0 \\ 0 & \frac{1}{L} \end{bmatrix}, C = \begin{bmatrix} 1 & 0 \\ 0 & 1 \end{bmatrix}, D = \begin{bmatrix} 0 & 0 \\ 0 & 0 \end{bmatrix}$$

3   Controller Design

The control of single phase PWM rectifier basically consist of two control loops, namely the outer voltage loop and the inner current loop. The control objective of voltage loop control is just to keep constant DC voltage, a PI controller can be used to achieve this purpose, but the inner current loop control is responsible of AC current control such that the effect of disturbance or parametric uncertainties should be eliminated in order to keep sinusoidal AC current with same grid frequency. Thus the inner current loop robust control for single phase PWM rectifier is very essential aspect. Figure 3 shows the control system block diagram for single phase pwm rectifier, H_∞ loop shaping controller is implemented for current loop control and PI controller is adopted to achieve dc-link voltage regulation.

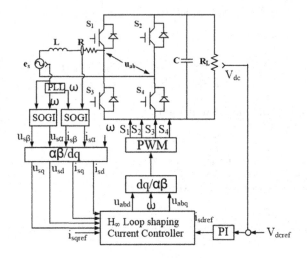

Fig. 3. Block diagram of proposed control scheme

3.1 H_∞ Loop Shaping

The most important item in H_∞ loop shaping design is the pre-compensator W_1 and post-compensator W_2 . These weights are used to shape the open loop frequency response of the nominal plant. For a given plant, the weighting functions are chosen to achieve sufficiently high loop gain at low frequency for disturbance attenuation and low loop gain at high frequency for good noise rejection as illustrated in Fig. 4.

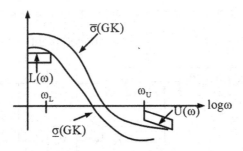

Fig. 4. Open loop singular value shaping

The H_∞ loop shaping design can be summarized as follow:

– Loop shaping: select pre-compensator W_1 and/or pos-compensator W_2, the singular value of the nominal plant are shaped to give a desired open-loop shape as shown in Fig. 5 . the shaped plant $G_s = W_2 G W_1$
– Robust Stabilization: Select $\varepsilon \leq \varepsilon_{\max}$ then synthesis a stabilizing controller K_s which satisfies:

$$\varepsilon_{\max} = \left\| \begin{bmatrix} I \\ K_s \end{bmatrix} (I - G_s K_s)^{-1} \tilde{M}_s^{-1} \right\|_\infty$$
$$= \sqrt{1 - \left\| [\tilde{M}_s \ \tilde{N}_s] \right\|_H} < 1$$

where $\tilde{M}_s$ and $\tilde{N}_s$ are coprime decomposition of G_s as follow:

$$G_s = \tilde{M}_s \tilde{N}_s^{-1}$$

The final feedback controller K is obtained by combination of the H_∞ controller K_s with shaping functions W_1 and W_2 such that $K = W_1 K_s W_2$ as illustrated in Fig. 6:

3.2 ν Gap Metric

ν gap metric is used to quantify the difference between two systems, it is powerful tool to study uncertainty and robustness in feedback systems.

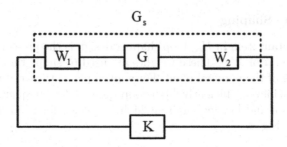

Fig. 5. Plant with pre- and post-compensators

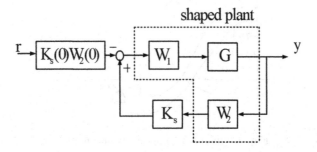

Fig. 6. Loop shaping controller

Suppose G and G_Δ are two systems, the ν gap distance between them can be expressed as:

$$\delta_\nu = \begin{cases} \psi(G, G_\Delta) \text{ if } \begin{cases} \det(I + G_\Delta^* G) \neq 0 \forall \omega \\ \text{wnodet}(I + G_\Delta^* G) + \eta(G) - \eta(G_\Delta) - \eta_0(G_\Delta) = 0 \\ 1 \qquad\qquad\qquad\qquad\qquad \text{otherwise} \end{cases} \end{cases} \quad (6)$$

where:

$$\psi(G, G_\Delta) = (I + G_\Delta G_\Delta^*)^{-1/2}(G - G_\Delta)(I + G^* G)^{-1/2}$$

η indicates the number of right half-plane (RHP) poles, η_0 indicates the number of poles on the imaginary axis, wno denotes the winding number. The H_∞ norm $\|\psi(G, G_\Delta)\|_\infty$ defined as:

$$\|\psi(G, G_\Delta)\|_\infty := \sup_\omega \bar\sigma \left(\psi(G(j\omega), G_\Delta(j\omega)) \right) \quad (7)$$

$\bar\sigma(\psi)$ is the maximum singular value of ψ and ω denotes angular frequency ($s = j\omega$)

The generalized stability margin of the closed loop system as shown in Fig. 7 $[G, C]$ is defined as:

$$b_{G,C} := \begin{cases} \|H(G, C)\|_\infty^{-1} \ if [G, C] is stable \\ 0 \qquad\qquad\qquad otherwise \end{cases} \quad (8)$$

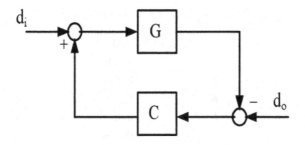

Fig. 7. Standard feedback configuration

Where $H(G, C) = \begin{bmatrix} I \\ C \end{bmatrix} (I - GC)^{-1} \begin{bmatrix} G & I \end{bmatrix}$

Theorem 1. *Suppose (G_0, C) is stable and $\delta_\nu(G_0, G_1) < 1$. Then (G_1, C) is stable if*

$$b_{G_0, C(\omega)} > \psi(G_0(j\omega), G_1(j\omega)), \forall \omega$$

Where $b_{G_0, C}(\omega)$ and $\psi(G_0(j\omega), G_1(j\omega))$ are defined as follows

$$\frac{1}{b_{G_0, C(\omega)}} := \bar{\sigma} \left(\begin{bmatrix} I \\ C(j\omega) \end{bmatrix} (I - G(j\omega)C(j\omega))^{-1} \begin{bmatrix} G(j\omega) & I \end{bmatrix} \right)$$

$$\psi(G_0(j\omega), G_1(j\omega)) := \bar{\sigma}(\psi(G_0(j\omega), G_1(j\omega)))$$

and $\psi(G_0(j\omega), G_1(j\omega)) := (1 + G_1(j\omega)G_0(j\omega))^{1/2}(G_0(j\omega) - G_1(j\omega))$
$(I + G_1(j\omega)G_0(j\omega))^{-1/2}$

4   Simulation Results

The simulation is done by MATLAB/SIMULINK environment and the parameters of the single phase PWM rectifier in this paper are given in Table 1.

From Fig. 8 the desired shape has high gain in low frequency and low gain in high frequency which guarantee a good tracking performance, ensure d axis current disturbance attenuation and robustness therefore W_1 is selected as:

$$W_1 = \begin{bmatrix} (1.3s + 1000)/(0.0001s + 30) & 0 \\ 0 & (1.3s + 1000)/(0.0001s + 30) \end{bmatrix}$$

and W_2 is chosen as identy matrix $W_2 = \begin{bmatrix} 1 & 0 \\ 0 & 1 \end{bmatrix}$

From Fig. 9 according to theorem 1 all perturbed plants $\Delta G_{L=1.2\,\mathrm{mH}}$ and $\Delta G_{L=2.7\,\mathrm{mH}}$ can be stabilized by the controller, the perturbed plant $\Delta G_{L=0.8\,\mathrm{mH}}$ will not be stabilized by the controller but for high frequencies the stability can be achieved.

Table 1. PWM rectifier parameters

PARAMETER	VALUE
V_s	1550V
V_{dc}	3000V
R	0.3Ω
L	5.4mH
C	0.009F
R_L	25Ω

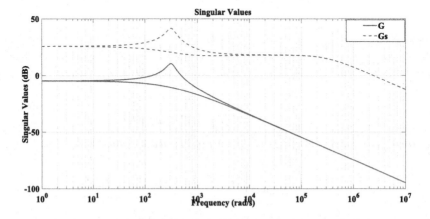

Fig. 8. The original plant and shaped plant

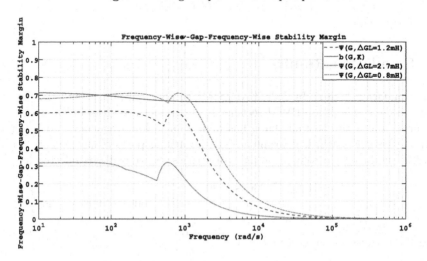

Fig. 9. Frequency response of stability margin $b_{G,C(\omega)}$ and $\psi(G, \Delta G)$

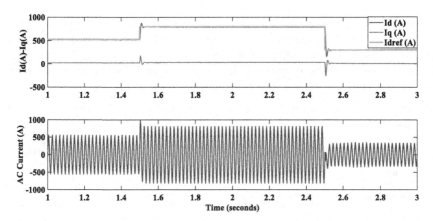

Fig. 10. dq axis currents ac current with nominal parameters under H_∞ loop shaping controller

Figure 10 shows when the reference current I_{dref} at 1.5 s changed to 800A the direct axis current I_d perform good tracking with very small maximum overshoot and settling time of 20 ms. At 2.5 s the reference current again changed to 300A also I_d is good tracking as the figure illustrate.

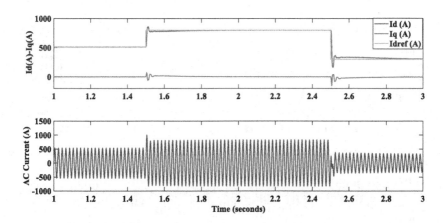

Fig. 11. dq axis currents ac current with nominal parameters under PI controller

PI controller is used for comparison with the proposed controller from Fig. 11 it is clear that tracking is not good as H_∞ loop shaping controller when the reference current changed to 800A the settling time is 108 msec. When the grid inductance is reduced to 2.7 mH and 1.2 mH From Fig. 12 and Fig. 13 respectively dq currents can be stabilized by the controller because the frequency response ν

gap metric for all frequencies is less than the stability margin as shown in Fig. 9, also good tracking performance is achieved. The inductance is reduced to 0.8mH refering to Fig. 9 the controller is not able to stabilize the system because ν gap metric exceeds the stability margin, the waveforms of dq currents and ac current show clearly that the system is not stable according to Fig. 14

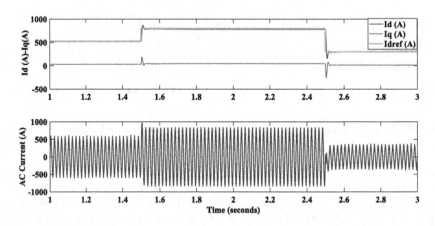

Fig. 12. dq axis currents ac current when $L = 2.7\,\text{mH}$ under H_∞ loop shaping controller

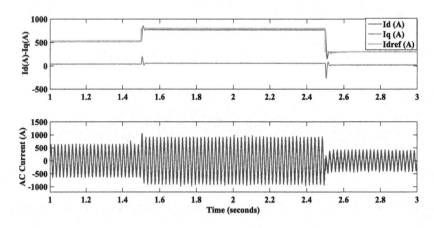

Fig. 13. dq axis currents ac current when $L = 1.2\,\text{mH}$ under H_∞ loop shaping controller

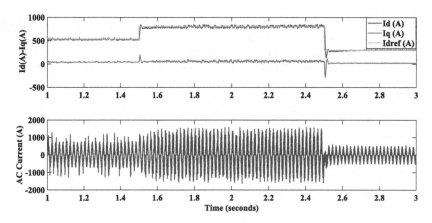

Fig. 14. dq axis currents ac current when $L = 0.8\,\mathrm{mH}$ under H_∞ loop shaping controller

5 Conclusion

H_∞ loop shaping is applied to inner current loop of single phase PWM rectifier. The simulation results shows that when the system encounter parametric uncertainties, H_∞ loop shaping delivers good performance and robustness. So we can conclude that H_∞ loop shaping can handle broad range of parametric uncertainties.The next step of research will be more specific control design according to different magnitude of the faults and parametric uncertainties. We shall investigate the variation of gap metric under these circumstances and carry out control design that place more emphasis on the performance. This could be approached by loosing the stability margin requirement for loop shaping.

References

1. Yao, S., Liu, Z., Chuan, X., Zhang, G.: H infinity control of line-side converter in electric multiple unit. In: 2016 12th World Congress on Intelligent Control and Automation (WCICA), pp. 1786–1790 (2016)
2. Geng, Z., Liu, Z., Xinxuan, H., Liu, J.: Low frequency oscillation suppression of the vehicle-grid system in high-speed railways based on h infinity control. Energies **1594** (2018)
3. Lv, M., Hu, Y., Liu, P.: Attitude control for unmanned helicopter using h-infinity loop-shaping method. In: 2011 International Conference on Mechatronic Science, Electric Engineering and Computer (MEC), pp. 1746–1749 (2011)
4. Li, Z., Zheng, C., Guo, F.: H infinity loop shaping control for quadruple tank system. In: 2014 6th International Conference on Intelligent Human-Machine Systems and Cybernetics (IHMSC) (2014)
5. Ahmed, K.: Elsakkary, the gap metric: robustness of stabilization of feedback systems. IEEE Trans. Autom. Control, pp. 240–247 (1985)
6. Georgiou Tryphon, T., Smith Malcolm, C.: Optimal robustness in the gap metric. In: Proceedings of the 28th IEEE Conference on Decision and Control, pp. 2331–2336 (1989)

7. Vinnicombe, G.: Frequency domain uncertainty and the graph topology. IEEE Trans. Autom. Control **38**, 1371–1383 (1993)
8. Wan, Z., Ren, J., Wan, T., Jing, S., Xiong, S., Guo, J.: A control strategy of single phase voltage source PWM rectifier under rotating coordinate. In; 2017 2nd International Conference on Power and Renewable Energy (ICPRE), pp. 162–165 (2017)

State Identification of Three-Trans Towers' Bolts Based on Quadratic Wavelet Transform and KNN

Yanhong Luo, Xuefang Liu[✉], Bowen Zhou, Dongsheng Yang, Qiubo Nie, and Songsheng Wang

College of Information Science and Engineering, Northeastern University, Shenyang 110819, China
1870487@stu.neu.edu.cn, 15940264143@163.com

Abstract. The environment of the three-trans towers is complicated, with the influence of strong noise and non-deterministic interference, the features of the vibration response signal of bolts are very weak, which is difficult to identify the bolts' state. In order to avoid man-made damage to the tower and realize the identification of bolt state under natural vibration conditions, a state identification method based on quadratic wavelet transform and KNN was proposed. First, the wavelet denoising method based on new threshold function was used to filter out the noise in original signal. Then, the dB10 wavelet function commonly used in fault diagnosis was selected to decompose and reconstruct the denoised signal, the dimensionless time domain characteristics of the low-frequency signal component and the information entropy of the high-frequency detail component were calculated respectively. Finally, the bolts state feature dataset was established. By comparing common classifiers, KNN with the best performance was selected. The experimental case analysis verified the effectiveness of the proposed method for bolts state identification. The overall research has certain practical and engineering value.

Keywords: Bolt state identification quadratic · Wavelet transform · Feature extraction · KNN

1 Introduction

The transmission lines crossing high-speed railways, highways, and important channels often have tower collapse accidents due to the destruction of external forces, which greatly threatens the safety of pedestrians and traffic below [1]. Bolts are important components that affect the safe operation of transmission towers, monitoring their connection state can effectively avoid the accidents. However, common maintenance methods, such as ground inspections, pole inspections and man-made knocking, have problems such as time-consuming and labor-intensive, untimely maintenance, dependence on subjective experience, damage to the towers, which are difficult to meet actual needs.

© Springer Nature Singapore Pte Ltd. 2020
M. Fei et al. (Eds.): LSMS 2020/ICSEE 2020 Workshops, CCIS 1303, pp. 169–183, 2020.
https://doi.org/10.1007/978-981-33-6378-6_13

Considering the three-trans towers are affected by natural vibration obviously, a bolt state identification method without the damage caused by man-made knocking is proposed. According to the different responses of vibration signal before and after bolt loosening, the state of the bolt is judged. But the characteristics of the original signal are weak and cannot be recognized intuitively. Therefore, the denoising and feature extraction of bolt state signals are mainly studied.

The bolts state identification based on vibration signal is similar to traditional fault diagnosis, focusing on the analysis of fault features. In [2], a coupled hidden markov method based on gaussian mixture model is proposed, which introduced whitening algorithm and stepping algorithm to deal with weak fault signals. In [3], a vibration signal filtering method based on Kullback-Leibler divergence genetic algorithm is proposed. Then various types of faults are identified by envelope demodulation with effective fault characteristic frequency. In [4], a method for extracting features of rolling bearings under noise background by wavelet singularity detection technology is proposed, which effectively reduce the influence of noise on signal characteristics. Yan et al. [5] proposed a multi-scale morphological analysis algorithm. Combined with entropy weight and grey relational analysis, the multi-domain fault features of rolling bearings are extracted. In [6], a method combined with integrated empirical mode decomposition, autoregressive and support vector regression is proposed to establish the state prediction model, which can provide a valid reference for abnormal detection of high-speed train gearbox. L. Bo et al. [7] proposed an intelligent analysis method for the fault diagnosis of element rolling bearing, vibration nonlinear features are extracted by recurrence quantification analysis based on the integrated interaction relationship, then the nonredundant feature subset is established. S. Maurya et al. [8] proposed a novel method for feature extraction, which includes fusion of handcrafted features derived from empirical mode decomposition and refined features extracted from deep neural network, then followed by feature selection on fused features. X. Zhang et al. [9] proposed a fault diagnosis method based on variational mode decomposition and modulus squared threshold. two main intrinsic modal function components were selected based on the kurtosis to reconstruct the signal, then the reconstructed signal is denoised by square-modulus threshold wavelet, finally, the fault characteristic frequency of the de-noised signal was extracted. In [10], a fault diagnosis method to optimize the whole process is proposed to improve separately the three stages of feature extraction, feature selection, and mode classification. Firstly, the signal is processed by non-local mean denoising, then the denoised signal is reconstructed by empirical mode decomposition. Next 9 time-domain characteristic indicators and 4 frequency domain characteristic indicators are calculated separately. Feature subsets are selected based on distance evaluation after two stages of feature selection. By comparing three common classification algorithms, the optimal classification model is selected. In [11], a fault diagnosis method based on multiple features is proposed, which includes 11 time-domain features and 5 types of entropy of the signal. Then these characteristic indicators are constructed the multi-dimensional optimal feature dataset. And the Light-GBM classification model is trained, finally the accurate identification of mechanical fault state of the high-voltage circuit breaker is achieved. F. Li et al. [12] proposed an effective analysis method with the optimal spectral mode separation. The mode separation based on wavelet transform and adaptive wavelet bank design can generate a better time-frequency resolution, which is helpful to extract the time-frequency information of the signal.

Summarizing the above methods, it can be known selecting the appropriate noise reduction method and feature extraction strategies are the key of fault recognition. Furthermore, a novel bolts state identification method is proposed. The remainder of this paper is organized as follows. Section 2 introduces the improved wavelet denoising method with new threshold function, extracted multi-dimensional time domain feature extraction, the concept of information entropy and KNN algorithm principle. Section 3 introduces the overall framework of the proposed method. Section 4 presents the result analysis of measured data. Finally, Sect. 5 concludes the paper.

2 Signal Denoising and Feature Extraction

2.1 Wavelet Denoising Based on New Threshold Function

In order to reduce the impact of noise mixed in original signal, and obtain the multidimensional characteristics of different states, the signal multi-level denoising process is studied. Signal processing method based on wavelet decomposition has good timefrequency focusing. The wavelet threshold can be selected at any level and the noise can be effectively suppressed for the decomposed signal [13]. However, the form of threshold function and the setting of thresholds are the key to wavelet denoising. There are four common forms: hard threshold, soft threshold, soft-hard threshold tradeoff and new threshold function.

For the wavelet coefficients outside the threshold range, the hard threshold function doesn't handle, but the coefficients within the threshold are set to zero. However, for the soft threshold function, the coefficients outside the threshold range gradually shrink toward zero, and the coefficients within the threshold range are set to zero. The soft and hard threshold trade-off method controls the threshold function by introducing an adjustable parameter, and changes the adjustment parameter to reduce the constant error between the actual coefficients and the estimated coefficients [14].

At the threshold, the hard threshold function is discontinuous, which causes the reconstructed signal to oscillate. For the soft threshold function, there is a constant deviation between the estimated coefficients and the real coefficients, which leads to severe distortion of the signal reconstructed from coefficients, although it's continuous. The soft and hard threshold tradeoff method overcomes the shortcomings of the above functions to a certain extent, but the above problems still exist. Therefore, consider introducing e-index and establish the new threshold function:

$$\hat{W}_{j,k} = \begin{cases} \text{sgn}(W_{j,k})\Big(\big|W_{j,k}\big| - \alpha\lambda e^{(1-\alpha)(\lambda - |W_{j,k}|)}\Big), \big|W_{j,k}\big| \geq \lambda \\ 0, \big|W_{j,k}\big| < \lambda \end{cases} \tag{1}$$

λ is the set threshold, $W_{j,k}$ is the decomposed wavelet coefficient, $\hat{W}_{j,k}$ is the estimated wavelet coefficient, j is the number of decomposition layers, $\text{sgn}(x)$ is the sign function, changing the value of α can adjust flexibly the denoising effect of the threshold function, $0 < \alpha < 1$.

The new threshold function avoids the shortcomings of the traditional function, its threshold can make adaptive adjustments as the scale of the wavelet decomposition changes, making it more flexible and practical. Adjustable parameters determine the degree of approximation of the wavelet threshold, which can retain some high-frequency signals when processing noisy signals, and better suppress signal oscillation.

The threshold is another key of wavelet denoising. Using the same threshold for wavelet coefficients of different decomposition layers may cause the distortion of the reconstructed signal. And the threshold should change with the number of layers. The greater the number of layers, the smaller the threshold. The layered threshold is:

$$\lambda_j = \frac{\sigma\sqrt{2\log_2 N}}{\ln(j+1)} \tag{2}$$

λ_j is the threshold of layer j, λ_j decreases with the increase of decomposition scale j, which is consistent with the variation characteristics of the coefficients of noise on each decomposition scale. N is the length of the signal, σ is the standard deviation of noise, in practice, σ is always unknown and its estimated value is used:

$$\sigma = \frac{median(|W_{1,k}|)}{0.6745} \tag{3}$$

$W_{1,k}$ is the wavelet coefficient of the first layer, $median(x)$ is the median operation.

2.2 Multi-feature Extraction Based on Quadratic Wavelet Transform

Extracting the feature in different original signals is the key to identify the bolts state. The above denoised signal is divided into multiple shorter periods, each short period is called a frame, and the signal processing for one frame is equivalent to continuous signal processing with fixed characteristics.

After wavelet decomposition, the high and low frequency components contain different characteristics. In order to extract features at different resolutions, db10 is selected as the wavelet basis function to decompose the denoised signal [15], and the high frequency information is separated, the low frequency information is retained. Then the feature calculation is performed on the high and low frequency components respectively.

Time domain feature analysis is a direct and effective method. When the bolt is loose, a small gap will be formed near the junction. With the excitation of external vibration, the gap collision contact will occur, resulting in a non-stationary and nonlinear change in the vibration response signal. At this time, the vibration response is complicated, which is difficult to be fully represented by a single time-domain feature. Therefore, consider constructing multi-dimensional time domain characteristics. Dimensional time domain characteristics are sensitive to changes in the original signal, but they are susceptible to environmental disturbances, and will change with changes in operating conditions. The dimensionless time-domain characteristics can eliminate the influence of disturbances and are widely used in feature extraction, mainly including: margin factor, form factor, pulse factor, crest factor, and kurtosis factor [16]. Assuming that the vibration signal is x, the mathematical description of each feature is shown in Table 1.

Table 1. Mathematical description of dimensionless time-domain features

Index	Formula	Index	Formula
Peak	$X_p = \max\{|x_i|\}$	Margin Factor	$L = \frac{X_p}{X_r}$
Mean	$\bar{x} = \frac{1}{N} \sum_{i=1}^{N} x_i$	Form Factor	$S = \frac{X_{rms}}{|\bar{x}|}$
Root Mean Square	$X_{rms} = \sqrt{\frac{1}{N} \sum_{i=1}^{N} x_i^2}$	Pulse Factor	$I = \frac{X_p}{\bar{x}}$
Square Root Amplitude	$X_r = \left(\frac{1}{N} \sum_{i=1}^{N} \sqrt{|x_i|} \right)^2$	Crest Factor	$C = \frac{X_p}{X_{rms}}$
Kurtosis	$\beta = \frac{1}{N} \sum_{i=1}^{N} x_i^4$	Kurtosis Factor	$K = \frac{\beta}{X_{rms}^4}$

In order to quantify the uncertainty and complexity of the high-frequency components of different signals, information entropy is introduced to assess the degree of irregularity in high-frequency components. The greater the number of components and states, the greater the information entropy.

The concept of entropy is commonly used to measure the degree of disorder of a thermodynamic system. And it is also used to measure the amount of information, which can be understood as the probability of occurrence of a piece of information or the degree of system regularity.

The more scattered the signal, the higher the randomness, and the larger information entropy. The more concentrated the distribution, the lower the randomness, and the smaller information entropy. The more ordered the system, the lower information entropy and vice versa.

Set a signal source $X = \{x_1, x_2, \ldots, x_n\}$, the probability of occurrence of $x_i (i = 1, 2, \ldots, n)$ is $P(x_i)$, and there is $\sum_{i=1}^{n} P(x_i)$. The information entropy can be expressed as:

$$H(x) = E(-\log P(x_i)) = -\sum_{i=1}^{n} P(x_i) \log(P(x_i)) \tag{4}$$

$P(x_i)$ represents the probability of occurrence of x_i, $E()$ represents the mathematical expectation. It can be known from (4) the information entropy is related to the probability of occurrence of the variable, which has nothing to do with its specific position. The information entropy can reduce the interference of random noise [17].

Therefore, the information entropy is used to evaluate the uncertainty and irregularity of the signal, effectively measure the unknown degree of the sequence, and estimate its complexity.

2.3 KNN Algorithm

The basic principle of KNN (K-Nearest Neighbor Classification) is: select k training samples that are nearest to the input sample in the feature space, then give the output results according to certain decision rules, which are the most of the k training samples [18]. The algorithm is simple and easy to implement, and the results obtained are accurate. There is no need to know the specific distribution of each feature, and it is not sensitive to outliers. The algorithm specific steps are as follows:

(1) Perform 0-1 operation on the input dataset.
(2) Input the pre-processed training set and test set.
(3) Set the K value and related distance, using the European distance:

$$dist(X_1, X_2) = \sqrt{\sum_{i=1}^{n} (x_{1i} - x_{2i})^2} \tag{5}$$

X_1, X_2 is the distance between two objects of the training sample, the closer the two objects are, the more similar they are, x_{1i}, x_{2i} is the i-th attribute value.
(4) Calculate the Euclidean distance between the test sample and the training sample.
(5) Calculate the weight between the test object and k neighbors, compare the weights between the classes, classify the test objects into the category with the largest weight. The weight is calculated as follows:

$$p(\bar{x}, C_i) = \sum_{d_i \in KNN} dist(\bar{x}, \bar{d}_i) y(\bar{d}_i, C_j) \tag{6}$$

$$y(d_i, C_j) = \begin{cases} 1, & d_i \in C_j \\ 0, & d_i \notin C_j \end{cases} \tag{7}$$

$\bar{x}$ is the feature vector of the test set, $\bar{d}_i$ is the sample in the training set, C_j is the sample set of j, $dist(\bar{x}, \bar{d}_i)$ is the mathematical description for calculating the similarity between samples, $y(\bar{d}_i, C_j)$ is the category attribute function.
(6) Output the class label of the sample with the largest weight among the training samples.

3 The Process of Bolts State Identification

Based on the above signal processing and analysis, the bolt state signals are analyzed, the overall framework of the proposed method is given, as shown in Fig. 1. And the specific steps are as follows:

Step 1. Data pre-processing: standardize the original signal and remove outliers, convert the voltage value to the corresponding speed value, analyze the normalized signal to improve the comparability between different state signals.

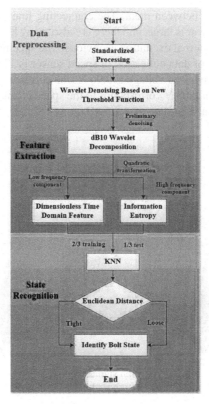

Fig. 1. Flow chart of bolt state identification

Step 2. Signal denoising: perform wavelet denoising on the pre-processed signal based on new threshold function, select dB3 as the wavelet basis function to decompose 3 layers and reconstruct the noise-reduced signal.

Step 3. Feature extraction: select dB10 as the wavelet basis function, and perform a 1-layer decomposition on the denoised signal to obtain the reconstructed low and high frequency component, then calculate the dimensionless time-domain features of low-frequency component and the information entropy of high-frequency component.

Step 4. State classification: establish the bolt state feature dataset based on the above calculation, 2/3 of the samples are used to train the network and parameter learning, other samples are used to test models and verify performance. Compared with some common classifiers, the KNN with the best performance is selected, the state of bolts is identified based on the Euclidean distance.

4 Experimental Case Analysis

4.1 Signal Acquisition

An actual 66 kV transmission tower in Northeast China is selected for experimental verification. The tower is 33 m high, about 100 m from expressway and about 250 m from railway, which is located between them. The connecting line crosses the expressway. The traffic close to the tower is large, and there are also many high-speed trains. Natural vibration generated by external excitation is obvious. The specific conditions are shown in Fig. 2. The frequency of data collection of speed sensor is 12.5 kHz, 150,000 points per signal acquisition, the initial voltage is 0.96 V (the corresponding speed is 0 mm/s), and the voltage change of 0.08 V corresponds to the speed change of 1 mm/s. Sampling every 5 points, signal analysis frequency 2.5 kHz.

Fig. 2. Vibration signal acquisition site

The original time-domain waveform when the bolts are in different states is shown in Fig. 3. We can see that the initial signal fluctuations in tight state are relatively stable, including some noise components, the overall amplitude change is small. The signal fluctuation in loose state is relatively more complex, the signal changes irregularly with time, which is a non-stationary signal.

4.2 Multi-feature Extraction

The original signal needs to be denoised so as to clarify the differential characteristics of different state signals. Figure 4 is the result of normalized processing of different state signals after wavelet denoising with new threshold function, in which the wavelet basis function is dB3 and the number of decomposition layers is 3. We can see that the new threshold function overcomes the problem, that is, the soft threshold and the hard threshold have many oscillation points at peaks and peaks-valleys, the noise is effectively filtered out, and the reconstructed signal waveform becomes smoother. The changes in the details of the signal become more clearly, the overall trend is closer to the real vibration, especially the tight state signal.

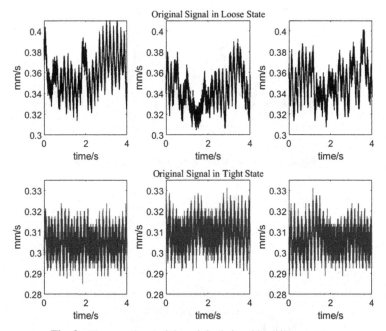

Fig. 3. The waveform of the original signal in different states

DB10 wavelet commonly used in fault diagnosis is selected as the basis function to perform a layer of decomposition on the denoised signal, so as to improve the resolution and refine the time-frequency change characteristics of the different state signals. The signal is decomposed into low and high frequency component. The decomposition results are shown in Fig. 5.

It can be seen that the high and low frequency signal component of different state signals respectively contain different features. The smoothness of the tight state signal is greater than the loose state signal, but the uncertainty pulses in the high frequency component are fewer. Calculate the dimensionless time domain characteristics of the reconstructed low frequency component and the information entropy of the high frequency component. Each time the number of input data is N = 10000, 34 groups of signals in different states are taken respectively. The corresponding characteristics of different state signals are shown Fig. 6.

4.3 Bolt State Identification

A 2040*6 feature dataset was established based on the above calculation. The number of features in each state accounted for 50%. And 1360 groups were selected as the training set, KNN, ELM, GRNN, PNN, RBF, BP, and SVM were trained respectively, the rest as the test set.

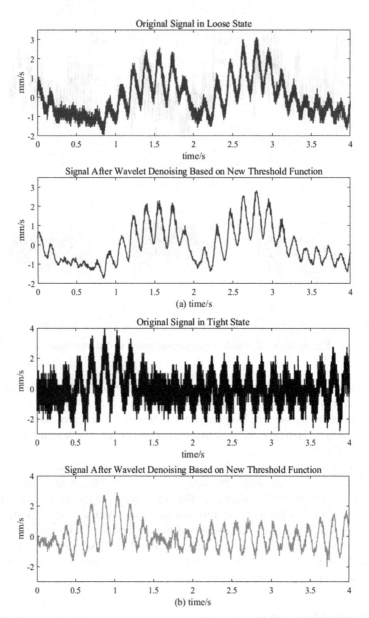

Fig. 4. Wavelet denoising based on new threshold function (a) Loose state (b) Tight state

Due to the small number of samples analyzed, by comparing common classifier, KNN with the best performance is selected. In order to better analyze and evaluate the performance of each classifier, the accuracy, precision, recall and F1 score are additionally introduced, the different metrics and index values are shown in Table 2.

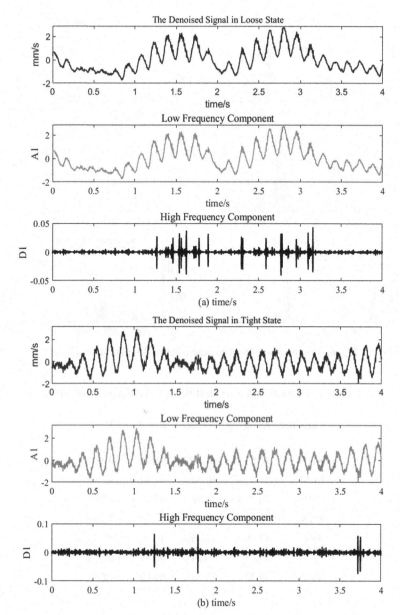

Fig. 5. High-frequency and low-frequency signal components based on dB10 wavelet decomposition (a) Loose state (b) Tight state

The accuracy of KNN is 90.85%, which has obvious advantages over BP, ELM, PNN, GRNN, RBF, and SVM, the changes of other evaluation values are consistent. The recall of RBF is 1, that is, the classifier classifies all states into one state, this classifier has no reference value.

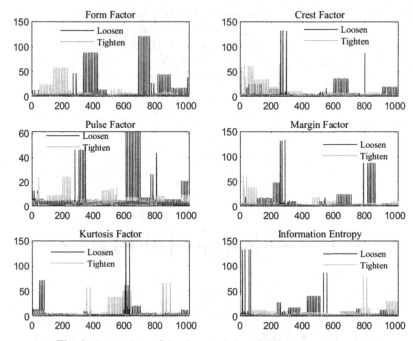

Fig. 6. Comparison of the characteristics of different state signals

Table 2. Evaluation of state identification of different classifiers

Classifier	Accuracy	Precision	Recall	F1 score
KNN	90.85%	90.15%	95%	90.81%
ELM	83.53%	80.32%	88.82%	84.36%
GRNN	70.29%	64.97%	97.65%	78.03%
PNN	73.53%	67.07%	98.24%	79.71%
RBF	47.06%	47.06%	1	64%
BP	86.47%	86.74%	90.83%	88.74%
SVM	59.26%	55.33%	69.69%	61.69%

The principle of KNN is to find the k points based on the Euclidean distance, which is closest to the training dataset according to the given test sample, then classify the state of the test sample by the k nearest sample data. Based on the above analysis, the distribution and difference of the bolts' vibration response features in different states is more obvious, which can further illustrate and explain the effectiveness of the proposed method. And the final classification results of KNN are shown in Fig. 7.

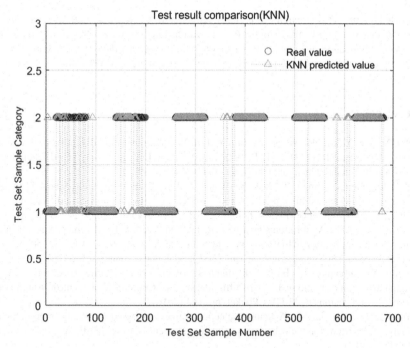

Fig. 7. KNN classification results

5 Conclusion

The operating conditions of the three-trans tower are complex, the bolt state signals are disturbed by the external interference. Without damage to the transmission tower, the method of feature extraction and bolt state identification under natural vibration was proposed. After the original signal was processed by wavelet denoising processing based on new threshold function and quadratic wavelet decomposition, the multi-dimensional time domain features of low-frequency component and the information entropy of high-frequency component were extracted respectively. After comparing common classifiers, KNN with the best performance was selected for bolt states classification. Finally, the experimental case analysis results based on measured data further proved the applicability and rationality of the proposed method. With the development of 5G and the construction of smart grid, the structure of the tower is more complicated, the proposed method is simple and reasonable, which can provide an exemplary application for the maintenance of the towers located in important sections of transmission lines. Considering that the state change of the bolt is not abrupt, the analysis when the bolt is critically loosened will be added in the future, and the feature selection and classification algorithm will be further optimized.

Acknowledgment. This research was partially supported by the National Natural Science Foundation of China (61703081), the Liaoning Revi-talization Talents Program (XLYC1801005), Natural Science Foundation of Liaoning Province (20170520113), and the State Key Laboratory of Alternate Electrical Power System with Renewable Energy Sources (LAPS19005).

References

1. Wu, X., Cai, Y., Shi, R., Wang, Y., Zuo, Y., Liu, X.: Intelligent fault diagnosis of three-trans towers' bolts based on EMD and LS-SVM. In: 2019 IEEE Innovative Smart Grid Technologies - Asia (ISGT Asia), Chengdu, China, pp. 3543–3548 (2019)
2. Cao, L., Shen, Y., Shan, T., Xia, Y., Wang, J., Lin, Z.: Bearing fault diagnosis method based on GMM and coupled hidden markov model. In: 2018 Prognostics and System Health Management Conference (PHM-Chongqing), Chongqing, China, pp. 932–936 (2018)
3. Liao, Z., Chen, P.: A vibration signal filtering method based on KL divergence genetic algorithm – with application to low speed bearing fault diagnosis. In: 2018 IEEE 23rd International Conference on Digital Signal Processing (DSP), Shanghai, China, pp. 1–5 (2018)
4. Qijun, X., Zhonghui, L.: Early fault diagnosis of the rolling bearing based on the weak signal detection technology. In: 2016 14th International Conference on Control, Automation, Robotics and Vision (ICARCV), Phuket, pp. 1–4 (2016)
5. Yan, X., Liu, Y., Jia, M.: A feature selection framework-based multiscale morphological analysis algorithm for fault diagnosis of rolling element bearing. IEEE Access **7**, 123436–123452 (2019)
6. Liu, Y., Qiao, N., Zhao, C., Zhuang, J.: Vibration signal prediction of gearbox in high-speed train based on monitoring data. IEEE Access **6**, 50709–50719 (2018)
7. Bo, L., Liu, X., Xu, G.: Intelligent diagnostics for bearing faults based on integrated interaction of nonlinear features. IEEE Trans. Ind. Inform. **16**(2), 1111–1119 (2020)
8. Maurya, S., Singh, V., Verma, N.K.: Condition monitoring of machines using fused features from EMD based local energy with DNN. IEEE Sens. J. **20**(15), 8316–8327 (2020)
9. Zhang, X., Luan, Z., Liu, X., et al.: Fault diagnosis of rolling bearing based on kurtosis criterion VMD and modulo square threshold. J. Eng. **2019**(23), 8685–8690 (2019)
10. Van, M., Kang, H.: Bearing fault diagnosis using nonlocal means algorithm and empirical mode decomposition based feature extraction and two-stage feature selection. IET Sci. Meas. Technol. **9**(6), 671–680 (2015)
11. Qi, J., Gao, X., Huang, N.: Mechanical fault diagnosis of a high voltage circuit breaker based on high-efficiency time-domain feature extraction with entropy features. Entropy **22**(4), 478–494 (2020)
12. Li, F., Wu, B., Liu, N., Hu, Y., Wu, H.: Seismic time-frequency analysis via adaptive mode separation-based wavelet transform. IEEE Geosci. Remote Sens. Lett. **17**(4), 696–700 (2020)
13. Wang, J.: A wavelet denoising method based on the improved threshold function. In: 2014 International Conference on Wavelet Analysis and Pattern Recognition, Lanzhou, China, pp. 70–74 (2014)
14. Zhan, Z., Qin, H.: A wavelet threshold denoising algorithm based on new threshold function. Comput. Technol. Dev. **29**(11), 47–51 (2019)
15. Zhang, J., Zhang, S., Guan, L., Yang, Y.: Pattern recognition of bearing defect severity based on multiwavelet packet sample entropy method. Zhendong Ceshi Yu Zhenduan/J. Vibr. Test. Diagn. **35**(01), 128–132 (2015)
16. Qi, J., Gao, X., Huang, N.: Mechanical fault diagnosis of a high voltage circuit breaker based on high-efficiency time-domain feature extraction with entropy features. Entropy **22**(4), 478–494 (2020)

17. Gou, X., Mu, S., Jin, W., Li, X.: Fault type recognition of high-speed train bogie based on dual-channel integration of information entropy. In: 2016 2nd IEEE International Conference on Computer and Communications (ICCC), Chengdu, China, pp. 1880–1884 (2016)
18. Wang, Z., Liu, S., Luo, Q.: KNN classification algorithm based on improved k-modes clustering. Comput. Eng. Des. **40**(08), 2228–2234 (2019)

Torque Ripple Minimization of PMSM Drives with Multiple Reference Frame and DNN Based High-Fidelity Torque Model

Chengli Jia[1], Tianfu Sun[1(✉)], Gang Yu[2], Jianing Liang[1], and Dingfang Lin[1]

[1] Shenzhen Institutes of Advanced Technology, Chinese Academy of Sciences, Shenzhen, China
tianfu.sun@foxmail.com
[2] Harbin Institute of Technology (Shenzhen), Shenzhen, China

Abstract. Torque ripple minimization is important to realize optimal operating state of permanent magnet synchronous machines (PMSMs). In order to suppress the torque ripples, this paper proposes a novel control technique for PMSM drives with a deep neural networks (DNN) based high-fidelity motor model and multiple reference frame (MRF). The high-fidelity model of PMSM is utilized to produce a reference current corresponding to a reversed torque harmonic to offset the torque harmonic inherent in the motor, and MRF is adopted to achieve follow-up control of high frequency signal. By optimizing the d-q axis current waveforms, the torque ripples in the resultant output torque will be minimized. The efficiency of the proposed scheme is explored through simulations. The simulation results of transient and steady state illustrate that the feasibility of the proposed control algorithm in torque ripple suppression.

Keywords: Torque ripple minimization · Multiple reference frame (MRF) · Motor models · Deep neural networks (DNNs) · Permanent magnet synchronous motor (PMSM)

1 Introduction

The permanent magnet synchronous machines (PMSMs) possess series of superiority such as high efficiency and dynamic performance, therefore, they have been widely used in electric vehicle and robot motion control [1, 2]. However due to several reasons inside and outside the motor, the resultant torque of PMSMs contain torque ripples which deteriorate the PMSM driving performance [3]. To avoid the torque ripples, the state-of-the-art work in torque ripple minimization can be divided into two research directions, i.e., to optimize motor mechanical structure or to improve the control algorithm. The former one including optimizing the winding distribution and structure of the stator [4, 5], however, such method may increase the machinery cost. Therefore, more attentions have been paid to improve motor control algorithms.

Although the torque ripple is induced by many reasons, the frequency of their resultant torque ripple is a multiple of the electrical angular frequency of the motor. An

© Springer Nature Singapore Pte Ltd. 2020
M. Fei et al. (Eds.): LSMS 2020/ICSEE 2020 Workshops, CCIS 1303, pp. 184–198, 2020.
https://doi.org/10.1007/978-981-33-6378-6_14

effective method to cancel these torque ripples is current harmonic injection [6]. Appropriate injection current will produce torque ripple with the same amplitude and opposite phase. Thus, the torque ripple of the resultant torque will be suppressed. In the existing literature, effective control algorithms have been designed base on this theory, and one of the most adopted approach is the feed forward-based method [7] which requires the accurate modeling of torque ripple for the design of the optimal reference current. However, due to the machine parameter nonlinearity and uncertainties, accurate modeling of the machine torque ripples is always difficult. Although the finite element analysis-based method can model the motor torque ripple with a relatively high accuracy, it is cumbersome to process the data and store them in look-up tables. The amplitude of speed harmonics is proportional to the amplitude of torque harmonics [8]. Therefore, in order to avoid the difficulties in modeling torque ripple, speed harmonics are worth considering However, in actual industrial applications, high load inertia will filter the torque ripple and make the speed ripple difficult to be detected. Thus, the application of such speed harmonic-based torque ripple minimization method is greatly limited. Recently, a novel high-fidelity model for PMSMs based on DNN was proposed in [9] and this high-fidelity modelling technique is utilized by this paper to design the corresponding harmonic component of injection current.

Due to the limitation of the controller bandwidth, it is necessary to select a suitable controller for high frequency harmonic components. Although the proportional resonant (PR) controller can be used to control harmonic current [10], its parameters should vary with the working state. Focusing on this problem, a current controller named multiple reference frame (MRF) which is based on the principle of synchronous rotating coordinate system is proposed in [6]. Therefore, PI controller can be used in harmonic current control after coordinate transformation.

In this paper, a torque ripple minimization control system combining multiple reference frame (MRF) with DNN based high-fidelity torque model is proposed. This paper is compiled in the following order. System mathematical equations and diagrams of the proposed control are detailed in Sect. 2. Simulation results are analyzed in detail in Sect. 3. Based on simulation results, the conclusion and discussion are drawn in Sect. 4. Simulation tests have shown that this method significantly minimize the torque ripple.

2   Control of PMSM

The proposed control scheme can be separated into four parts, i.e., ① the DNN based high-fidelity torque model, ② the method of extracting torque harmonics from the DNN based torque model, ③ the method of obtaining the reference current with the injected harmonics from torque harmonics, and ④ the compensation of the torque ripple with MRF method.

2.1   The DNN Based High-Fidelity Torque Model

The relationship between the parameters of a three-phase PMSM in d-q coordinate system is shown as follow:

$$v_d = L_d \frac{di_d}{dt} + Ri_d - p\omega_m L_q i_q. \tag{1}$$

$$v_q = L_q \frac{di_q}{dt} + Ri_q + p\omega_m L_d i_d + p\omega_m \Psi_m. \tag{2}$$

$$T_e = \frac{3p}{2} \left[\Psi_m i_q + \left(L_d - L_q \right) i_d i_q \right]. \tag{3}$$

Where, v_d, v_q, i_d, i_q, L_d, L_q are the d-q axis voltages, currents and inductances, respectively. Ψ_m is the permanent magnet flux linkage. p is pole pairs of PMSM. ω_m is mechanical angular velocity.

According to (3), a set of i_d and i_q can be used to calculated the value of resultant torque. Since L_d, L_q and Ψ_m are nonlinear and vary with rotor positions, therefore, it is difficult to accurately model these motor parameters.

On the other hand, deep neural networks have outstanding ability in fitting nonlinear systems, which make it suit for modeling PMSM torque ripples. The structure of the high-fidelity PMSM torque model based on DNN proposed in [9] is exhibited in Fig. 1.

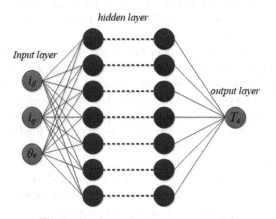

Fig. 1. The form of PMSM torque model

Although it is difficult to model the motor torque through accurate mathematical formula due to the nonlinearity, the precise motor torque can be obtained from a given set of inputs by using neural network according to Fig. 1.

The activation function of the DNN in this paper is the tanh function given in (4). Owing to solving the problem of none zero centered, its convergence speed is faster.

$$f(x) = \frac{1 - e^{-2x}}{1 + e^{-2x}}. \tag{4}$$

In the process of training neural network, the target mean squared error (MSE) in training the DNN was set to 0.0001 to ensure the output accuracy of the network. In order to avoid the situation that the neural network does not converge, the training iteration will be stopped when the number of iterations reaches 80000. By training different network topologies, several networks that meet the requirements are available. Here, the form of the network of torque model is expressed as $\left[x, h_1, h_2, \cdots, h_i, y \right]$, where x is number of neurons in input layer, h_i is the number of neurons in ith hidden layer and y is the number of neurons in output layer. The comparison of the training result of different DNN topologies is given in Table 1 [9].

Table 1. Comparison of the training result

No.	Network structure	Epochs	MSE
1	[3,25,30,1]	30121	0.0001
2	[3,25,25,1]	71419	0.0001
3	[3,20,20,1]	80000	0.000151
4	[3,15,20,1]	80000	0.000410
5	[3,15,15,1]	80000	0.000455
6	[3,30,30,30,1]	8358	0.0001
7	[3,20,20,20,1]	18728	0.0001
8	[3,20,15,10,1]	42173	0.0001
9	[3,20,10,10,1]	80000	0.000127
10	[3,15,15,15,1]	80000	0.00011
11	[3,20,10,10,1]	80000	0.000217
12	[3,15,10,5,1]	80000	0.000758
13	[3,20,20,20,20,1]	10014	0.0001
14	[3,15,15,15,15,1]	30246	0.0001
15	[3,15,15,1510,1]	43361	0.0001
16	[3,15,15,10,10,1]	80000	0.00013

The neural network with the minimum number of layers and nodes should be selected in the actual application. Comparing the network structures that meet the requirements in Table 1, the second group of the network is chosen for subsequent algorithm design.

Finite element analysis (FEA) can ensure that the simulation is the same as the actual working state of the motor to the greatest extent. Therefore, the results of FEA are used to training the neural work. The parameters of the motor used in this paper are shown in detail in Table 2.

Table 2. Parameters of IPMSM

Parameter	Value
Pole pairs of motor	3
Phase resistance	51.2 mΩ
Continuous/Maximum current	58.5/96 A

(continued)

Table 2. (*continued*)

Parameter	Value
DC bus voltage	120 V
Number of slots	36
Base/maximum rotating speed	1350/4500 r/min
Continuous/peak output torque	35.5/70 N m
Peak power at base/maximum speed	10/7 kW

Here, the fitting effect of neural network for torque ripple is given under different working condition. Figure 2 (a) show that the torque command was set to 50 N · m in steady state at speed of 1000 r/min. Figure 2 (b) show that the torque command was set to different torque values.

In PMSM, the torque harmonics is caused by both internal and external factors, such as:

1) Measurement errors of current: This will generate torque harmonics of fundamental frequency and second harmonic frequency.
2) Flux harmonic and cogging torque: This will generate torque harmonics with the order of integer frequency multiplication of six.
3) Inverter nonlinearity: This will also generate torque harmonics with the order of multiples of six;

According to the above analysis, the frequency of torque harmonic is mainly 6 times of the fundamental frequency and integer multiples of 6. Therefore, this article only focuses on the amplitude of harmonic components whose orders are integer multiples of 6.

For the purpose of guaranteeing the correctness when the DNN based model is used later, the actual torque and the torque estimated by DNN are analyzed by Fourier analysis in Fig. 3.

According to Fig. 3, the torque harmonics estimated by DNN have a good performance of fitting the actual output torque harmonics of the motor.

2.2 Torque Harmonic Extraction from the DNN Based Torque Model

The torque harmonics of order k can be canceled by injecting harmonic currents of the same order. To obtain current harmonics, it is necessary to obtain the magnitude and phase angle of the torque harmonics from the output torque in advance. In [8], a method of extracting harmonic signal from composite signal is proposed. The detailed flow chart of the harmonic extraction module is shown Fig. 4.

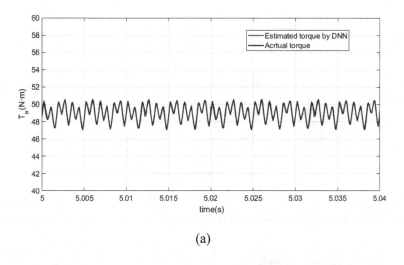

(a)

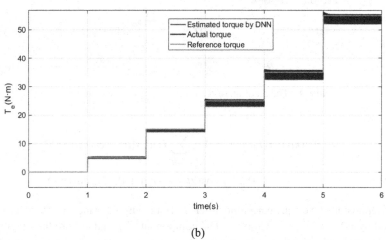

(b)

Fig. 2. Torque estimated by DNN model and the actual motor torque: (a) the torque ripple at the given reference torque is 50 N · m (b) The modeling effect at the given reference torque increasing from 0 to 55 N · m

If the torque estimated in Fig. 1 is multiplied by $sink\theta$ and $cosk\theta$, respectively, the result can be expressed as follow:

$$T_e \sin(k\theta) = T_{e0} \sin(k\theta) + \sum_{l \neq k} T_{el} cos(l\theta - \phi_{tl}) \sin(k\theta)$$
$$+ 0.5T_{ek} \sin(2k\theta - \phi_{tk}) + 0.5T_{ek} \sin(\phi_{tk}). \quad (5)$$

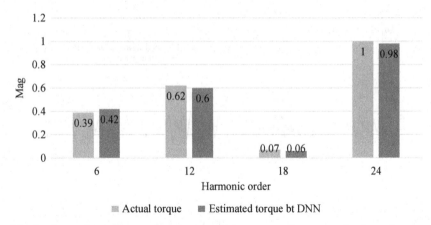

Fig. 3. Comparison of harmonic components between actual torque and estimated torque

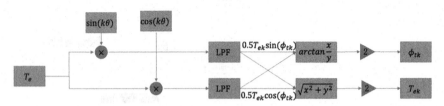

Fig. 4. The diagram of the torque harmonic detection

$$T_e \cos(k\theta) = T_{e0} \cos(k\theta) + \sum_{l \neq k} T_{el} cos(l\theta - \phi_{tl}) \cos(k\theta)$$

$$+ 0.5T_{ek} \cos(2k\theta - \phi_{tk}) + 0.5T_{ek} \cos(\phi_{tk}). \quad (6)$$

where k represents the torque harmonic order; T_e is the output torque with the harmonics estimated by DNN in Fig. 1; T_{e0} is the DC component; T_{ek} and ϕ_{tk} are the magnitude and phase angle of the kth torque harmonic component, respectively. θ is the rotor electrical position. From (5) and (6), the last terms of the right side of the equations are constant which will pass the low-pass-filters (LPF). Therefore, the values of T_{ek} and ϕ_{tk} can be calculated using the mathematical operations. After obtaining T_{ek} and ϕ_{tk}, the corresponding compensation current can be designed according to them.

2.3   Generation of Reference Current With Harmonics Injection

According to the above analysis of torque harmonic components, the output torque can be considered as the sum of the dc component and harmonic components. The formula is described as (7):

$$T_e = T_{e0} + \sum_k t_{ek} = T_{e0} + \sum_k T_{ek} \sin(k\theta - \phi_{tk}). \quad (7)$$

According to this torque model, torque ripple is a combination of harmonics of several orders and this paper only minimizes the dominant torque harmonic for simplicity. In this paper, it is assumed as kth order harmonic. This torque harmonic can be cancelled by injecting a kth current component into the base current vector reversely. The injected kth current component can be expressed as follows:

$$i_{dk} = I_{dk} \cos(k\theta - \phi_{dk}). \tag{8}$$

$$i_{qk} = I_{qk} \sin(k\theta - \phi_{qk}). \tag{9}$$

where i_{dk} and i_{qk} are d- and q-axis component of the kth harmonic current. I_{dk} and ϕ_{dk} are the magnitude and phase angle of i_{dk}, respectively; I_{qk} and ϕ_{qk} are the magnitude and phase angle of i_{qk}, respectively.

In the study of [8], it has been proved that when the relationship between i_{dk} and i_{qk} satisfy the Eq. (10) and (11), the kth torque harmonic will be cancelled out after injecting i_{dk} and i_{qk}.

$$I_{dk} = \beta I_{qk}. \tag{10}$$

$$\phi_{dk} = \phi_{qk} + \frac{\pi}{2}. \tag{11}$$

where $\beta = L_\Delta I_{q0}/(L_\Delta I_{d0} + \Lambda_0)$ and $L_\Delta = L_d - L_q$, Λ_0 is the dc bias of the magnet flux.

Meanwhile, according to the relationship between torque harmonics and current harmonics [11], torque harmonic caused by injection current is

$$t_{ek\_i} = K_P(L_\Delta I_{d0} + \Lambda_0)I_{qk} \sin(k\theta - \phi_{qk}) + K_P L_\Delta I_{q0}I_{dk} \cos(k\theta - \phi_{dk}). \tag{12}$$

Here, the harmonics generated due to the interaction of flux harmonic and current harmonic account for a small proportion of the total harmonics. Therefore, this proportion is ignored in (12).

Substitution of (10) and (11) into (12), the torque harmonic t_{extra_k}. caused by the injected harmonic current can be obtained in (13).

$$t_{extra\_k} = K_P \gamma I_{qk} \sin(k\theta - \phi_{qk}). \tag{13}$$

where $\gamma = (L_\Delta I_{d0} + \Lambda_0 + L_\Delta I_{q0}\beta)$.

Therefore, after injecting the reference current, the sum of t_{ek} and t_{extra_k} is the resultant kth torque harmonic, t_{total_k}. The sum of the (7) and (13) is:

$$t_{total_k} = t_{ek} + t_{extra\_k} = T_{ek} \sin(k\theta - \phi_{tk}) + K_P \gamma I_{qk} \sin(k\theta - \phi_{qk}). \tag{14}$$

Since the torque ripple minimization control target is to make the $t_{total\_k} = 0$, base on (14) the following equation can be obtained:

$$I_{qk} = \frac{\alpha T_{ek}}{\gamma K_P}. \tag{15}$$

$$\phi_{qk} = \phi_{tk} + \pi. \tag{16}$$

From (15) and (16), the i_{qk} can be obtained according the torque harmonic extracted from the torque model. According to (10) and (11), the i_{dk} can be directly obtained.

The α in (15) is a factor to compensate the gain caused by aforementioned signal processing and the neglected torque mentioned earlier.

2.4 Compensation of the Torque Ripple with MRF

As a result of the limited bandwidth of the PI controller, it cannot regulate the high frequency harmonic currents to realize no-error control. The MRF introduces a new reference frame which transforms the harmonic current into direct current. Therefore, the PI controller can be used to realize the no-error control of high frequency harmonic current [6]. The framework of the MRF controller is depicted in Fig. 5.

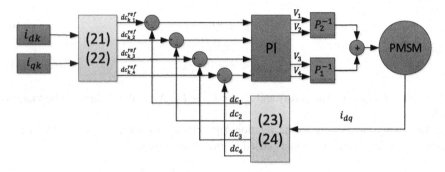

Fig. 5. The diagram of the MRF controller

According to Fig. 5, the two rotational transformations P_1 and P_2 can be defined as follow:

$$P_1 = \begin{bmatrix} \cos k\theta & -\sin k\theta \\ \sin k\theta & \cos k\theta \end{bmatrix}, \quad P_2 = \begin{bmatrix} \cos k\theta & \sin k\theta \\ -\sin k\theta & \cos k\theta \end{bmatrix}. \tag{17}$$

And they satisfy the following relationship:

$$P_1^{-1} = P_1^T, \ P_2^{-1} = P_2^T. \tag{18}$$

i_{dk} and i_{qk} multiply P_1 and P_2, respectively, gives:

$$P_1 \begin{bmatrix} i_{dk} \\ i_{qk} \end{bmatrix} = \begin{bmatrix} I_{dk}\cos k\theta \ \cos(k\theta - \phi_{dk}) - I_{qk}\sin k\theta \ \cos(k\theta - \phi_{qk}) \\ I_{dk}\sin k\theta \ \cos(k\theta - \phi_{dk}) + I_{qk}\cos k\theta \ \cos(k\theta - \phi_{qk}) \end{bmatrix}. \tag{19}$$

$$P_2\begin{bmatrix} i_{dk} \\ i_{qk} \end{bmatrix} = \begin{bmatrix} I_{dk}cosk\theta\ cos(k\theta - \phi_{dk}) + I_{qk}sink\theta\ cos(k\theta - \phi_{qk}) \\ -I_{dk}sink\theta\ cos(k\theta - \phi_{dk}) + I_{qk}cosk\theta\ cos(k\theta - \phi_{qk}) \end{bmatrix}. \tag{20}$$

The average values of (19) and (20) can be expressed as:

$$\text{mean}\left(P_1\begin{bmatrix} i_{dk} \\ i_{qk} \end{bmatrix}\right) = \frac{1}{2}\begin{bmatrix} I_{dk}cos\phi_{dk} - I_{qk}sin\phi_{qk} \\ I_{dk}sin\phi_{dk} + I_{qk}cos\phi_{qk} \end{bmatrix} = \begin{bmatrix} dc_{k\_1} \\ dc_{k\_2} \end{bmatrix}. \tag{21}$$

$$\text{mean}\left(P_2\begin{bmatrix} i_{dk} \\ i_{qk} \end{bmatrix}\right) = \frac{1}{2}\begin{bmatrix} I_{dk}cos\phi_{dk} + I_{qk}sin\phi_{qk} \\ -I_{dk}sin\phi_{dk} + I_{qk}cos\phi_{qk} \end{bmatrix} = \begin{bmatrix} dc_{k\_3} \\ dc_{k\_4} \end{bmatrix}. \tag{22}$$

Where, dc_{k_i} is the dc components of the harmonics in the new reference frame.

The inputs of PI Fig. 5 in controller are the difference between $dc_{k\_i}^{ref}$ minus dc_{k_i}. Where $dc_{k\_i}^{ref}$ can be directly obtained from the reference current i_{dk} and i_{qk}, and dc_{k_i} are calculated from the kth harmonic current. The following equations have been proved in [6].

$$\text{mean}\left(P_1\begin{bmatrix} i_d\ i_q \end{bmatrix}^T\right) = \text{mean}\left(P_1\begin{bmatrix} I_{d0}\ I_{q0} \end{bmatrix}^T\right) + \text{mean}\left(P_1\begin{bmatrix} i_{dk}\ i_{qk} \end{bmatrix}^T\right)$$

$$= \text{mean}\left(P_1\begin{bmatrix} i_{dk}\ i_{qk} \end{bmatrix}^T\right). \tag{23}$$

$$\text{mean}\left(P_2\begin{bmatrix} i_d\ i_q \end{bmatrix}^T\right) = \text{mean}\left(P_2\begin{bmatrix} I_{d0}\ I_{q0} \end{bmatrix}^T\right) + \text{mean}\left(P_2\begin{bmatrix} i_{dk}\ i_{qk} \end{bmatrix}^T\right)$$

$$= \text{mean}\left(P_2\begin{bmatrix} i_{dk}\ i_{qk} \end{bmatrix}^T\right). \tag{24}$$

where I_{d0} and I_{q0} are the dc components.

Therefore, according to (24) and (25), dc_{k_i} is equal to dc_i. The input of PI controller become the difference between $dc_{k\_i}^{ref}$ minus dc_i, where dc_i is calculated according to the actual measured currents. The resultant voltage harmonics injected to the motor are:

$$\begin{bmatrix} v_{dk}\ v_{qk} \end{bmatrix}^T = P_1^{-1}\begin{bmatrix} V_1\ V_2 \end{bmatrix}^T + P_2^{-1}\begin{bmatrix} V_3\ V_4 \end{bmatrix}^T. \tag{25}$$

v_{dk} and v_{qk} are the voltage injected into the d-q axis. V_i is obtained by the output of PI controller.

Considering the above four parts, the overall control framework of the proposed for suppressing torque ripple is showed in Fig. 6.

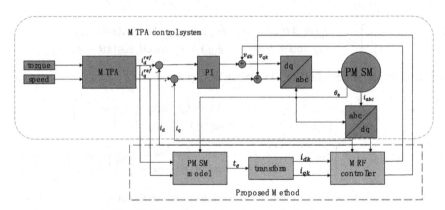

Fig. 6. Flow chart of the whole control system

3 Simulation Study

In order to verify the feasibility of the proposed control algorithm, simulations were carried out based on a 10 kW nonlinear motor drive system. The parameters of motor prototype are also the data of finite element analysis.

In the simulation, torque command was first set to 50 N · m in steady state, and the speed command was set to 1000 r/min. After injecting the compensation harmonic current designed above, the harmonic components of measured motor currents are significantly improved. The comparison of their current waveform can be observed in Fig. 7.

The additional harmonic current will produce additional torque harmonic, which will offset the original torque harmonic. The comparison of resultant torque waveform is depicted in Fig. 8.

Due to the proposed torque compensation scheme, the torque ripple was significantly reduced. The torque ripple is not completely suppressed because this paper focuses on the dominant order of harmonics. The harmonics of other orders are not processed.

The resultant torques were also compared in frequency domain by Fourier transform, the amplitude comparison of the torque harmonic with and without compensation scheme is shown in Fig. 9.

According to Fig. 9, the dominant torque harmonic is 24th. After applying the proposed strategy to the PMSM, the amplitude of 24th torque harmonic was significantly reduced. But the 24th torque harmonics still exist to a certain extent. That is because the compensation coefficient α in the simulation is set as a fixed constant, so the system can't make the corresponding compensation in time for the change of the mechanical parameters of the motor. Therefore, both the overall and single order torque ripples cannot be completely suppressed.

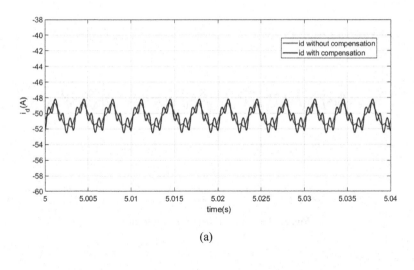

(a)

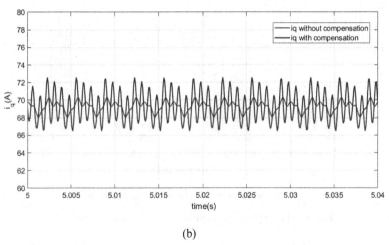

(b)

Fig. 7. The comparison of d-q axis current waveform

The suppression effect of total torque ripple in different working state can be obtained in Fig. 10.

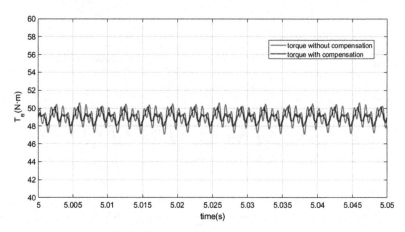

Fig. 8. The comparison of output torque

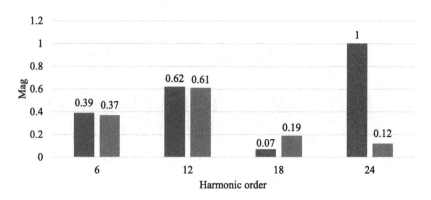

■ torque without compensation ■ torque with compensation

Fig. 9. The amplitude comparison of the torque harmonic

According the result of Fig. 10, the simulation results have proved that the proposed compensation method is feasible in both steady state and transient state to some extent.

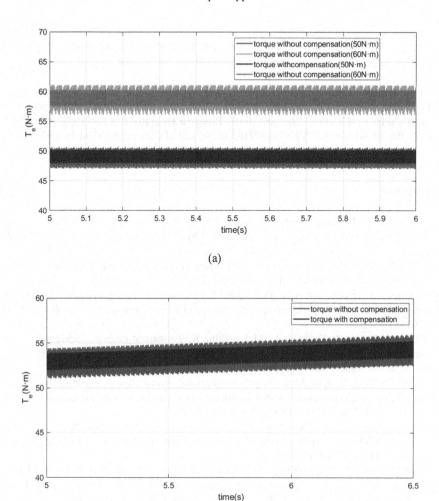

(a)

(b)

Fig. 10. The effect of minimizing torque ripple (a) 50 N · m, 60 N · m, (b) 50 N · m, to 60 N · m,

4 Conclusion

This paper presents a control algorithm of torque ripple suppression of PMSM relying on multiple reference frame and high-fidelity model. A high-fidelity PMSM model provides a method to calculate the optimized current. MRF makes the current harmonic follow the optimized current in steady and transient state. When the output v_{dk} and v_{qk} of MRF controller are added to the system, torque ripple will be suppressed. However, this method can only suppress single order harmonic. The design of motor will affect the content of harmonic components. If the dominant order harmonic of the motor is much larger than the harmonics of the other orders, the control theory proposed in this paper will have a satisfactory suppression result on the overall torque ripple. However, if the harmonic amplitudes of each order are similar, the effect of this method will be limited. Therefore, future research efforts are required to get a better overall suppression effect even if the amplitudes of the harmonics of each order are similar.

Acknowledgments. This work was supported by the Key-Area R&D Program of Guangdong Province (2020B090925002, 2019B090917001), Chinese National Natural Science Foundation of China (NSFC) (51707191), Shenzhen Basic Research Program (JCYJ20180507182619669, JCYJ20170818164527303, JCYJ20180507182628567), SIAT Outstanding Youth Fund (Y8G020)

References

1. Sun, T., Koc, M., Wang, J.: MTPA control of IPMSM drives based on virtual signal injection considering machine parameter variations. IEEE Trans. Ind. Electron. **65**(8), 6089–6098 (2018)
2. Yi, P., Sun, Z., Wang, X.: Research on PMSM harmonic coupling models based on magnetic co-energy. IET Electr. Power Appl. **13**(4), 571–579 (2019)
3. Güemes, J.A., Iraolagoitia, A.M., Del Hoyo, J.I., Fernández, P.: Torque analysis in permanent-magnet synchronous motors: a comparative study. IEEE Trans. Energy Convers. **26**(1), 55–63 (2011)
4. Islam, R., Husain, I., Fardoun, A., McLaughlin, K.: Permanent-magnet synchronous motor magnet designs with skewing for torque ripple and cogging torque reduction. In: IEEE Industry Applications Annual Meeting, pp. 1552–1559. IEEE (2009)
5. Wang, D., Wang, X., Kim, M.K., Jung, S.Y.: Integrated optimization of two design techniques for cogging torque reduction combined with analytical method by a simple gradient descent method. IEEE Trans. Magn. **48**(8), 2265–2276 (2012)
6. Feng, G., Lai, C., Tian, J., Kar, N.C.: Multiple reference frame based torque ripple minimization for PMSM drive under both steady-state and transient conditions. IEEE Trans. Power Electron. **34**(7), 6685–6696 (2019)
7. Flieller, D., Nguyen, N.K., Wira, P., Sturtzer, G., Abdeslam, D.O., Mercklé, J.: A self-learning solution for torque ripple reduction for nonsinusoidal permanent-magnet motor drives based on artificial neural networks. IEEE Trans. Ind. Electron. **61**(2), 655–666 (2014)
8. Feng, G., Lai, C., Kar, N.C.: Practical testing solutions to optimal stator harmonic current design for PMSM torque ripple minimization using speed harmonics. IEEE Trans. Power Electron. **33**(6), 5181–5191 (2018)
9. Zhang, K., et al.: High-fidelity model for interior permanent magnet synchronous machines considering the magnet saturation and spatial harmonics based on deep neural network. In: 22nd International Conference on Electrical Machines and Systems (ICEMS), pp. 1–5 (2019)
10. Gao, J., Wu, X., Huang, S., Zhang, W., Xiao, L.: Torque ripple minimisation of permanent magnet synchronous motor using a new proportional resonant controller. IET Power Electron. **10**(2), 208–214 (2017)
11. Feng, G., Lai, C., Kar, N.C.: An analytical solution to optimal stator current design for PMSM torque ripple minimization with minimal machine losses. IEEE Trans. Industr. Electron. **64**(10), 7655–7665 (2017)

Intelligent Manufacturing and System

Demand Forecasting of a Fused Magnesia Smelting Process Based on LSTM and FRA

Jingwen Zhang[1], Kang Li[2(✉)], and Tianyou Chai[1]

[1] State Key Laboratory of Synthetical Automation for Process Industries College of Information Science and Engineering, Northeast University, Shenyang 110819, China
[2] School of Electronic and Electric Engineering, University of Leeds, Leeds, UK
K.Li1@leeds.ac.uk

Abstract. In a Fused Magnesia Smelting Process(FMSP), its electricity demand is defined as the average electric power consumption over a fixed period of time and often used to calculate the electricity cost. The power supply has to be switched off once the demand value exceeds one specific threshold for safety and economic reasons. However, it has been shown that through appropriate current control of the FMSP, the demand can be reduced hence avoiding the shut-down of the process. A key issue to adopt the control strategy to avoid switch-off of electricity is to forecast the power demand and its trend However, this is technically challenging given the complexity and unknown dynamics of the process. In this paper, a hybrid approach combining a linear model with an unknown high order function is proposed. The linear model is used to capture the priori information from the domain knowledge and historic data, while the unknown dynamics in FMSP embedded in the error of the linear model are approximated with a high order nonlinear function. The Recursive Least Square algorithm (RLS) is used for identifying the unknown parameters in the linear model. A Long-Short Term Memory (LSTM) trained by the Fast Recursive Algorithm (FRA) is proposed to fit the unknown high-order function. Finally, the output weights of LSTM is updated by the RLS again. Experimental studies reveal that compared with other hybrid models such as a linear model combined with Radial Basis Function Neural Network (RBF), the proposed model offers the better performance.

Keywords: Demand forecasting · LSTM · FRA

1 Introduction

Fused magnesia is a key refractory matter used in many fields like the chemical industry, metallurgy and aerospace industry, etc. To produce the fused magnesia, magnesia is smelted by a Fused Magnesium Furnace(FMF) which is an energy intensive process. The energy cost accounts for over 60% of the total production

© Springer Nature Singapore Pte Ltd. 2020
M. Fei et al. (Eds.): LSMS 2020/ICSEE 2020 Workshops, CCIS 1303, pp. 201–215, 2020.
https://doi.org/10.1007/978-981-33-6378-6_15

cost. The electricity demand is used to calculate the electric energy cost during the Fused Magnesia Smelting Process(FMSP) and it can not exceed a certain limit $\bar{P}_l$ namely the maximum demand which is specified by State Grid Corporation of China(SGCC), otherwise the factory will be penalized with a huge fine. For this reason if the demand exceeds this limit for 4 sampling periods(28 s) continuously, one or more FMFs have to be turned off to ensure the total demand is below the maximum limit at the 5th sampling time. But turning off the power frequently, the product quality will be reduce, leading to unexpected economic loss. Fortunately, the current controller designed for the FMF can adjust the length of the arc to reduce the current when the demand exceeds its limit, hence reducing the overall demand. But some unknown effects, such as unknown process dynamics and external disturbances, etc., may affect the performance of the current controller, hence the demand may not necessarily accordingly. For these aforementioned reasons, it is crucial to have an accurate forecasting of the demand and its trend to allow the controller take appropriate control actions, and to determine if it is necessary to turn off FMFs. The FMF uses submerged arc mode to achieve the melting point of the raw material about 2850 °C, which is much higher than 1700 °C - the temperature required by the steel-making electric arc furnace [1]. Using this mode we can only measure the electric power and demand during the smelting process. However, some unavailable states and information such as the length of the arc and the height of the smelting pool also affect the model accuracy. But fortunately, experimental studies show that the past demand and electric power consumption of the process are the single largest contributing factor to forecasting the future demands.

Over the past decades time series models [2], regression models [3], Artificial Neural Networks(ANN) [4] or some hybrid models have been used to predict the electricity demand and its trend. Time series models are the simplest model used for energy demand forecasting and some statistical methods are usually used for trend analysis. Regression models with threshold is proved to have better performance than standard linear models and it can be used for both long term and short term forecasting. In [5] ANN is used to forecast peak load planning. In [6] a hybrid model with linear and non-linear parts is proposed to forecast the electric demand and its trend. In [7] the electric demand and its peak are predicted by a hybrid model with regression models and ANNs. To forecast the demand in FMSP, in [8] a Radial Basis Function Neural Network(RBFNN) has been used, and in [1,9] the linear model with RBFNN produced by different identification methods are used. By comparing with [1,8,9], we find hybrid model with a linear model combined with a neural network is more suitable for the demand forecasting in FMSP, but the demand forecasting accuracy is affected by the design of the neural network and the statistic method used for selecting the order of the input [1,8,9].

In [1,9] a mechanism model which combines with a linear model and an unknown high-order function is proposed to capture the dynamics of FMSP. The linear model uses past electric power consumption as its inputs, the process dynamics embedded in the error of the linear model in the smelting process can be written as an unknown high-order function with the past modelling errors and electric power consumptions as the model inputs. The biggest challenge for forecasting the demand and its trend is that there are overwhelmingly unknown information during the FMSP due to the submerged arc mode. The single RBFNN structure also limits the accuracy of the model. So there is a need to build a more accurate demand neural network with better forecasting performance.

In this paper we propose a hybrid model with a linear model and an unknown high order function to forecast the demand value and its trend. The parameters of the linear model will be identified by Recursive Least Squares(RLS) [10], and the Long-Short Term Memory(LSTM) [11] is chosen to fit the unknown high order function. A Fast Recursive Algorithm(FRA) [12] is used to build LSTM in identifying the output weights. The RLS is used for updating the output weights online. Comparing with some ANNs like RBFNN, the hybrid model with LSTM proposed in this paper has the best performance in our experimental studies using the real data from the FMSP.

2   Demand Forecasting Model

2.1   Introduction to FMSP

Figure 1 gives a brief illustration of the FMSP, which comprises a power supply system, current control systems, power systems, multiple FMFs and a demand monitoring system. Due to the high melting point of the raw materials, the FMF has to use the submerged arc mode (the electrodes are embedded into the raw material). The current control system will adjust the distance between the three-phase electrodes and the surface of the molten pool for generating the arc to melt the raw material and form a magnesium oxide bath. The smelting process is to add raw material into the furnace until it is full.

A number of uncertainties are associated with the FMSP which stop the current controller acting appropriately to timely adjust the set point. For the cases when the current controller does not perform, we can turn off the FMF will be switched off manually to stop the demand exceeding its limit. The power consumptions of FMFs can be collected by the power transducer and the demand can be calculated timely, i.e. the current demand is calculated every seven seconds. With these information, tthe current demand and its future prediction can be obtained. If the demand exceeds its limitation for 4 successive sampling points, the power of one or more FMFs will be cut off. Vice versa, the FMFs can be turned on when the demand is below the limit.

However if the power of FMFs are switched off and on frequently, the quality of the fused magnesia will be significantly decreased, and the energy waste can be huge as well. Therefore it is crucial to model the demand accurately for the current controller such that it can action properly to avoid unnecessary power cut-off and economic losses.

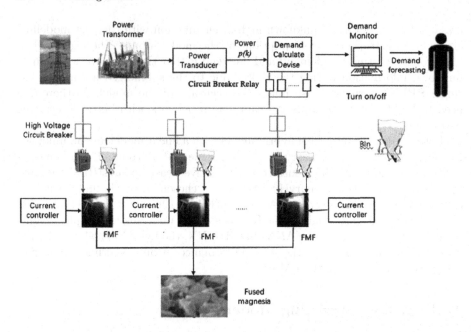

Fig. 1. A sketch of demand monitoring for FMSP.

2.2   Demand Forecasting Model

In FMSP, each FMF may have different values such as the charging current, smelting state, etc., therefore ideally the demand of each FMF can be modelled separately. But since numerous contributing factors to the demand are unknown about each FMF, such as the length of arc, distance between the three-phase electrodes, the surface area of the molten pool and its height, etc., therefore it is difficult to build accurate demand forecasting model for each FMF. A practical approach would be to built an aggregated model to predict the overall demand of all FMFs. The demand $\bar{P}(k)$ of all FMFs at time instant k is defined as the average value of electric power consumptions $p(k), p(k-1), \cdots, p(k-n+1)$ in Eq. (1).

$$\bar{P}(k) = \frac{1}{n} \sum_{i=1}^{n} p(k-i+1) \tag{1}$$

where $p(k) = \sum_{i=1}^{m} \sqrt{3} U I_i(k) \cos \varphi$ is the electric power of m FMFs at time instant k, U is the voltage of each FMF, I_i is the current of ith FMF, and $\cos \varphi$ is the power factor. According to [1], the electric power forecasting model can be formulated as in Eq. (2),

$$p(k+1) = \boldsymbol{\psi}(k)\boldsymbol{\theta} + v(k) \tag{2}$$

where $\boldsymbol{\psi}(k) = (p(k), p(k-1), p(k-2), p^*)$, $p^* = m\sqrt{3}U I^* \cos \varphi, \boldsymbol{\theta} = (\theta_1, \theta_2, \theta_3, \theta_4)^T$, I^* is the set value of current and $v(k)$ is an unknown high order function which will be fitted by LSTM trained with FRA. We can rewrite Eq. (1) as follows

$$\bar{P}(k+1) = \bar{P}(k) + \frac{p(k+1) - p(k+1-n)}{n} \tag{3}$$

We can see that if we want to predict the demand $\bar{P}(k+1)$ at time instnat $k+1$, we have to predict the electric power consumption $p(k+1)$ first. So our target in this paper is to forecast the electric power consumption $\hat{p}(k+1)$ firstly and then calculate the forecasting demand $\hat{\bar{P}}(k+1)$ using Eq. (3).

3 Demand Forecasting Algorithm

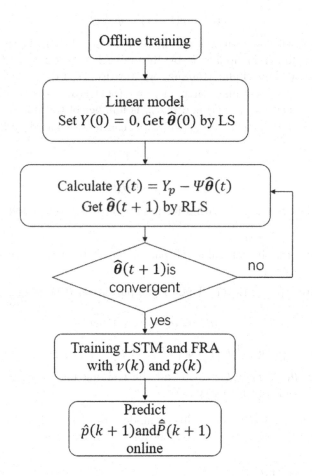

Fig. 2. The process for demand forecasting.

3.1   Offline Parameter Identification for Linear Model by RLS

In this section we use Recursive Least Square(RLS) algorithm to identify the parameters θ of linear model in (2) and the unknown high order nonlinear function $v(k)$ is fitted by LSTM trained with FRA as show in Fig. 2. The initial value of θ is can be identified by the Least Square(LS) algorithm. According to Eq. (2), we write linear model as Eq. (4) and we set $v(k) = 0$ for $k = 1, 2, \cdots$ at first. Then the parameter θ will be identified by RLS until it is convergent and then it will be fixed as $\hat{\theta}$. Afterwards, we fit $v(k), k = 1, 2, \cdots$ by LSTM trained with FRA and its value can be calculated by Eq. (5). Finally, an online parameter update algorithm will be given.

$$p(k + 1) - \hat{v}(k) = \psi(k)\theta \tag{4}$$

$$v(k) = p(k + 1) - \psi(k)\hat{\theta} \tag{5}$$

We need to identify linear model parameter θ in Eq. (4) with Least Square(LS) [10] algorithm first and update it by Recursive Least Square(RLS) [10] until it is convergent. For offline training, we rewrite Eq. (4) in matrix form as Eq. (6), where t is training time, $Y_p = (p(1), p(5), \cdots, p(n_{train}))^T$, $\Psi = (\psi(0); \psi(1); \cdots ; \psi(n_{train} - 1))$ and n_{train} is number of training data, $p(k) = 0$ for $k \leqslant 0$. $Y(t) = (v(1), v(2), \cdots, v(n_{train}))^T$ can be calculated by Eq. (7).

$$Y_p - Y(t) = \Psi\theta(t) \tag{6}$$

$$Y(t) = Y_p - \Psi\hat{\theta}(t) \tag{7}$$

The offline identification algorithm for parameter θ is given as follows:

(1) Set $t = 0, Y(0) = \mathbf{0}$;
(2) Calculate the initial value of parameter θ by LS algorithm $\hat{\theta}(0) = \left(\Psi^T\Psi\right)^{-1}\Psi^T Y_p$;
(3) Calculate $Y(t)$ by Eq. (7). Update the parameter $\hat{\theta}(t)$ using Eq. (8)

$$\hat{\theta}(t) = \hat{\theta}(t - 1) + Y(t)G \tag{8}$$

where $G = \left(\Psi^T\Psi\right)^{-1}\Psi\left[I + \Psi\left(\Psi^T\Psi\right)^{-1}\Psi\right]^{-1}$ and I is an identity matrix;

(4) If $\hat{\theta}(t) - \hat{\theta}(t - 1) = \xi, \xi \to 0$ end the training for linear model, else set $t = t + 1$ and return to step (3).

Once the parameters $\hat{\theta}(t)$ in the linear model are identified, we calculate new error $v(k)$ by Eq. (7) as the target values for LSTM training.

3.2   LSTM Model

The Recurrent Neural Network(RNN)(Fig. 3, left) in Eq. (9), (10) trained by back-propagation through time (BPTT) is usually used for short-term time series

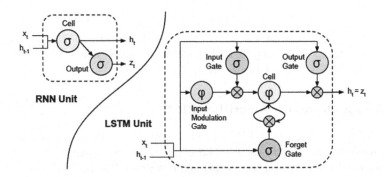

Fig. 3. A diagram of a basic RNN cell (left) and an LSTM memory cell (right) [13].

forecasting [11], but the error signals accumulating over time tend to blow up or vanish in long-term time series forecasting.

$$h_t = g\left(W_{xh}x_t + W_{hh}h_{t-1} + b_h\right) \tag{9}$$

$$z_t = g\left(W_{hz}h_t + b_z\right) \tag{10}$$

where x_t is input.

The LSTM (Fig. 3, right) which is a variant of RNN with gates can be used for both long-term and short-term time series forecasting [11]. The LSTM updates for timestep t given inputs x_t, h_t, c_{t-1} are :

$$i_t = \sigma\left(W_{xi}x_t + W_{hi}h_{t-1} + b_i\right) \tag{11}$$

$$f_t = \sigma\left(W_{xf}x_t + W_{hf}h_{t-1} + b_f\right) \tag{12}$$

$$o_t = \sigma\left(W_{xo}x_t + W_{ho}h_{t-1} + b_o\right) \tag{13}$$

$$g_t = \phi\left(W_{xc}x_t + W_{hc}h_{t-1} + b_c\right) \tag{14}$$

$$c_t = f_t \odot c_{t-1} + i_t \odot g_t \tag{15}$$

$$h_t = o_t \odot \phi\left(c_t\right) \tag{16}$$

where $\sigma(x) = \frac{1}{1+e^{-x}}$ is a sigmoid function in range $[0,1]$, $\phi(x) = \frac{e^x - e^{-x}}{e^x + e^{-x}}$ is a tanh function in range $[-1,1]$.

The LSTM with gates can avoid gradient blow up or vanish with over 1000 time steps, therefore the LSTM is chosen to fit unknown high order nonlinear function. The electric demand and electric power consumption updates ever 7 s, so it is difficult to update all parameters in LSTM online. To tackle this problem, weights linking the hidden notes of the last layer and the output are trained by FRA and then the weights can be updated online by RLS, so that the electric demand can be forecasted in 7 s.

The inputs at time instant k of the unknown high order nonlinear function is $\boldsymbol{x}(k) = (p(k), \cdots, p(k-n_p), v(k-1), \cdots, v(k-n_v))$, the output $v(k)$ can be written as

$$v(k) = \boldsymbol{\beta}\boldsymbol{h}_t(k) + b \tag{17}$$

where $\boldsymbol{\beta} = (\beta_1, \cdots, \beta_m)$ is trained by FRA, $\boldsymbol{h}_t(k) = (h_t^1(k), \cdots, h_t^m(k))^T$, $h_t^i(k) = tanh\left(h_{t-1}^i(k)\right)$, for $i = 1, \cdots, m$, m is the number of the hidden notes of the last layer, t is the number of the hidden layers, and b is the bias. The hidden notes in $1, \cdots, t-1$ layers are LSTM units.

3.3   A Fast Recursive Algorithm(FRA)

A Fast Recursive Algorithm(FRA) proposed in [12] is an alternative nonlinear system modeling method to the popular orthogonal least square(OLS) algorithm with improved stability and computational efficiency. We use FRA instead of LS to avoid calculating the pseudo inverse of the information matrix and FRA is described as follows.

Assume that there are m hidden nodes in the last hidden layer, define R_h as Eq. (18), where $\phi = (\phi_1, \cdots, \phi_m)$ is the information matrix and I is an identity matrix.

$$R_h = \begin{cases} I, & h = 0 \\ I - \phi\left[(\phi)^T \phi\right]^{-1}(\phi)^T, & 1 < h < H_l \end{cases} \tag{18}$$

According to [12] R_h can update iteratively by (19) for $h = 1, 2, \cdots, m-1$ and it avoids calculating the pseudo inverse of $\left[(\phi)^T \phi\right]$.

$$R_{h+1} = R_h - \frac{R_h\phi_{h+1}\left(R_h\phi_{h+1}\right)^T}{(\phi_{h+1})^T R_h\phi_{h+1}} \tag{19}$$

In [12] there are some propositions of R_h listed as (20)–(23) and the proofs can be found in [12].

$$\left(R_h\right)^T = R_h \tag{20}$$

$$\left(R_h\right)^2 = R_h \tag{21}$$

$$R_h R_j = R_j R_h = R_h, h \geq j \tag{22}$$

$$R_h\phi_i = 0, \forall i \in \{1, 2, 3, \cdots, m\} \tag{23}$$

As for traditional LS algorithm the parameters can be obtained by (24). Multiply ϕ on both sides of Eq. (24), yielding (25)

$$\hat{\beta}_i = \left[(\phi)^T \phi\right]^{-1} \phi Y \tag{24}$$

$$\phi\hat{\beta} = \phi\left[(\phi)^T \phi\right]^{-1} \phi Y$$
$$= (I - R_h) Y \tag{25}$$

where $Y = (v(1), \cdots, v(k))$.

Then define an upper triangular matrix as $A = [a_{h,p}]_{h \times H_i}$ as Eq. (26), where $h = 1, 2, \cdots, m, p = h, h+1, \cdots, m$.

$$a_{h,p} = (\phi_h)^T \phi_- \sum_{j=1}^{h-1} \frac{a_{j,h} a_{j,p}}{a_{j,j}} \tag{26}$$

And $a_{h,y}$ as Eq. (27)

$$a_{h,y} = (\phi_h)^T Y - \sum_{j=1}^{h-1} \frac{a_{j,h} a_{j,y}}{a_{j,j}} \tag{27}$$

Multiply $(\phi)^T$ to both sides of Eq. (25), yielding (28).

$$(\phi)^T \phi \hat{\beta} = (\phi)^T (I - R_h) Y \tag{28}$$

According to proposition (23), we can reformulate Eq. (25) as Eq. (29).

$$a_{h,h} \beta_h + \sum_{p=h+1}^{m} a_{j,p} \beta_p = a_{h,y} \tag{29}$$

Based on Eq. (29), the output weights β_h can be calculated as Eq. (30), which can avoid calculating the pseudo inverse of the information matrix.

$$\hat{\beta}_h = \frac{a_{h,y} - \sum_{p=h+1}^{m} a_{h,p} \hat{\beta}_p}{a_{h,h}} \tag{30}$$

3.4   Demand Forecasting in FMSP Online

The $v(k)$ in Eq. (17) has to be written as follows for online forecasting

$$v(k) = \beta(k) h_t(k) + b(k) \tag{31}$$

where $\beta(k), b(k)$ are updated online.

When new electric power $p(k)$ is measured and demand $\bar{P}(k)$ is calculated at time instant k, we fix linear model parameter θ and update the output $v(k-1)$ of LSTM by Eq. (5). Output weights β of LSTM can be updated by RLS as follows.

(1) Set a sliding window with width n_{train};
(2) Measure new electric power consumption $p(k)$ and update demand $\bar{P}(k)$;
(3) Update input matrix as

$$X(k) = [x(k - n_{train} + 1), \cdots, x(k)], l = 1, 2, \cdots, L$$

and compute the output of LSTM as

$$Y(k) = [v(k - n_{train}), \cdots, v(k-1)]^T$$

(4) Update the information matrix $\Phi(k) = (\boldsymbol{h}_t(k - n_{train} + 1), \cdots, \boldsymbol{h}_t(k))$ for LSTM with X;

(5) Update $\boldsymbol{\beta}(k)$ and $b(k)$ by RLS as Eq. (32)

$$\left(\hat{\boldsymbol{\beta}}(k), \hat{b}(k)\right) = \left(\hat{\boldsymbol{\beta}}(k-1), \hat{b}(k-1)\right) + Y(k)G(k) \tag{32}$$

where

$$G(k) = P(k-1)\Phi(k)\left[I + \Phi^T(k)P(k-1)\Phi(k)\right]^{-1}$$
$$P(k) = P(k-1) - G(k)\Phi^T(k)P(k-1)$$

and $P(0) = \left(\Phi^T(k)\Phi(k)\right)^{-1}$;

(6) Forecasting $\hat{v}(k)$ by LSTM with $\hat{\boldsymbol{\beta}}(k), \hat{b}(k)$ as Eq. (31), calculate $\hat{p}(k+1)$ by (2) and then calculate the predictive value of demand $\hat{\bar{P}}(k+1)$ by (3);

(7) Set $k = k+1$ at $k+1$ and return to step (2).

In summary, we can predict the demand $\hat{\bar{P}}(k+1)$ in FMSP and the next electric power consumption $\hat{p}(k+1)$ at time instant k by the aforementioned method. Furthermore, we can analyse the trend of demand for FMSP and evaluate the accuracy of the predictions detailed in the next section.

4   Accuracy and Trend of Demand Forecasting Analysis

We use Root Mean Squared Error(RMSE) [1] to evaluate the accuracy of demand forecasting by our method proposed in this paper. RMSE is described by Eq. (33), where N is the number of observations.

$$RMSE = \sqrt{\frac{1}{N}\sum_{k=1}^{N}\left(\bar{P}(k) - \hat{\bar{P}}(k+1)\right)^2} \tag{33}$$

We choose Mean Absolute Scaled Error(MASE) [?] in Eq. (34) as another evaluation for demand forecasting accuracy.

$$MASE = \frac{1}{N}\sum_{k=1}^{N}\frac{2\left|\bar{P}(k) - \hat{\bar{P}}(k+1)\right|}{\bar{P}(k) + \hat{\bar{P}}(k+1)} \tag{34}$$

The Direction Symmetry(DS) in Eq. (35) is used for analysing the trend of demand.

$$DS = \frac{1}{N}\sum_{k=1}^{N}d(k) \tag{35}$$

where $d(k)$ is defined as follows.

$$d(k) = \begin{cases} 1, & \left[\bar{P}(k) - \bar{P}(k-1)\right]\left[\hat{\bar{P}}(k) - \hat{\bar{P}}(k-1)\right] \geq 0 \\ 0, & \left[\bar{P}(k) - \bar{P}(k-1)\right]\left[\hat{\bar{P}}(k) - \hat{\bar{P}}(k-1)\right] < 0 \end{cases} \tag{36}$$

The DS does not indicate the upward or downhill tendency of demand in a continuous period, and the True Positive Rate(TPR) in Eq. (37) and True Negative Rate(TNR) in Eq. (38) are used to solve this deficiency.

$$TPR = \frac{TP}{TP + FN} \tag{37}$$

$$TNR = \frac{TN}{TN + FP} \tag{38}$$

The relationship between TPR and TNR are listed in Table 1.

Table 1. Relationship table

$\bar{P}$	$\hat{\bar{P}}$	
	Up	Down
Up	TP	FP
Down	TN	FN

5 Experimental Results

We used real data from a fused magnesia factory within a particular day to test the demand forecasting algorithm in this paper. The limitation of demand is $\bar{P}_l = 22100\,\text{kVA}$ and a total of 4 FMFs were operating in full capacity. $U = 190\,\text{V}$ was the voltage of each FMF, $I_i = 15000\,\text{A}$ was the set value for the ith FMF'current and $\cos\varphi = 0.92$ was the power factor. We sampled the continuous electric power consumption $p(1), \cdots, p(N)$ and calculated demand $\bar{P}(1), \cdots, \bar{P}(N)$, where the total number of samples is $N = 4000$, among which half of the samples are used for training, i.e. $n_{train} = 2000$ samples are for training offline and the remainder is for testing online.

5.1 Identification of Linear Model Parameters

As discussed above, we first set $v(k) = 0$ for $k = 1, 2, \cdots, n_{train}$ and then identifying parameters $\theta = (0.0234, 0.1586, -0.4169, 1.2344)^T$ by LS. Update θ by RLS in Eq. (8) until θ is convergent which lead to the set of value $\theta = (-1.2082, 0.3686, -0.0702, 0.3003)^T$. Then we fix parameters θ and fit the unknown high order function in Eq. (7) by LSTM trained with FRA.

5.2 Results of Demand Forecasting

We train LSTM by the method mentioned earlier in this paper with training data set and then test the online forecasting performance of the trained model

Page 212, J. Zhang et al.

on the test data set. In Fig. 4 we can see that the error between actual demand and its prediction is in [−200, 400] kVA, and we can forecast the accurate trend when the actual demand exceeds its limitation via the method developed in this paper. Figure 5 shows the electric power consumption prediction results and which are acceptable. Both the prediction errors of demand and electric power consumptions are Gaussian white noise.

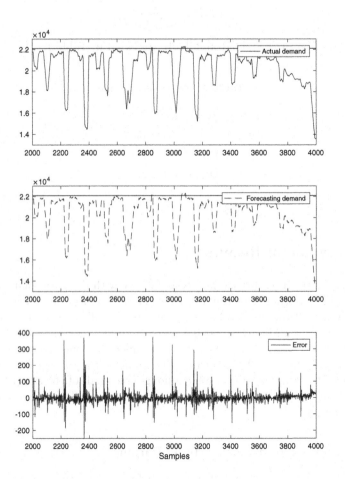

Fig. 4. Results of demand forecasting

Results of RMSE, MASE, DS, TPR and TNR for LSTM and RBF to model the unknown high order function are shown in Table 2. From Table 2 we can see that the unknown high order function $v(k)$ fitted by LSTM trained by FRA has the best performance for demand forecasting in this paper. The LSTM model has the trend forecasting accuracy is up to almost 90% and it can be used to judge whether it is necessary to turn off or turn on the FMF.

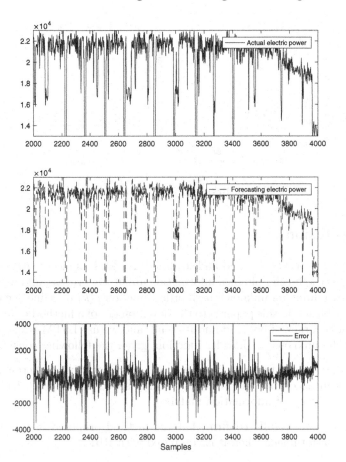

Fig. 5. Results of power forecasting

Table 2. Results of demand forecasting with LSTM and RBF

	RMSE	MASE	DS	TPR	TNR
LSTM	26.1457	0.016	93.76	88.69	91.75
RBF	33.9494	0.028	89.39	84.61	85.21

In Fig. 6 the probability distribution of the predictive error of electric power consumption and auto-correlation coefficient curve of the predictive error of demand are illustrated. From the probability distribution curve it is shown that the probability of the error equals to zero is 2.5% and the auto-correlation coefficient curve of the predictive error of demand shows the errors are least correlated.

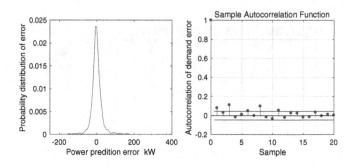

Fig. 6. Results of demand forecasting

6 Conclusion

Forecasting the power demand and its trend in FMSP is faced with a challenge that there is too much unknown information about the process, and we solve this problem by adding an unknown high order function $v(k)$ to a linear model. The main contribution in this paper is to the development of a method to improve the prediction accuracy of demand and its tend, and we use LSTM trained by FRA to model the unknown high order function, given the following considerations:

(1) LSTM is a kind of deep network and it can improve the prediction accuracy;
(2) FRA can accelerate the network training speed and avoid calculating the pseudo-inverse of information matrix.

The experimental results confirm that the method proposed in this paper has the better performance when it is compared with RBF.

References

1. Yang, J., Chai, T., Luo, C., Yu, W.: Intelligent demand forecasting of smelting process using data-driven and mechanism model. IEEE Trans. Ind. Electron. **66**(12), 9745–9755 (2018)
2. Hagan, M.T., Behr, S.M.: The time series approach to short term load forecasting. IEEE Trans. Power Syst. **2**(3), 785–791 (1987)
3. Moghram, I., Rahman, S.: Analysis and evaluation of five short-term load forecasting techniques. IEEE Trans. Power Syst. **4**(4), 1484–1491 (1989)
4. Al-Hamadi, H.M., Soliman, S.A.: Long-term/mid-term electric load forecasting based on short-term correlation and annual growth. Electr. Power Syst. Res. **74**(3), 353–361 (2005)
5. Hsu, C.-C., Chen, C.-Y.: Regional load forecasting in Taiwan–applications of artificial neural networks. Energy Convers. Manag. **44**(12), 1941–1949 (2003)
6. Pao, H.-T.: Comparing linear and nonlinear forecasts for Taiwan's electricity consumption. Energy **31**(12), 2129–2141 (2006)
7. Maia, A.L.S., de Carvalho, F.D.A., Ludermir, T.B.: Symbolic interval time series forecasting using a hybrid model. In: 2006 Ninth Brazilian Symposium on Neural Networks (SBRN 2006), pp. 202–207. IEEE (2006)

8. Yang, J., Chai, T.: Data-driven demand forecasting method for fused magnesium furnaces. In: 2016 12th World Congress on Intelligent Control and Automation (WCICA), pp. 2015–2022. IEEE (2016)
9. Yang, J., Chai, T., Zhang, Y., Zhiwei, W.: Data and model driven demand forecasting method for fused magnesium furnaces. Acta Autom. Sin. **44**(8), 1460–1474 (2018)
10. Badoni, M., Singh, A., Singh, B.: Variable forgetting factor recursive least square control algorithm for dstatcom. IEEE Trans. Power Delivery **30**(5), 2353–2361 (2015)
11. Hochreiter, S., Schmidhuber, J.: Long short-term memory. Neural Comput. **9**(8), 1735–1780 (1997)
12. Li, K., Peng, J.X., Irwin, G.W.: A fast nonlinear model identification method. IEEE Trans. Autom. Control **50**(8), 1211–1216 (2005)
13. Donahue, J., et al.: Long-term recurrent convolutional networks for visual recognition and description. In: Proceedings of the IEEE Conference on Computer Vision and Pattern Recognition, pp. 2625–2634 (2015)

Bearing Fault Diagnosis Based on Variational Mode Decomposition and Modified CNN

Guolian Hou[(✉)], Chen Yao, Linjuan Gong, and Jianhua Zhang

School of Control and Computer Engineering, North China Electric Power University, Beijing 102206, China
hgl@ncepu.edu.cn

Abstract. In order to solve the problem that it is difficult to extract slight fault features in the process of fault diagnosis of rolling bearing, this paper proposes a fault diagnosis method based on variational mode decomposition (VMD) and modified convolution neural network (CNN). Firstly, in the process of eigenvalue extraction, VMD decomposition is used to extract more fault feature details of rolling bearing vibration signals; then, dense block and other methods are applied in the network with fewer layers; finally, global average pooling is used instead of full connection layer for the complex calculation of full connection layer. Through the diagnosis experiments of different fault conditions of rolling bearing, it is proved that the proposed method can improve the fault recognition rate and has good feasibility.

Keywords: Variational mode decomposition · Convolution neural network · DenseNet · Intelligent fault diagnosis

1 Introduction

Rolling bearing is most crucial part among rotating machinery field, and one of the most easily damaged parts. Research has shown that more than 45% of the accidents of this field impute to the failure of rolling bearings [1]. The fault signals of modern rotating machinery are often accompanied by nonlinear, high-dimensional, non-stationary and non-Gaussian distribution. Therefore, it is difficult for the traditional fault diagnosis methods to meet the fault diagnosis requirements of rolling bearing in contemporary rotating machinery. The advanced technology has been developed to guarantee that the production of rotating machinery is efficient and the operation is safe.

Recently, deep learning technology has been widely concerned, because of its strong ability in solving the problem caused by the failure of traditional artificial intelligence methods [2–4]. Literally, there are many levels in the deep learning algorithms, and each layer is designed simply so that new representations of inputs can be developed automatically for detection or classification. Although these methods may not be as fast as those based on physical models, They are well suited for complex systems that are difficult to establish explicit models or signal symptoms. Deep learning networks usually start with raw data. Layers of the identical kind are stacked together and the input can

© Springer Nature Singapore Pte Ltd. 2020
M. Fei et al. (Eds.): LSMS 2020/ICSEE 2020 Workshops, CCIS 1303, pp. 216–224, 2020.
https://doi.org/10.1007/978-981-33-6378-6_16

be adaptively converted to a high-level and more unknowable representation layer for suppressing irrelevant changes and maintaining the discrimination information [5]. On account of the design above, the network just mentioned has been proved to have greater functionality and is adaptive in dealing with different data [6].

In [7], a standard convolutional neural network was reconstructed according to the actual object, thus the constructed CNN neural network conformed to the actual physical meaning of the actual object. At the same time, it relied on the excellent classification ability of the neural network to classify the fault. Then, the causes and duration time of the fault were predicted through the proposed method. In [8], a novel method based on lenet-5 is adopted for actual fault diagnosis in production and early warning. The introduced method, which could transform signal into two-dimensional image, was effective for data processing. As for the method presented in [9], the features were learned adaptively from the original mechanical data without prior knowledge through the deep learning network. In [10], the vibration signals were sorted and combined to produce the gray value, and the processed gray value image was used for fault diagnosis. In [11], a novel network which used many complementary sensor dates was proposed and worked well in domain of fault diagnosis. The adopted network can be simply developed to a structure of more profundity for intricate diagnosis assignment as well as alleviated the wastage of the features and the wastage of gradient. However, the original network mentioned above is designed for image recognition, the depth and complexity of the network far exceed the actual needs of fault diagnosis, which only need to diagnose a few fault conditions. Too complex structure cannot show the advantage of diagnosis accuracy when dealing with relatively simple problems, but will lead to longer calculation time. It is difficult to get better diagnosis effect when the network structure is shallow.

In addition, data processing has a very important impact on subsequent network learning and helps to achieve better diagnosis results. In reference [12], a method for feature extraction and fault diagnosis of planetary gears was displayed. The pattern matrix was divided into several sub-matrices. This algorithm could extract the subtle and easily overlooked feature information successfully and diagnosis different faults accurately. As in [13], VMD was used to decompose non-stationary signals into multiple stationary sub signals, and many feature selection algorithms were implemented for further selection of sub-signals. The prediction results of each subsequence were further integrated nonlinearly.

Inspired by the DenseNet network in the field of computational vision, a model using VMD and modified CNN is used in this treatise to achieve higher diagnosis accuracy in the case of lower network layers. VMD can decompose the original mechanical fault signal into sub-signals of different scales, and decompose the original signal without destroying the overall structure and function In addition, because the dense connection structure is adopted in the model, the problems of fault feature recognition errors and omissions are largely avoided, and the gradient is also improved. for the sake of optimizing the structure of DenseNet network, information mining and fault identification can be better acquired without hundreds of network layers. Furthermore, the structure of full convolution layer is used instead of full connection layer to make the output more flexible and reduce the calculation burden.

2 Theory Background

2.1 VMD

VMD is an adaptive, completely non recursive method famous for its self-adaption without recurrence for modal variation and signal processing. This technology shows excellent advantage in determining the number of modal decompositions of the given sequence according to the actual situation. In the subsequent search and solution process, realize the effective separation of the intrinsic mode component. VMD overcomes the problem of end-point effect and model component overlapping in EMD method, and has a more solid mathematical theoretical basis. VMD can reduce the non-stationary of time series with high complexity and strong nonlinearity. It can decompose multiple subsequences with different frequency scales and relative stationary. It is applicable series that is not a stationary sequence. The core idea of VMD is to construct and solve the difficult task of variation.

Assuming the original signal f is decomposed into k components. In order to ensure that the decomposition sequence is a finite bandwidth modal component with center frequency and after estimating all the bandwidth, the bandwidth is taken as the lower limit on the premise that the sum of the components is equal to the total value. The corresponding constraint variation expression is obtained as follows.

$$\begin{cases} \min\limits_{\{u_k\},\{a_k\}} \left\{ \sum\limits_k \left\| \partial_t \left[(\delta(t) + j/\pi t) * u_k(t) \right] e^{-j\omega_k t} \right\|_2^2 \right\} \\ \text{s.t.} \sum\limits_{k=1}^{K} u_k = f \end{cases} \tag{1}$$

in which, K is the number of modes to be decomposed while $\{u_k\}$ and $\{a_k\}$ correspond to the kth modal component and center frequency after decomposition. Besides, $\delta(t)$ stands for Dirac function while $*$ is convolution operator.

By introducing the Lagrange multiplication operator λ, the expression with constraints is converted to a formula without constraints. The converted Formula is as follows is expressed as follows.

$$L(\{u_k\}, \{\omega_k\}, \lambda) = \\ \alpha \sum\limits_k \left\| \partial_t \left[(\delta(t) + j/\pi t) * u_k(t) \right] e^{-j\omega_z t} \right\|_2^2 + \\ \left\| f(t) - \sum\limits_k u_k(t) \right\|_2^2 + \left\langle \lambda(t), f(t) - \sum\limits_k u_k(t) \right\rangle \tag{2}$$

where α is the second penalty factor, which is used to reduce the dry reactance of Gaussian noise.

Using the alternating direction multiplier iterative algorithm, the modal components and center frequencies are optimized, and searching saddle points of the augmented Lagrange function. The expressions of u_k, ω_k and λ after the alternating optimization iteration are shown as follows.

$$\hat{u}_k^{n+1}(\omega) = \frac{\hat{f}(\omega) - \sum\limits_{i/k} \hat{u}_i(\omega) + \hat{\lambda}(\omega)/2}{1 + 2\alpha(\omega - \omega_k)^2} \tag{3}$$

$$\omega_k^{n+1} = \frac{\int_0^\infty \omega \left| \hat{u}_k^{n+1}(\omega) \right|^2 d\omega}{\int_0^\infty \left| \hat{u}_k^{n+1}(\omega) \right|^2 d\omega} \tag{4}$$

$$\hat{\lambda}^{n+1}(\omega) = \hat{\lambda}^n(\omega) + \gamma \left(\hat{f}(\omega) - \sum\limits_k \hat{u}_k^{n+1}(\omega) \right) \tag{5}$$

Repeat the iteration until $\sum\limits_k \left\| \hat{u}_k^{n+1} - \hat{u}_k^n \right\|_2^2 / \left\| \hat{u}_k^n \right\|_2^2 < \varepsilon$ and $n < N$, where ε represent convergence accuracy. By iterating the above formulas, the time domain mode $u_k(t)$ can be calculated.

2.2 Standard Convolutional Neural Networks Module

The standard convolutional neural networks include the convolutional layer and the pooling layer. The first one is composed of convolution filter with information mining ability, and all filter parameters are equal, which reduces the computational burden to some extent. Multiple filters work together to get the output. Calculation of each convolution layer is formulated as

$$z_j^{l+1} = \sigma \left(\sum\limits_i x_i^l * w_{ij}^l + b_j^{l+1} \right) \tag{6}$$

The second kind is pooling layer. Its function is to aggregate the previous convolution layer, so that the size of the input feature graph of the upper convolution layer will be reduced after the aggregation of the sub sampling layer, so as to reduce the number of features and parameters. What the sub sampling layer does is to scan the upper layer of the convolution layer, scan a specific area each time, and then calculate the maximum value (maximum pooling) or average value (mean pooling) of the area characteristics as the representation of the region characteristics. Denoting the pooling function as $p(\cdot)$, the mathematical formula can be expressed as

$$a_j^l = \mathop{p}\limits_{n \in R} \left(z_j^l(n) \right) \tag{7}$$

R represents the pooling region, n is the position coordinates, and a_j^l denotes the output. Average pooling and max pooling are commonly used pooling operations.

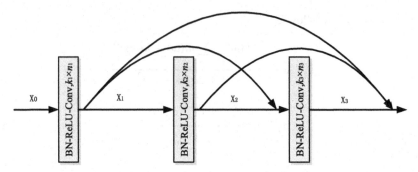

Fig. 1. Structure of the dense block.

2.3 Dense Network Module

A simple example is shown in Fig. 1. Even if there are only a few simple dense blocks, the information it receives is far more than that of other network structures on the same layer, because each layer starting from the second layer receives all the previous information.

$$\mathbf{x}_l = H_l([\mathbf{x}_0, \mathbf{x}_1, \ldots, \mathbf{x}_{l-1}]) \tag{8}$$

where $x_0, x_1, \ldots, x_{l-1}$ are input variables, respectively. H_l represents a composite function of three consecutive operations: batch normalization (BN), rectified linear unit (ReLU) and a convolution operation. One of the advantages of DenseNet is that the network is narrower and the parameters are less, as a result of the dense block design. In the dense block, the number of output feature maps of each volume layer is very small, rather than thousands of widths like other networks. Furthermore, this connection makes the transfer of features and gradients to be more effective, and network training easier.

A significant difference between DenseNet and conventional network structure is that the layer of DenseNet can be very narrow. Define the super parameter K as the growth rate of the network. The growth rate determines how much new information each layer contributes to the overall state, which is different from the traditional network needing to copy all information layer by layer.

In order to reduce the number of feature mapping and improve the compactness of the model, the transition block is designed. The transition block consists of a batch normalization layer, an activation function layer, a 1×1 convolution layer and a 2×1 average pool layer.

3 Framework of the Proposed Method

The adopted VMD-Modified CNN intelligent fault diagnosis is presented in this section,. Firstly, using VMD to break up the signal to three scales. Then, the signals obtained in the previous step are normalized, and 400 sampling points of a sample are transformed from vector form to 20 * 20 matrix form, which is composed of numerous rows. Then, the matrixes will be converted to grayscale images.

As shown in Fig. 2, the proposed model includes three same subnetworks and a concatenate layer.

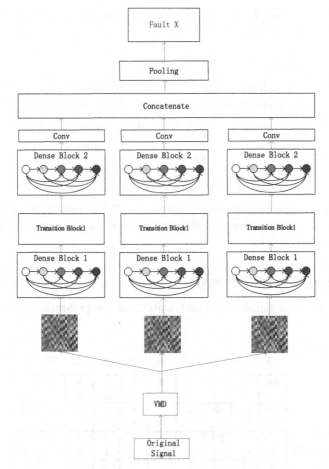

Fig. 2. Structure of proposed model.

The three signals of different scales are input into three modified models, respectively. Besides, the finial outputs of three models are concreted, and then the fully connected layer was replaced by convolution layer. The growth rate is set as 4 and as shown in Table 1, where the $13 - [20 \times 20]$ denotes that there are thirteen feature maps of size 20×20; the conv represents the convolution operation.

4 Experimental Validation

4.1 Data Description

To validate the effectiveness of the proposed method, BDCS Fault Test Data is selected. The data used in the simulation are selected from bearing data: normal vibration data at 1797 rpm and vibration data at three fault levels of rolling element, inner ring and outer ring, with a total of nine faults, which is numbered #1 to #9 and normal, numbered #0.

Table 1. Architecture for intelligent fault diagnosis

Layers	Output size	Parameters
Input	$1 - [20 \times 20]$	/
Dense Block1	$13 - [20 \times 20]$	3×3conv, stride 1
Transition Block1	$8 - [10 \times 10]$	2×2conv, stride 2
Dense Block2	$20 - [10 \times 10]$	3×3conv, stride 1
Conv Pooling	$10 - [1 \times 1]$ 10	10×10conv/

The bearing rotates one circle to sample 400 points. 120000 sampling points are selected from each fault time series data. The sliding window is used to sample 400 points, and the sliding step size is 80.

4.2 Results Analysis and Discussion

In this section, the detailed results and analysis of comprehensive experiments are presented.

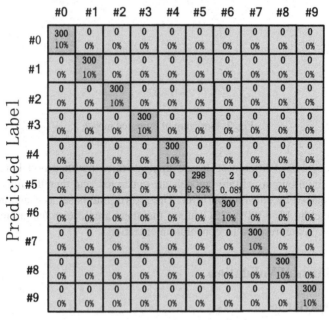

Fig. 3. Confusion matrix of the proposed model.

After normalizing and reorganizing the original data into the format of 20 × 20, a total of 15000 samples are obtained. Of these, 12000 were used for training and the remaining 3000 for testing. The 15000 samples were composed of nine kinds of failure and normal, a total of ten health conditions, each accounting for one tenth.

As shown in Fig. 3, the test accuracy of each of the ten health conditions obtained by this method has been given. #0 represents normal, and #1−#9 represents nine different faults as mentioned above. The row label represents the health status corresponding to the predicted result, and the column represents the real health status. The number and proportion of the corresponding prediction results are given in the lattice. The accuracy of each health condition is higher than 99.2% and overall diagnosis accuracy achieves 99.93%. The result of standard convolutional neural networks is 92.8% while LSTM network is 93.4%, which fully proves the effectiveness and robustness of the proposed method for each category.

5 Conclusion

This paper presents a network model based on VMD and modified CNN for intelligent fault diagnosis. In this method, VMD is used to decompose the signals and obtain more abundant state information in parallel. The advantages of DenseNet network are migrated and applied in fault diagnosis. In addition, the model can be easily extended for both simple and complex diagnostic tasks. In particular, the model combines information mining, feature extraction and fault classification. Compared with the shallow classification model relying on manual feature extraction, the model has a great improvement in training speed and accuracy, and is more intelligent. Finally, the comprehensive experiments on bearing data of Laboratory of Case Western Reserve University were carried out for performance evaluation, and the results demonstrated the effectiveness and superiority of the proposed method.

Acknowledgements. This work is supported by the National Natural Science Foundation of China (Grant No. 61973116). The reviewers' insightful comments and valuable suggestions are also greatly appreciated.

References

1. Zhang, Q., Gao, J., Dong, H.: WPD and DE/BBO-RBFNN for solution of rolling bearing fault diagnosis. Neurocomputing **312**, 27–33 (2018). https://doi.org/10.1016/j.neucom.2018.05.014
2. Pan, J., Zi, Y., Chen, J., et al.: LiftingNet: LiftingNet: a novel deep learning network with layerwise feature learning from noisy mechanical data for fault classification. IEEE Trans. Ind. Electron. **65**(6), 4973–4982 (2017). https://doi.org/10.1109/tie.2017.2767540
3. Wen, L., Li, X., Gao, L., Zhang, Y.: A new convolutional neural network-based data-driven fault diagnosis method. IEEE Trans. Ind. Electron. **65**(7), 5990–5998 (2017). https://doi.org/10.1109/tie.2017.2774777
4. Zhao, H., Liu, H., Hu, W., Yan, X.: Anomaly detection and fault analysis of wind turbine components based on deep learning network. Renew. Energy **127**, 825–834 (2018). https://doi.org/10.1016/j.renene.2018.05.024

5. Lee, K.B., Cheon, S., Kim, C.O.: A convolutional neural network for fault classification and diagnosis in semiconductor manufacturing processes. IEEE Trans. Semicond. Manufact. **30**(2), 135–142 (2017). https://doi.org/10.1109/tsm.2017.2676245
6. Liu, C., Cheng, G., Chen, X., Pang, Y.: Planetary gears feature extraction and fault diagnosis method based on VMD and CNN. Sensors **18**(5), 1523 (2018). https://doi.org/10.3390/s18051523
7. Glowacz, A., Glowacz, Z.: Diagnosis of stator faults of the single-phase induction motor using acoustic signals. Appl. Acoust. **117**, 20–27 (2016). https://doi.org/10.1016/j.apacoust.2016.10.012
8. Glowacz, A., Glowacz, W., Glowacz, Z., et al.: Early fault diagnosis of bearing and stator faults of the single-phase induction motor using acoustic signals. Measurement **113**, 1–9 (2018). https://doi.org/10.1016/j.measurement.2017.08.036
9. Lin, P., Wang, C., Chen, T.: A stall warning scheme for aircraft engines with inlet distortion via deterministic learning. IEEE Trans. Control Syst. Technol. **26**(4), 1468–1474 (2017). https://doi.org/10.1109/tcst.2017.2709273
10. Wen, L., Li, X., Gao, L., Zhang, Y.: A new convolutional neural network-based data-driven fault diagnosis method. IEEE Trans. Ind. Electron. **65**(7), 5990–5998 (2018). https://doi.org/10.1109/tie.2017.2774777
11. Jiao, J., Zhao, M., Lin, J., Ding, C.: Deep coupled dense convolutional network with complementary data for intelligent fault diagnosis. IEEE Trans. Ind. Electron. **66**(12), 9858–9867 (2019). https://doi.org/10.1109/tie.2019.2902817
12. Liu, C., Cheng, G., Chen, X., Pang, Y.: Planetary gears feature extraction and fault diagnosis method based on VMD and CNN. Sensors **18**(5), 1523 (2018). https://doi.org/10.3390/s18051523
13. Wang, J., Li, Y.: Multi-step ahead wind speed prediction based on optimal feature extraction, long short term memory neural network and error correction strategy. Appl. Energy **230**, 429–443 (2018). https://doi.org/10.1016/j.apenergy.2018.08.11

Analysis of Suspension Characteristics of a New High-Temperature Superconducting Magnetic Levitation Bearing Based on Frozen-Image Model

Si-yu Wang, Nan Wu$^{(\boxtimes)}$, Dongsheng Yang, and Ye-sheng Zhu

Northeastern University, College of Information Science and Engineering,
Shenyang 110819, China
2954216938@qq.com

Abstract. The high temperature superconducting magnetic levitation bearing (HTSMB) uses the special magnetic flux pinning properties of the high temperature superconductor and the full diamagnetic effect of the Meissner effect to make the rotor achieve self-stable suspension. Compared with electromagnetic bearings (AMB), superconducting magnetic suspension bearings do not require a control system and have lower energy losses. This paper analyzes the levitation characteristics of a new high-temperature superconducting magnetic levitation bearing using a frozen-image model. The results show that the levitation force and stiffness of the bearing change with the changes of the initial cooling gap, axial displacement and radial displacement of the bearing. Under the same conditions, the radial suspension force is an order of magnitude greater than the axial suspension force; Within a certain range, the smaller the initial cooling gap, the better the suspension performance of the bearing, and the hysteresis characteristics have little effect on the suspension characteristics of the bearing.

Keywords: High-temperature superconducting magnetic levitation bearing ·
Magnetic flux freezing mirror model · Levitation force · Stiffness · Hysteresis characteristics

1 Introduction

In order to overcome the deficiency caused by mechanical friction in bearings, various non-contact magnetic bearings have been developed to replace mechanical bearings. Typical examples are active magnetic suspension bearings (AMB), hybrid magnetic suspension bearings (HMB), and passive magnetic suspension bearings (PMB) [1]. High temperature superconducting magnetic suspension bearing (HTSMB) is a new type of passive suspension bearing. Depending on the special internal structure of high temperature superconducting material, it has a unique magnetic flux nailing effect and the complete diamagnetism of meissner effect in the mixed state, and has self-stabilizing suspension characteristics [2]. Compared with active magnetic bearing, HTSMB does

© Springer Nature Singapore Pte Ltd. 2020
M. Fei et al. (Eds.): LSMS 2020/ICSEE 2020 Workshops, CCIS 1303, pp. 225–235, 2020.
https://doi.org/10.1007/978-981-33-6378-6_17

not need high-precision automatic control system, and compared with low-temperature superconductor, it does not need complex low-temperature supply system, which greatly reduces energy consumption and use cost [3–5].

High temperature superconducting magnetic levitation bearings have excellent levitation bearing capacity. Foreign research institutions have conducted long-term research and application [6, 7]. Related products of German ATZ company have achieved a levitation load ratio of 1:200 [8]. Rigney and Trivedi had achieved 520,000 rpm ultra-high-speed superconducting magnetic levitation bearing experiments, with good anti-interference and self-recovery capabilities [9]. At the Argonne National Laboratory in the United States, a 2.25 kW hFESS was developed using high-temperature superconducting magnetic levitation bearings [10]. China is still mostly in the theoretical and laboratory research stage, and there are few researches on the practical application of superconducting magnetic suspension bearings.

In this paper, aiming at the design optimization requirements of a new high temperature superconducting magnetic suspension bearing, the frozen-image model is used to calculate the stiffness and suspension force of the superconducting magnetic suspension bearing theoretically [11]. The qualitative research on the initial cooling clearance and hysteresis characteristics of the bearing provides a more comprehensive theoretical basis for the design optimization of the bearing.

2   System Modeling and Theoretical Calculation

2.1   Structural Model of a New High Temperature Superconducting Magnetic Suspension Bearing

In this paper, a new high temperature superconducting magnetic suspension bearing composed of six superconducting blocks of the bearing stator is studied and the structure is shown in Fig. 1.

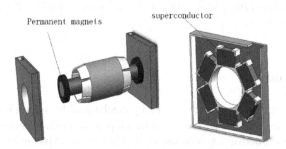

Fig. 1. Schematic diagram of the new HTSMB

Permanent magnetic rings are nested on both sides of the bearing rotor, and six high-temperature superconducting blocks are fixed around the permanent magnetic ring. The dimensions are shown in Table 1. Under the environment of liquid nitrogen, due to the high temperature superconductor pinning characteristics of superconductor in the initial cooling process has frozen the phenomenon of magnetic flux, and frozen flux and the external magnetic field between the permanent magnet and superconducting appeal, make the permanent magnet in the middle of the superconducting piece stable suspension, if there are the relative position of the permanent magnets at this time, the magnetic flux pinning force and loren magnetic interaction, forcing the rotor once again return to balance position, so that the HTS maglev system has a good stability of axial and radial.

Table 1. Structural parameters of permanent magnet and superconducting

Parameter	Value
Permanent magnet remanence/T	0.25
Inner diameter of permanent magnet/mm	30
Outer diameter of permanent magnet/mm	50
Permanent magnet height/mm	10
Superconductor length/mm	30
Superconductor width/mm	20
Superconductor height/mm	10

2.2 Frozen-Image Model

The magnetic flux freezing mirror model is shown in Fig. 2. The yOz plane is the surface of the high temperature superconductor opposite to the permanent magnet. The permanent magnet is simplified as a magnetic dipole M_1, whose direction is consistent with the negative direction of the x-axis, and the magnetic pole moment is m_1, when the bearing rotor is in the initial cooling position, the coordinates of the magnetic dipole are $(-x_0, 0, 0)$. The diamagnetic image magnetic dipole M_2 with a magnetic pole moment m_2 is equivalent to the shielding current of the surface of a high-temperature superconductor, M_2 and M_1 is always symmetrical about the surface of the superconductor, that is, M_2 follows the M_1's movement and moves in the opposite direction. Once the cooling is completed, a frozen magnetic flux will be formed at the initial cooling position of the high-temperature superconductor, which is simplified as a frozen mirror magnetic dipole M_3 and M_3 fixed at a position where the permanent magnet is symmetrical about the yOz plane. When the permanent magnet moves in the x-axis direction, as the external magnetic field is strengthened or weakened, some of the magnetic flux will enter and exit the superconductor. Magnetic field, its pole moment m_4 is related to the movement of the permanent magnet in the x direction. The position of M_4 is fixed to a position symmetrical to yOz when the gap between the permanent magnet and the superconductor is the smallest. When the permanent magnet moves in the z-axis axial direction, the change in the internal magnetic flux of the superconductor will be described by moving the mirror magnetic dipole M_5 in the z-direction. The position of M_5 is also fixed at the position of M_4, the direction of M_5 and the magnitude of the magnetic pole moment m_5 are related to the movement of the permanent magnet in the z direction. M_4 and M_5 are related to the movement history of permanent magnets, so they can reflect the hysteresis characteristics of superconducting magnetic levitation systems to a certain extent.

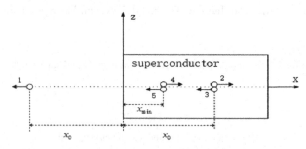

Fig. 2. Frozen-image model

According to magnetism, the interaction force between two magnetic dipoles is given by the following formula:

$$F = (m \cdot \nabla)B \tag{1}$$

m is the magnetic moment of one of the magnetic dipoles, ∇ is the differential operator, and B is the magnetic field generated by the other magnetic dipole. The magnetic field B generated by a magnetic dipole with a magnetic moment m at an arbitrary point P $(x, 0, z)$ is:

$$B = \frac{\mu_0 m}{4\pi}\left(\frac{3xz}{r^5}i + \frac{2x^2 - z^2}{r^5}j\right) \tag{2}$$

μ_0 is the vacuum permeability, $r = \sqrt{x^2 + z^2}$, i is the unit vector in the z direction, and j is the unit vector in the x direction.

After the equivalent of the magnetic dipole, the force on the permanent magnet at any point p $(x, 0, z)$ is the vector sum of the forces of the magnetic dipoles M_2 and M_3 acting on the permanent magnet equivalent magnetic dipole M_1, respectively:

$$
F_z(x, z) = A\left\{\frac{z^3 - 4z(x + x_0)^2}{\left[z^2 + (x + x_0)^2\right]^{7/2}}m_3\right.
$$
$$
\left.+ \frac{z^3 - 4z(x + x_{\min})^2}{\left[z^2 + (x + x_{\min})^2\right]^{7/2}}(m_4 + m_5)\right\} \tag{3}
$$
$$
F_x(x, z) = A\left\{\frac{2}{(2x)^4}m_2\right.
$$
$$
+ \frac{3(x + x_0)z^2 - 2(x + x_0)^3}{\left[z^2 + (x + x_0)^2\right]^{7/2}}m_3
$$

$$+ \frac{3z^2(x+x_{min})^2 - 2(x+x_{min})^3}{\left[z^2 + (x+x_{min})^2\right]^{7/2}}(m_4 + m_5)\Bigg\} \tag{4}$$

Among them $A = (3\mu_0 m_1)/4\pi$, x is the distance between the magnetic dipole M_1 and the surface of the superconductor, z is the displacement of M_1 in the z-axis direction, x_0 is the distance between the magnetic dipole M_1 and the surface of the superconductor during initial cooling, and x_{min} is the distance between the superconductor and M_1 during the movement of m1 Minimum clearance of the surface.

2.3 Suspension Force and Stiffness Calculation

It can be seen from Fig. 3 that the permanent magnet and the superconducting block are symmetrical about the z-axis. When qualitatively analyzing the axial and radial suspension forces and stiffness characteristics, the force on the rotor can be viewed as the vector sum of the force of six superconductor magnetic fields on the permanent magnet magnetic dipole.

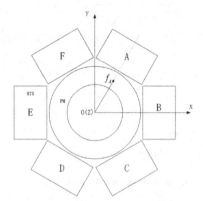

Fig. 3. The structure and the suspension force of HTSMB

When the rotor moves in the z-axis direction at the equilibrium position, the rotor receives the sum of the forces of six superconductors in the z-axis direction, and the force of each superconductor is the same. At this time, $x = x_0$, the force on the rotor in the z direction is:

$$F_z(0, z) = 6A\Bigg\{ \frac{z^3 - 16zx_0^2}{\left[z^2 + 4x_0^2\right]^{7/2}}m_3$$

$$+ \frac{z^3 - 4z(x_0 + x_{min})^2}{\left[z^2 + (x_0 + x_{min})^2\right]^{7/2}}(m_4 + m_5)\Bigg\} \tag{5}$$

The axial stiffness is:

$$K_{zz} = \frac{dF_z}{dz} \tag{6}$$

Because superconductors are all about axis symmetry, the force in the x direction is analyzed radially. Assuming that the bearing rotor moves in the x axis direction, the force on the rotor is the sum of the forces of the six superconductors on the permanent magnets in the x axis direction, that is:

$$F_x(x, z) = f_{Ax} + f_{Bx} + f_{Cx} + f_{Dx} + f_{Ex} + f_{Fx} \qquad (7)$$

Obviously, $f_{Ax} + f_{Dx} = f_{Cx} + f_{Fx}$, and the characteristics of the levitation force generated by the superconducting block A and the superconducting block D in the x-axis direction are similar to those of the superconducting block B and the superconducting block E. The qualitative analysis of bearing suspension force characteristics can be simplified. In the formula, β is used to represent the acting force of superconducting blocks in other directions on the rotor in the x-axis direction

$$
\begin{aligned}
F_x(x, z) = A\beta \Big\{ & \frac{2}{(2x)^4} m_2 \\
+ & \frac{3(x + x_0)z^2 - 2(x + x_0)^3}{\left[z^2 + (x + x_0)^2\right]^{7/2}} m_3 + \frac{2}{(4x_0 - 2x)^4} m_2 \\
& + \frac{3(3x_0 - x)z^2 - 2(3x_0 - x)^3}{\left[z^2 + (3x_0 - x)^2\right]^{7/2}} m_3 \\
& + \frac{3z^2(x + x_{min})^2 - 2(x + x_{min})^3}{\left[z^2 + (x + x_{min})^2\right]^{7/2}} (m_4 + m_5) \\
+ & \frac{3z^2(2x_0 + x_{min} - x)^2 - 2(2x_0 + x_{min} - x)^3}{\left[z^2 + (2x_0 + x_{min} - x)^2\right]^{7/2}} (m_4 + m_5) \Big\}
\end{aligned}
\qquad (8)
$$

Similarly, the stiffness in the x-axis direction is

$$K_{xx} = \frac{dF_x}{dx} \qquad (9)$$

3 Levitation Force and Stiffness Analysis

3.1 Axial Suspension Force and Stiffness

According to the results calculated in the previous section, the levitation force and stiffness characteristics of the HTSMB at different displacements can be obtained. Without considering the hysteresis characteristics, $m_1 = m_2 = m_3 = 5.0 \times 10^{-2} A \cdot m^2$, $m_4 = m_5 = 0$, the initial radial cooling gap x_0 takes different values, and the moving direction is the negative z-axis direction. The calculation results are shown in Figs. 4 and 5.

It can be seen from Fig. 4 that when the rotor moves in the negative direction of the z-axis, the permanent magnet receives a force in the positive direction of the z-axis. Within a certain range, the larger the axial displacement of the rotor, the greater the

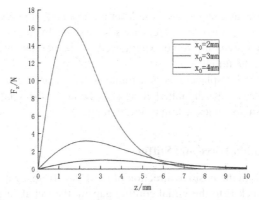

Fig. 4. Axial levitation forces at different initial radial clearances

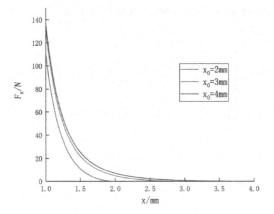

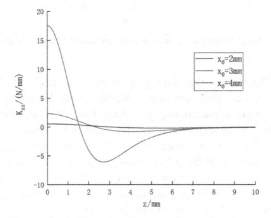

Fig. 5. Axial stiffness at different initial radial clearances

force that the permanent magnet receives. After a certain position, the axial force starts to decrease, but the direction of the force is always the same. In the case of different initial radial gaps, the axial levitation force experienced by the permanent magnet decreases with the increase of the initial gap.

The effect of axial displacement and radial initial clearance on axial stiffness is shown in Fig. 5. The smaller the initial clearance, the larger the axial stiffness value is, and they are all positive values within a certain range.

3.2   Radial Levitation Force and Stiffness

By moving the rotor in the positive direction of the x-axis from a gap of 1 mm from the superconducting block E to the initial cooling gap x_0, the radial suspension force and stiffness characteristics of the bearing were analyzed. In this process, x increases from 1 mm to x_0, and the force on the permanent magnet is always the positive direction of the x-axis. The calculation results are shown in Figs. 6 and 7.

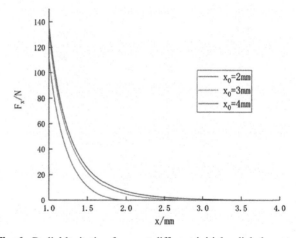

Fig. 6. Radial levitation forces at different initial radial clearances

It can be seen from Fig. 6 that during the process of increasing the gap between the permanent magnet and the superconducting block, the levitation force decreases rapidly, and the different initial cooling gaps have little effect on the magnitude and change characteristics of the levitation force at the gap of $1-2$ mm.

By moving the rotor in the positive direction of the x-axis from a gap of 1 mm from the superconducting block E to the initial cooling gap, the radial suspension force and stiffness characteristics of the bearing were analyzed. In this process, a increases from 1 mm to c, and the force on the permanent magnet is always the positive direction of the x-axis. The calculation results are shown in Figs. 6 and 7.

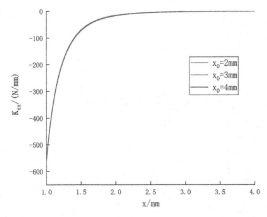

Fig. 7. Radial stiffness at different initial radial clearances

3.3 Analysis of Hysteresis Characteristics of HTSMB

In the case of considering hysteresis characteristics, $m_4 \neq 0$ and $m_5 \neq 0$. Taking superconducting block E as an example, when the permanent magnet moves from the initial cooling gap to the negative direction of the x-axis for the first time $m_4 = m_{41}$, the permanent magnet moves from the minimum gap back to the initial cooling gap $m_4 = m_{42}$; when the permanent magnet moves from the initial cooling gap to the positive direction of the x-axis $m_4 = m_{43}$, the permanent magnet returns from the maximum gap $m_4 = m_{44}$. If there is no axial movement, then $m_5 = 0$, when the permanent magnet first moves in the z-axis direction $m_5 = m_{51}$, and when it moves in the reverse direction $m_5 = m_{52}$.

The magnetic pole moments of magnetic dipoles M_4 and M_5 are

$$m_{41} = -a_1(x_0 - x) \tag{10}$$

$$m_{42} = -a_1(x_0 - x_{min}) + a_2(x - x_{min}) \tag{11}$$

$$m_{43} = b_1(x - x_{min}) \tag{12}$$

$$m_{44} = b_1(x_{max} - x_{min}) - b_2(x_{max} - x) \tag{13}$$

$$m_{51} = c_1 z \tag{14}$$

$$m_{52} = c_1 z_{max} - c_2(z_{max} - z) \tag{15}$$

In the formula, x_{max} and z_{max} are the radial maximum clearance and axial maximum displacement, a_1, a_2, b_1, b_2, c_1, c_2 are different proportional constants, $a_1 = 1.0A \cdot m$, $a_2 = 5.0A \cdot m$, $b_1 = 0.5A \cdot m$, $b_2 = 1.5A \cdot m$, $c_1 = 6.0A \cdot m$, $c_2 = 4.0A \cdot m$.

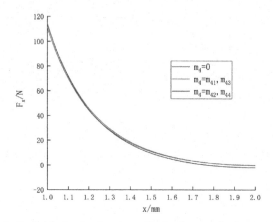

Fig. 8. Radial levitation force considering hysteresis characteristics

Assume that x_{min} is 1 mm, x_{max} is 3 mm, z_{max} is 4 mm, and x_0 is 2 mm. When the permanent magnets only move in the radial direction, the radial suspension force of the high-temperature superconducting magnetic levitation bearing is shown in Fig. 8.

It can be seen from the figure that the hysteresis characteristics of the HTSMB have a small effect on the radial levitation force, and the levitation force and levitation characteristics of the system have not changed significantly.

Suppose z_{max} is 4 mm, x_{min} is 1 mm, and $x = x_0 = 2$mm. The permanent magnets move only in the z-axis direction. The axial suspension force of the high-temperature superconducting magnetic levitation bearing is shown in Fig. 9.

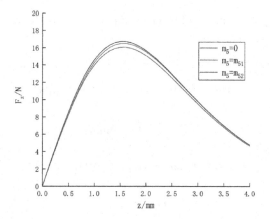

Fig. 9. Axial levitation force considering hysteresis characteristics

Similar to the radial suspension force, the effect of hysteresis on axial suspension force is not significant in HTSMB. Therefore, during the design and optimization of HTSMB, hysteresis is not a key factor affecting the bearing performance.

4   Conclusion

In this paper, the frozen-image method is used to calculate and analyze the suspension characteristics of a new HTSMB. Firstly, the HTSMB and its corresponding frozen-image model are briefly introduced, the suspension force and stiffness of the bearing are obtained according to the force formula between magnetic dipoles, and the calculated results are analyzed. The following conclusions can be drawn from the calculation results: the smaller the initial radial cooling clearance, the greater the axial suspension force. The radial suspension force and its stiffness within a certain radial clearance are not significantly affected by the initial cooling clearance. In the case of the same initial cooling clearance, the axial suspension force and its stiffness are both smaller than the radial suspension force and its stiffness, with a difference of about one order of magnitude. Under the displacement in a certain range, the bearing suspension is opposite to its motion direction, that is, the rotor can achieve self-stable suspension in both axial and radial directions. The hysteresis characteristics of the bearing will affect the axial and radial levitation forces, but its levitation characteristics have not changed significantly.

References

1. Tang, J., Zhang, Y., Fang, J.: Superconducting levitation styles for superconducting energy storage flywheel. In: International Conference on Mechatronics and Automation, Harbin, pp. 2889–2893 (2007)
2. Basaran, S., Sivrioglu, S.: Radial stiffness improvement of a flywheel system using multi–surface superconducting levitation. Supercond. Sci. Technol. **30**(3), 035008 (2017)
3. Pesch, A.H., Smirnov, A., Pyrhönen, O., Sawicki, J.T.: Magnetic bearing spindle tool tracking through μ -synthesis robust control. IEEE/ASME Trans. Mechatron. **20**(3), 1448–1457 (2015)
4. Hull, J.R.: Superconducting bearings. Supercond. Sci. Technol. **13**(2), 1 (2000)
5. Ma, K.B., Postrekhin, Y.V., Chu, W.K.: Superconductor and magnet levitation devices. Rev. Sci. Instrum. **74**(12), 4989–5017 (2003)
6. Nagaya, S., Komura, K., Kashima, N., Kawashima, H., Kakiuchi, Y., Minami, M.: Improvement and enlarging of the CFRP flywheel with superconducting magnetic bearings. Physica C **392**, 769–772 (2003)
7. Floegel-Delor, U., Rothfeld, R., Wippich, D., Goebel, B., Riedel, T., Werfel, F.N.: Fabrication of HTS bearings with ton load performance. IEEE Trans. Appl. Supercond. **17**(2), 2142–2145 (2007)
8. Wolsky, M.A.: An overview of flywheel energy systems with HTS bearings. Supercond. Sci. Technol. **15**(5), 836–837 (2002)
9. Ma, K.B., Postrekhin, Y.V., Chu, W.K.: Superconductor and magnet levitation devices. Rev. Sci. Instr. **74**(12), 4989–5017 (2003)
10. Werfel, F.N., Floegel-Delor, U., Rothfeld, R., et al.: Modelling and construction of a compact 500 kg HTS magnetic bearing. Supercond. Sci. Technol. **18**(2), S19–S23 (2005)
11. Hull, J.R.: Effect of permanent-magnet irregularities in levitation force measurements. Supercond. Sci. Technol. **13**(6), 854–856 (2000)

3D Curve Planning Algorithm of Aircraft Under Multiple Constraints

Yang Zhou[1], Wenju Zhou[1(✉)], Minrui Fei[1], and Sen Wang[2]

[1] Shanghai Key Laboratory of Power Station Automation Technology, School of Mechatronic Engineering and Automation, Shanghai University, Shanghai, China
zhouwenju@shu.edu.cn

[2] UK National Robotarium, School of Engineering and Physical Sciences, Heriot-Watt University, Edinburgh EH14 4AS, UK

Abstract. The trajectory planning of the aircraft is generally based on different mission requirements, under certain constraints to find an available optimal mission route. The traditional 3D trajectory planning algorithm is easy to fall into the local optimum. The search speed is full when the algorithm is searched. Some algorithms can only perform polyline search and fail to fully consider the physical reality of the aircraft. Aiming at the above problems, this paper proposes a path planning algorithm that combines A* algorithm and Dubins curve comprehensive optimization. The algorithm in this paper adopts heuristic search algorithm, performs two pruning by setting parameters, and through reasonable parameter settings, in a short time, the UAV's three-dimensional curve trajectory planning is quickly performed, which is greatly improved compared with other current algorithms.

Keywords: Heuristic search · 3D track planning · Curve planning

1 Introduction

Aircraft trajectory planning is generally based on different mission requirements, the aircraft's own physical limitations (maximum turning radius) [1] and other constraints to find an available optimal mission route. However, the general 3D trajectory planning area is wide and there are many alternative points. It is impossible to complete the exhaustive selection of the optimal trajectory for all paths in a short time in practical problems. The research of UAV trajectory planning algorithm is an important subject in this field. Due to the error of the aircraft's own positioning system, it is difficult to accurately locate in the flight route with complex environment and many influencing factors. Especially for the long-distance flight process, any error will be cumulatively amplified, eventually leading to the failure of the mission. Therefore, the timely correction of system errors and the accurate judgment of the navigation route to select the optimal route is a problem to be solved in the current field of aircraft.

In order to shorten the search time, a large number of different search algorithms have also been proposed. R. J. Szczerba proposed a method called sparse algorithm (SAS)

© Springer Nature Singapore Pte Ltd. 2020
M. Fei et al. (Eds.): LSMS 2020/ICSEE 2020 Workshops, CCIS 1303, pp. 236–249, 2020.
https://doi.org/10.1007/978-981-33-6378-6_18

[2]. The algorithm greatly reduces the search time by combining path optimization conditions, but this is a two-dimensional search algorithm that cannot be solved well when solving three-dimensional problems. Unresolved this problem Yan Jiangjiang proposed a 3D trajectory planning method based on feasible priority [3], which uses the feasible priority criterion to effectively cut out the search space so that 3D trajectory planning can be applied to real-time Track planning, but this method cannot be used for track planning. In order to solve the physical limitation of the maximum turning radius, the earliest Dubins discussed the Dubins curve in the case of the initial motion state and the end state through mathematical geometric methods. The shortest curve problem [4]. This also allows the straight line path planning before, to get a smooth flight path. However, the Dubins curve is not the shortest path under the condition that the initial motion state and termination are uncertain.

Some scholars have also proposed some intelligent search algorithms such as artificial potential field method [5], genetic algorithm [6], ant colony algorithm [7–9] and particle swarm algorithm [10], but these algorithms have long search time, can not meet the real-time requirements. In view of the above problems, this paper proposes a path planning algorithm that combines the algorithm and the Dubins curve comprehensive optimization for the UAV's path planning. The algorithm in this paper uses a heuristic search algorithm, and the search space is pruned twice by setting parameters. Through reasonable parameter settings, the UAV's three-dimensional curve trajectory planning can be quickly performed in a short time, compared with other current The algorithm has been greatly improved.

2 Model

In this question, the positioning error of the aircraft is divided into vertical error and horizontal error. For every 1 m of flight, the vertical error and horizontal error increase by unit each. Only when the vertical error and horizontal error are less than B, the aircraft. In order to overcome the accumulated error of the aircraft positioning system, it is necessary to correct the positioning error regularly and fixedly. In this problem scenario, hundreds of correction points with known positions are provided. These coordinate points have been determined before the flight and meet the constraints (see in the case described later), the error can be corrected by these correction points. During the flight process, how to select the correction point and then plan the global optimal flight route requires modeling analysis.

Aircraft trajectory planning needs to comprehensively consider factors such as horizontal correction, vertical correction, and minimum turning radius to plan the shortest path and minimum node path from the departure point to the target point for the aircraft.

If in the flight area, the set of correction nodes is S:

$$S = \{s_1, s_2, \ldots, s_m\} \tag{1}$$

The set of all possible tracks from the starting point A to the ending point B in the flight area is E:

$$E = \{e_1, e_2, \cdots, e_n\} \tag{2}$$

If s_i and $s_j(s_i, s_j \in S)$ are two adjacent correction points of $e_k(e_k \in E)$ on one of the paths, the edge connecting the two correction points is denoted by $V(s_i, s_j)$ and d_{ij} is the cost of the edge:

$$\begin{cases} \min f(e_k) = \sum_{(s_i,s_j) \in e_k} d_{ij} \\ s.t. e_k \in E, s_i \in S, s_j \in S \end{cases} \tag{3}$$

The constraints for aircraft path planning are as follows:

(1) The vertical error and horizontal error will increase by δ unit each time the aircraft flies 1 m. Only when both the vertical error and the horizontal error are less than θ, the aircraft can fly normally according to the planned route, and record the current vertical error as σ_1 and the horizontal error as σ_2.

$$\sigma_1 < \theta \tag{4}$$

$$\sigma_2 < \theta \tag{5}$$

(2) The aircraft needs to correct the positioning error during the flight. There are some safe positions (correction points) in the flight area for error correction. If the vertical and horizontal errors can be corrected in time, the aircraft can fly according to the predetermined route, correct the errors through several correction points and finally reach the destination.

(3) At point A of departure. The vertical and horizontal errors of the aircraft are both 0:

$$\sigma_1 = 0 \tag{6}$$

$$\sigma_2 = 0 \tag{7}$$

(4) After the aircraft performs vertical error correction at the vertical error correction point, its vertical error will become 0, and its horizontal error will remain unchanged:

$$\sigma_1 = 0 \tag{8}$$

(5) After the aircraft performs horizontal error correction at the horizontal error correction point, its horizontal error will become 0, and the vertical error will remain unchanged.

$$\sigma_2 = 0 \tag{9}$$

(6) The vertical error correction can only be performed when the vertical error of the aircraft is not greater than α_1 units and the horizontal error is not greater than α_2 units:

$$\sigma_1 < \alpha_1 \tag{10}$$

$$\sigma_2 < \alpha_2 \tag{11}$$

(7) The horizontal error correction can only be performed when the vertical error of the aircraft is not greater than β_1 units and the horizontal error is not greater than β_2 units.

$$\sigma_1 < \beta_1 \tag{12}$$

$$\sigma_2 < \beta_2 \tag{13}$$

(8) The aircraft is restricted by the structure and control system when turning, and it cannot complete the instant turn (the aircraft's direction of advance cannot be changed suddenly), assuming that the minimum turning radius of the aircraft is r [11]:

$$r_i > r \tag{14}$$

3 Multi-constraint 3D Trajectory Planning Algorithm

3.1 Heuristic 3D Track Search Algorithm

The heuristic search algorithm [12] is to evaluate all search positions for each search position, and when the current best position is obtained, then search from this position until the target. According to the above, under the constraints of the minimum turning radius, horizontal and vertical error correction, etc., in order to quickly and accurately search for the shortest path, first of all artificially specify the search direction always towards the end point, as shown in Fig. 1, where point A It is the current correction point, point B is the end point, l is the plane passing point A with normal vector AB, point C is the next correction point, and θ is the angle between vector AC and AB π > θ > 0.When performing heuristic search, it is stipulated that θ is greater than a certain initial threshold, the range θ is π > θ > 0,The larger θ is, the more likely it is to find the global optimal solution, but the lower the search efficiency, the smaller the search The higher the efficiency, the greater the possibility of finding no feasible solution. In this paper,θ uses π/2. When θ equals π/2, the search point is always in front of the l-plane.

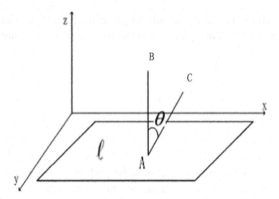

Fig. 1. The relationship between the target search node and the current node

Each flight is an "effective" flight and can be closer to the end. In this way, not only can the shortest path be obtained, but also the search time can be greatly reduced.

In this paper, a tree structure is used, and the correction point that satisfies the constraint conditions is traversed from the root node as the next slave node of the tree to form a new tree, and then the slave nodes of the new tree are traversed to add the next node that meets the requirements to the next layer of slave nodes. In order to prevent the constraint conditions from being broad, the searched tree is too large and the time complexity is too high. In this paper, two hyperparameters a and b are set. When the total number of branches of the tree reaches the threshold a, the tree is pruned at this time There are two ways of pruning. The first one only prunes the last layer of the tree. When a slave node branch tree in the last layer of the tree exceeds b, the current b branches are sorted using a path sorting algorithm (in the The algorithm will be described in detail later), the first b optimal path branches of the node are retained, and the other branches of the node are all pruned. The second method is to prun all the total branches. When the first pruning operation cannot satisfy the total score is less than or equal to a, then continue to use the second pruning operation, otherwise do not perform the second pruning The branch operation uses a path sorting algorithm to sort all the branches, retaining the first a optimal path branch. This loop repeats iteratively until all trees search to the end or reach our maximum number of iterations.

The path sorting algorithm in this paper iterates based on the cost of the existing path and the cost estimate required to extend the path all the way to the target.

$$f(n) = g(n) + h(n) \tag{15}$$

Where $g(n)$ is the actual distance traveled, $h(n)$ is the estimated distance to be traveled in the future, here c is multiplied by the straight line distance from the current node to the end point, c is determined according to the different complexity of the problem and related experience of different problems, generally taken as The number between $-$1.5, 1.05 used in this article, $f(n)$ is the estimated total flight distance. In this paper, the candidate paths are ranked in order with $f(n)$ as the criterion.

The algorithm can be described as:

Generate the starting node node;
Set the maximum number of iterations n, the number of branches from the node b, the total number of branches a, the current number of iterations i = 0, the path length list l;
repeat:
Search for all correction points that can be reached under the constraints of the current node as the next-level slave node as a new tree;
if current total number of branches> a then:
Use the first pruning method to prun the current tree;
if current total number of branches> a then:
Use the second pruning method to continue pruning the current tree;
    end if
   end if
  Add 1 to the current iteration number i;
   Until the current iteration i is greater than or equal to the maximum number of iterations n or the last node of all branches is the end point.
for all branches of the tree species c
  Calculate the path length of each branch c and add it to the list l
end for
Calculate the path length of each branch c and add it to the list l

According to the above search algorithm, we can get a tree network graph, as shown in Fig. 2 below, which can be expressed as a graph in graph theory, where each branch is an alternative path.

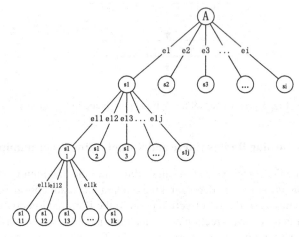

Fig. 2. Schematic diagram of node search

Based on the above algorithm description, the flow chart of this question's trajectory planning algorithm is shown in Fig. 3 below.

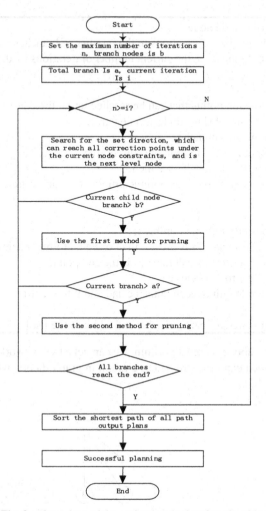

Fig. 3. Flow chart of the optimal path planning algorithm

3.2   Minimum Turning Radius Curve and Straight Line Programming Algorithm

In order to reduce the complexity of the algorithm, only the maximum turning angle in the horizontal direction is considered, and the pitch angle is not considered. Assuming that the aircraft flies from the A correction point to the B correction point and changes the flight direction to the D correction point, its flight model can be described as shown in Fig. 4 below, and the flight trajectory in Fig. 5 is the enlarged steering in the actual simulation model Figure.

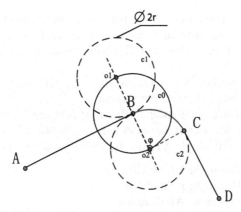

Fig. 4. Schematic diagram of aircraft steering model analysis

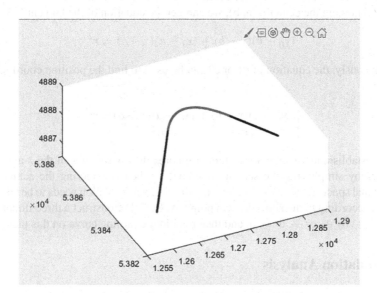

Fig. 5. Partial enlarged view of steering track

Suppose the coordinates of point A are $(x1, y1)$ and the coordinates of point B are $(x2, y2)$. Using point B as the center of the circle to make a courtyard $c0$ with r radius, the equation of $c0$ is:

$$(x - x2)^2 + (y - y2)^2 = r^2 \tag{16}$$

Make a vertical line of AB through point B:

$$x - x2 = -\frac{x2 - x1}{y2 - y1}(y - y2) \tag{17}$$

Simultaneous Eqs. (1) and (2) can get the intersection points o1 and o2, and use o1 and o2 as the center to make circles c1 and c2 with radius r Obviously, the next correction node can not be in the range of c1, c2, otherwise it can not meet the requirement of turning radius r in the case of flying around not far away. Then any node D outside or on the circle must be tangent CD to c1 or c2. Obviously, the arc BC meets the minimum turning radius.

Given the coordinate data of the three points A, B, and D, the positional relationship between point D and AB can be determined according to the positive and negative vector cross product and the sine relationship of the angle between the vectors, and then the flight steering can be determined. Given A $(x1, y1)$, B $(x2, y2)$, D $(x0, y0)$, let a $= (x2-x1, y2-y1)$, b $= (x0-x1, y0-y1)$, $a \times b = |a||b| \sin \phi$ lallbl, then:

$a \times b > 0$, point D is to the left of AB;
$a \times b > 0$, point D is to the right of AB;
$a \times b = 0$, point D is in the AB direction.

After determining the steering direction, that is, after determining the steering circle c1 or c2, knowing the center o (a, b), we can get the equation of the tangent CD:

$$(y_0 - b)(y - b) + (x_0 - a)(x - a) = r^2 \tag{18}$$

Simultaneously, the equation of c1 or c2 can be used to find the position coordinates of point C:

$$\begin{cases} (y_0 - b)(y - b) + (x_0 - a)(x - a) = r^2 \\ (x - a)^2 + (y - b)^2 = r^2 \end{cases} \tag{19}$$

The establishment of the above model ignores the coordinates in the z-axis direction, thereby simplifying the search complexity. When considering the actual three-dimensional space trajectory planning, only the z-axis information needs to be taken into account, according to the above known points A, B, C, D Construct a three-dimensional plane $ax + by + cz + d = 0$, and then rebuild the turning curve on this plane.

4 Simulation Analysis

Data simulation adopts the online public data including data 1 and data 2 to carry out the simulation test. The data dot plots are shown in Figs. 6 and 7, where point A is the start point, point B is the end point, and blue and yellow hollow points are different correction points.

Figure 8 is the optimal path effect of the algorithm of this paper under the data one data set, and Fig. 9 is the optimal path effect of the algorithm of this paper under the data two data set. Table 1 is the relevant parameters of the optimal path under the data-data set of the 0–1 integer mixed two-objective function and the multi-constraint programming algorithm, including the circle center coordinates, tangent point coordinates, arc and direct track distance, etc., The 0–1 integer mixed bi-objective function and multi-constrained programming algorithm use selective exhaustive method to find the optimal polyline path. Aiming at the requirement of turning radius, it is proved that the path connecting the tangent points of the two turning circles cannot be a straight line, so the method of connecting by quadratic curve is adopted. The optimal path can

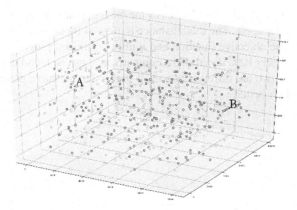

Fig. 6. Data 1 data dot plot (Color figure online)

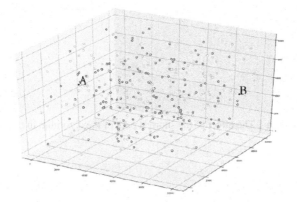

Fig. 7. Data 2 data dot plot (Color figure online)

be selected better and faster. The total planned path length is 105471.97 m. Table 2 is the correlation between the optimal path under data 2 of the 0–1 integer mixed two objective function and the multi-constrained programming algorithm. Parameters, including the coordinates of the center of the circle, the coordinates of the tangent point, the distance between the arc and the direct track, etc. The optimal path obtained by this algorithm is 110619.49 m. Table 3 is the relevant parameters of the optimal path of the algorithm under the data set of data 1 in this paper, including the coordinates of the center of the circle, the coordinates of the tangent point, the distance between the arc and the direct track, etc. The total path is 103567.76 m compared to 0–1 Integer mixed bi-objective function and multi-constrained programming algorithm, the algorithm of this paper is better than it, the two algorithms result in different paths when searching, and the algorithm of this paper shortens the navigation distance of nearly 2000 m to obtain a better solution, and the algorithm speed of this algorithm is only The search can be completed within 2 s to meet real-time planning requirements. Table 4 is the relevant parameters of the optimal path under the data set of Data 2 in this paper, including the coordinates of the center of the circle, the coordinates of the tangent point, the distance between the arc and the direct track, etc. The total distance is 109464.04 m compared to 0–1 integer mixed type two The objective function and multi-constrained programming algorithm are shortened by nearly 1000 m. From the comparison, it can be found that the path is the same node, but the two use different curves, it shows that the curve planning algorithm used in this article is better.

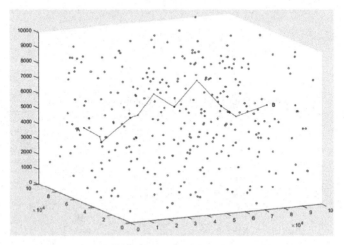

Fig. 8. Data 1 The effect of the shortest path of the algorithm in this paper

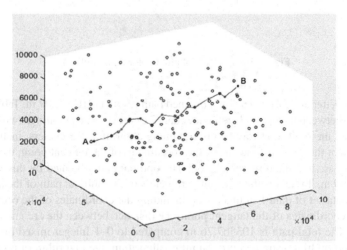

Fig. 9. Data 2 The effect of the shortest path of the algorithm in this paper

Table 1. Data 1 0–1 integer mixed two objective function and optimal path of multi-constrained programming algorithm

Path	Center coordinates	Tangent coordinate 1	Arc length	Distance
0–503	(0, 50000)	(0, 50000)	0	13387.92
503–69	(11461, 56814)	(11364, 56958)	63.58	8741.75
69–237	(19859, 59152)	(19802, 59340)	67.80	12447.1
237–155	(32805, 58421)	(32291,58620)	37.3	11142.3
155–540	(43139 55930)	(43193, 55919)	80.238	6630.63
540–250	(49229,54679)	(49141,54565)	92.241	11363.3
250–340	(60537, 54861)	(60526,55059)	50.765	12696.9
340–277	(73023,52606)	(72982, 52435)	93.209	11945.8
277–612	(84569,55035)	(84613,54840)	19.388	17116.2

Table 2. Data 2 0–1 integer mixed two objective function and optimal path of multi-constrained programming algorithm

Path	Center coordinates	Tangent coordinate 1	Arc length	Distance
0–163	(0, 50000)	(0, 50000)	0	13287.90
163–114	(12710,53668)	(12653,53856)	121.66	5168.67
114–8	(17738,52482)	(17677,52292)	208	13696.21
8–309	(28203,61363)	(28070, 61512)	215.44	5404.4
309–305	(33353,59679)	(33359, 59715)	23.95	5873.92
305–123	(38926.,57859)	(38854 57677)	163.53	9110.59
123–45	(47262, 61502)	(47226 61599)	20.87	9365.98
45-160	(55896,64974)	(55817, 65128)	118.6	8604.23
160–92	(64395, 64320)	(64367, 64125)	107.68	5697.07
92–93	(69759, 66259)	(69592, 66424)	48.18	9436.55
93–61	(78945, 68436)	(78992, 68242)	48.9	9788.05
61–292	(87839, 72524)	(87913, 72363)	43.4	6442.64
292–326	(93259, 76000)	(93141, 76154)	170.46	8743.28

Table 3. Data 1 The shortest curve path planning of this algorithm

Path	Center coordinates	Tangent coordinate 1	Arc length	Distance
0–503	(0, 50000)	(0, 50000)	0	13387.92
503–200	(11497, 56802)	(11577, 56985)	191.79	907.69
200–80	(12201, 56931)	(12211, 56731)	70.60	15751.62
80–237	(27799, 57743)	(27846, 57549)	36.79	4627.95
237–26	(32356, 58424)	(32374, 58623)	64.80	17735.36
26–375	(49762, 57263)	(49795, 57065)	51.51	4616.99
375–448	(54261, 58029)	(54322, 57838)	28.45	18587.05
448–485	(71539, 63133)	(71529, 63333)	51.66	5736.72
485–612	(76774, 63401)	(76807, 63598)	44.16	23571.29

Table 4. Data 2 The optimal curve path planning of this algorithm

Path	Center coordinates	Tangent coordinate 1	Arc length	Distance
0–163	(0, 50000)	(0, 50000)	0	13287.90
163–114	(12771,17839)	(12831 17973)	120.71	5341.81
114-8	(17839,52478)	(17973, 52330)	207.52	13956.93
8–309	(28298,61410)	(28369, 61597)	217.89	5563.89
309–305	(33304,59522)	(33373, 59710)	2.24	5968.73
305–123	(39003.,57852)	(39090, 57672)	160.42	9221.09
123–45	(47322, 61423)	(47309, 61622)	75.92	10008.10
45–160	(57188,62450)	(57238, 62256)	37.95	7485.49
160–92	(64369, 64319)	(64450, 64136)	31.59	5776.34
92–93	(69784, 66249)	(69746, 66446)	43.87	9485.10
93–61	(78967, 68445)	(79051, 68263)	48.12	9834.67
61–292	(87849, 72555)	(87966, 72393)	38.58	6554.35
292–326	(93335, 76026)	(93375, 76222)	165.59	6979.57

5   Conclusion

Aiming at the three-dimensional path planning method of the aircraft, this paper proposes A* combined path optimization algorithm with Dubins curve. This algorithm uses a heuristic search algorithm, and the search space is pruned twice by setting parameters. Parameter setting, which can quickly carry out the three-dimensional curve trajectory planning of the UAV in a short time. Compared with other algorithms at present, the search speed and search optimal solution are greatly improved, and can be used in real-time planning systems. The planned path is very close to the optimal path.

Acknowledgements. This research is financially supported by Key Project of Science and Technology Commission of Shanghai Municipality under Grant (16010500300) and Natural Science Foundation of China (61877065).

References

1. Zhou, L., Qian, W., Cao, G., Zhang, M.: Research on path planning algorithm based on 3D terrain. J Comput. Appl. Softw. **35**, 8 (2018)
2. Szczerba, R.J., Galkowski, P., Glicktein, I.S., Ternullo, N.: Robust algorithm for real-time route planning. IEEE Trans. Aerosp. Electron. Syst. **36**, 869–878 (2000)
3. Yan, J., Ding, M., Zhou, C., et al.: 3D route planning based on feasible first search. J. Astronaut. **30**, 139–144 (2009)
4. Dubins, L.E.: On curves of minimal length with a constraint on average curvature, and with prescribed initial and terminal positions and tangents. J. Am. J. Math. **79**, 497–516 (1957)
5. Gao, S., Xu, F., Guo, H.: Research on mobile robots' path planning based on a spring model. J. Chin. J. Sci. Instrum. **37**(4), 796–803 (2016)
6. Li, Y., Yan, J., Song, Z., et al.: Research on algorithm of UAV monitoring coverage path planning based on genetic algorithm. J. Comput. Sci. Appl. **9**(6), 1208–1215 (2019)
7. Gao, Z., Li, J., Yue, C.: Three-dimensional path planning based on improved ant colony algorithm. J. Inf. Eng. Univ. **15**, 599–602 (2014)
8. Nan, J., Zang, J., Gao, M.: Reverse modeling method for BRBP neural network power amplifier bsed on improved ant colony algorithm. J. Laser Optoelectron. Prog. **57**, 012001 (2020)
9. Xianlun, T., Guangdan, C., Zhang, P., et al.: Ant colony optimization based on maximum selection probability for path planning in unknown environment. J. Comput. Inf. Syst. **8**, 10325–10332 (2012)
10. Chen, Q., Lu, Y., Jia, G., et al.: Path planning for UAVs formation reconfiguration based on Dubins trajectory. J. Central South Univ. **25**, 2664–2676 (2018)
11. Zhou, H., Xiong, H.L., Liu, Y., et al.: Trajectory planning algorithm of uav based on system positioning accuracy constraints. Electron. **9**, 250 (2020)
12. Lv, Z., Yang, L., He, Y., et al.: 3D environment modeling with height dimension reduction and path planning for UAV. In: 2017 9th International Conference on Modelling, Identification and Control (ICMIC), pp. 734–739 (2017)

A Real-Time Dynamic Watermarking Detection Method of Networked Inverted Pendulum Servo Systems

Jingfan Zhang, Dajun Du$^{(\boxtimes)}$, Changda Zhang, and Lisi Yang

School of Mechanical Engineering and Automation,
Shanghai University, Shanghai 200444, China
smxzjf7@163.com, ddj@i.shu.edu.cn

Abstract. The paper proposes a real-time dynamic watermarking (DWM) detection method for networked inverted pendulum servo systems (NIPSSs). Firstly, considering network-induced delay, the discrete model of NIPSSs is established, which may be changed as cyber attacks occur. Secondly, according to DWM technology, an active detection mechanism is constructed to detect cyber attacks that distort the measurements from the sensor fed to the NIPSSs. Thirdly, to conform security and real-time requirements of NIPSSs, a real-time window-based DWM detection algorithm is designed to detect cyber attacks, where two indicators are defined to judge whether the measurement has been distorted. The proposed detection algorithm can detect cyber attacks in time and is easily implementable in the NIPSSs since there is no hardware update requirements. Finally, the effectiveness and feasibility of the DWM detection algorithm are confirmed by experimental analysis of the NIPSSs.

Keywords: Networked inverted pendulum servo systems · Cyber attacks · Dynamic watermarking · Real-time detection

1 Introduction

Due to the nonlinear, fast moving and instable characteristics [1], the inverted pendulum system is usually treated as an idea experimental platform to validate the theory and algorithm. These validated theory and algorithm are further employed to a lot of practical control fields such as unmanned aerial vehicle control, missile interception control and so on [2]. The network was integrated into the inverted pendulum system, leading to networked inverted pendulum servo systems (NIPSSs), which were employed to validate theory and method of networked control systems (NCSs) [3,4]. However, due to the interaction of open networks, the NIPSSs suffer inevitably from cyber attacks. For example, an unmanned aerial vehicle was attacked so that it arrived at a wrong location [5]. A offshore drilling platform suffered from a series of malware attacks so that

© Springer Nature Singapore Pte Ltd. 2020
M. Fei et al. (Eds.): LSMS 2020/ICSEE 2020 Workshops, CCIS 1303, pp. 250–263, 2020.
https://doi.org/10.1007/978-981-33-6378-6_19

it was shut down for 19 days in South America. Therefore, it is very important topic to investigate the attack detection techniques on the NIPSSs.

The detection techniques for cyber attacks can be roughly divided into passive detection and active detection. Passive detection aims at the detection of the attack based on some statistical hypothesis tests. For example, the most powerful and commonly used method is classical chi-squared test [6,7], which can detect the attack signal with Gaussian distribution. However, the current research has two defects: 1) The statistical hypothesis tests is based on a certain known of the attack signal such as the attack method of attack signal, statistical properties of attack signal and so on. However, some malicious cyber attacks may not be followed the known attack templates, and passive detection can no longer applies to unknown types of attacks. Consequently, the malicious cyber attacks can bypass the passive detection [8,9]. 2) The information of the system model might be exposed to the attackers [10]. For example, the statistical model of system can be learned based on the leaked data, and the malicious attacker can design a stealthy attack easily to bypass the passive detection. Therefore, these lead to the new research on the active detection.

Active detection requires not only the known system information, but also adds additional excitation signals to system. The dynamic watermarking (DWM) [11,12] technique is an active detection technology, which actively superimposes a watermark signal upon the control signals. Consequently, the watermark signals feed the measurements so that cyber attacks can be detected. Meanwhile, due to confidentiality of the watermark signals, DWM technique increases the cost for the attackers to obtain information, thus which makes the malicious attackers construct attack more difficult.

According to the above analysis, by using DWM technique to address the security and real-time requirements of the NIPSSs. The main contributions are: 1) According to the discrete model of NIPSSs and DWM technique, an active detection mechanism is constructed to detect cyber attacks that distorts the measurements from the sensor fed to the NIPSSs. 2) A real-time window-based DWM detection algorithm is designed, which conforms to the security and real-time requirements of the NIPSSs and ensures that any distortion of the measurements can be detected. 3) No hardware update is needed for the detection algorithm, which reduces the upgrade costs of NIPSSs.

The rest of this paper is organized as follows. In Sect. 2, we describe the discrete model of NIPSSs and formulate the problem of detection. In Sect. 3, the DWM technique is analyzed for establishing an active detection mechanism and a real-time window-based DWM detection algorithm is presented. Afterward, Sect. 4 confirms the effectiveness, feasibility and real-time performance of the proposed detection algorithm via experimental analysis. The conclusion and future work are given in Sect. 5.

2 Problem Formulation

2.1 The Model of NIPSSs

The framework of NIPSSs is shown in Fig. 1, where the inverted pendulum, sensor, actuator, and controller can exchange information via network. Firstly, after obtaining the state information, the sensor transmits real-time motion data to the controller via network. Secondly, the controller calculates control signal according to the algorithm designed for stability control of the NIPSSs, which is then transmitted to the motion control card via network. Furthermore, the motion control card generates control pulse to electric cabinet to drive servo motor and the cart of inverted pendulum [13]. Finally, the actuator achieves the closed-loop control for the inverted pendulum.

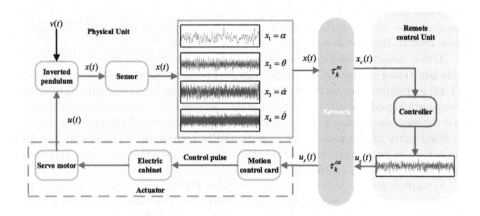

Fig. 1. Framework of the NIPSSs.

According to the Newton's kinematic theorem and linearization within the range ($|\theta| \leq 0.2 rad$), the nominal model of NIPSSs is modeled as follows:

$$\dot{x}(t) = Ax(t) + Bu(t) + v(t) \tag{1}$$

where $x(t) \in \mathbb{R}^4$ is the state vector, $x_1 = \alpha$, $x_2 = \theta$, $x_3 = \dot{\alpha}$ and $x_4 = \dot{\theta}$ represent cart position, pendulum angle, cart velocity, and angular velocity, respectively; $u(t) \in \mathbb{R}$ represents the actuator output signal and the process noise $v(t) \in \mathbb{R}^4$ is being zero-mean white Gaussian noise with Σ_v; A and B are as follow

$$A = \begin{bmatrix} 0 & 0 & 1 & 0 \\ 0 & 0 & 0 & 1 \\ 0 & 0 & 0 & 0 \\ 0 & \frac{lmg}{J} & 0 & 0 \end{bmatrix}, B = \begin{bmatrix} 0 \\ 0 \\ 1 \\ \frac{ml}{J} \end{bmatrix}$$

where l is the length of the pendulum bar from the rotation point to the core; m is the mass of the pendulum bar; J is the moment of inertia of the pendulum bar based on the point of rotation; g is the acceleration of gravity.

The NIPSSs adopts time trigger mechanism, i.e., the sensor, controller and actuator are set as the same sampling period h. Furthermore, the signal's timing diagram of NIPSSs is shown in Fig. 2

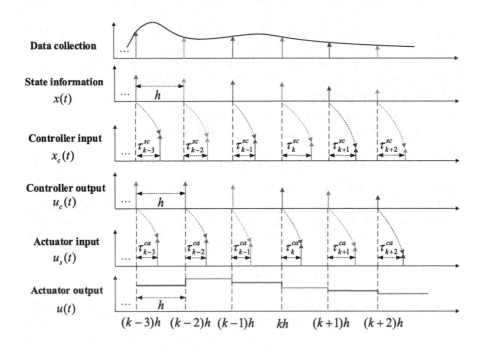

Fig. 2. Signal's timing diagram of the NIPSSs.

In Fig. 1, $x(t)$ is acquired by the sensor at the kth sampling instant, i.e.,

$$x(t) = x(kh), t \in \{kh, k = 0, 1, 2, ...\} \tag{2}$$

The sensor transmits $x(t)$ to the remote control unit via network, and networked-induced delay τ_k^{sc} between the sensor and the controller needs to be considered. Consequently, $x_c(t)$ is expressed by

$$x_c(t) = x(kh), t \in \{kh + \tau_k^{sc}, k = 0, 1, 2, ...\} \tag{3}$$

Next, the controller is designed as

$$u_c(t) = Kx(kh), t \in \{kh + \tau_k^{sc}, k = 0, 1, 2, ...\} \tag{4}$$

where K is the feedback gain.

The controller can not complete the control within a sampling period. Therefore, $x[k]$ has not been transferred to the controller at the kth sampling instant, so the controller is set to calculate the control signal according to the state variables obtained at the previous sampling instant. Then, the output of the controller is $u_c[k] = Kx[k-1]$.

Similarly, due to network-induced delay between the controller and the actuator, $u_c[k+1]$ has not been transferred to the actuator at the $(k+1)^{th}$ sampling instant. Therefore, actuator input $u_s(t)$ is received at $(k+1)h + \tau^{ca}_{k+1}$. So the actuator is set to execute the control signal calculated at the previous sampling period. As a result, at the kth sampling instant, the actuator executes the control signal $u_c[k-1]$. Hence, $u[k] = Kx[k-2]$ is the output of the actuator, which is operated. It is worth noting that the sampling period must greater than networked-induced delay $\tau_{\max} = \max_{k \in N}(\tau^{sc}_k, \tau^{ca}_k)$. So h is selected after $\tau_{\max}$ is statistically estimated by using experimental data.

Finally, the discrete close-loop system model of NIPSSs is

$$x[k+1] = A_d x[k] + B_d Kx[k-2] + v[k+1] \tag{5}$$

where $A_d = e^{AH}$, $B_d = \int_0^H e^{AH} dt B$.

Moreover, the noise was produced from system measurement, etc., and the experimental statistics show that $v(t) \in \mathbb{R}^4$ conforms to $\Sigma_v = diag(2.7 \times 10^{-7}, 5.5 \times 10^{-6}, 6 \times 10^{-4}, 1.33 \times 10^{-2})$.

2.2   Cyber Attack to the NIPSSs

For the NIPSSs control process, $x[k]$ is transmitted to remote control unit via the network. In general, $z[k]$ are supposed as the actual value measured, $z[k] = x[k]$. However, cyber attack may distort $z[k]$, leading to $z[k] \neq x[k]$. Cyber attack can tamper with the transmitted state information including:

1) Replay Attack: The attacker records a period of normal data, which is further transferred to the controller by replacing the actual data. Then, the controller will calculate the corresponding control command based on these malicious data, leading to inverted pendulum to unstable. Because these malicious data are a period of past data in the relatively smooth operation of the system, which may not be identified as the abnormal data by the traditional detection methods [14]. If the operator can only monitor the inverted pendulum state measurements $z[k]$, the actual state of the inverted pendulum can not be monitored.

2) Noise-Injection Attack: The attacker adds a bounded random value signal to the $z[k]$. As a result, data will have some extra noise such as impulse noise, Gaussian noise, background noise and so on. However, the system itself is subject to noise interference. Therefore, the traditional detection methods can not identify [15] noise and noise-injection attack signal. If the controller calculates the corresponding control command based on these data, the inverted pendulum control performance will degrade.

The above introduced attack methods may evade the existing attack detection and change the state variable information to damage the NIPSSs regardless of the attack means and form. Therefore, the objectiveness of this paper is to design a new detection method to detect cyber attacks in the NIPSSs.

3 Dynamic Watermarking Detection of the NIPSSs

The whole detection framework of NIPSSs is shown in Fig. 3. The measurement $z[k]$ is transmitted to remote control unit and detect unit. At the same time, the buffer transmits the measurement $z[k]$, $z[k-1]$, the watermark signal $w[k-2]$ and controller output signal $u_c[k-1]$ to the detector to check whether the measurement $z[k]$ is correct. The detector can detect any distortion of the measurements from the sensor and the alarm will be triggered.

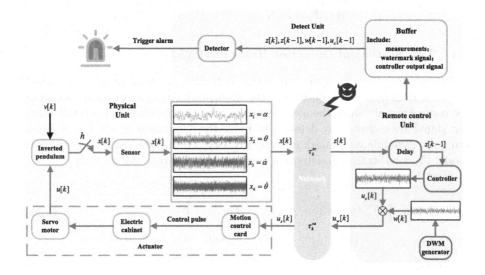

Fig. 3. Detection framework of the NIPSSs.

The DWM is an active defense strategy to solve the problem of cyber attack detection. It will be introduced in the NIPSSs to enhance the security of the system. The DWM is the superposition of a random signal on the controller output, the watermark signal can be used as the authentication signal. In this paper, the specific excitation value of the authentication signal is only known by the system operator while the watermark signal can not be separated by the malicious attacker. Therefore, the difficulty of obtaining system information is enhanced, as a result, the malicious attacker can not fabricate signals optionally. Therefore, the detection mechanism is designed to judge whether the output signal is tampered based on the statistical characteristics of the watermark.

For the NIPSSs, the addition of watermark signal and the use of attack detection can be completed directly without changing the hardware of the whole system, which will facilitate the security upgrade of the system. As an authentication signal, $w[k]$ is being zero-mean Gaussian signal with variance σ_w^2. Meanwhile, $w[k]$ and process noise signal $v[k+1]$ are independent. Controller output signal is $u_c[k]$ and the actual remote control unit output signal is

$$u_w[k] = u_c[k] + w[k] = Kz[k-1] + w[k]$$

and the actual actuator output signal is

$$u[k] = u_w[k-1] = Kz[k-2] + w[k-1]$$

Simultaneously, the discrete state-space model (5) becomes

$$\begin{aligned} x[k+1] &= A_d x[k] + B_d u_w[k-1] + B_d w[k-1] + v[k+1] \\ &= A_d x[k] + B_d K z[k-2] + B_d w[k-1] + v[k+1] \end{aligned} \tag{6}$$

3.1 Two Statistical Detection Tests of DWM

According to above analysis, an active detection mechanism is constructed. Under normal operating conditions, the watermark signal is superimposed upon the control signals, while the measurements, watermark signal and controller output signal are collected by detector to judge the correctness of $z[k+1]$. Considering the dynamic watermark detection theory [11] and the discrete state-space model (5), two statistical detection tests are calculated in the detector:

Test 1:

$$\begin{aligned} \lim_{T\to\infty} \frac{1}{T} \sum_{k=1}^{T} & (z[k+1] - A_d z[k] - B_d u_c[k-1]) \\ & (z[k+1] - A_d z[k] - B_d u_c[k-1])^T = \sigma_w^2 B_d B_d^T + \Sigma_v^2 \end{aligned} \tag{7}$$

Test 2:

$$\begin{aligned} \lim_{T\to\infty} \frac{1}{T} \sum_{k=1}^{T} & (z[k+1] - A_d z[k] - B_d u_c[k-1] - B_d w[k-1]) \\ & (z[k+1] - A_d z[k] - B_d u_c[k-1] - B_d w[k-1])^T = \Sigma_v^2 \end{aligned} \tag{8}$$

If the cyber attack occurs and the measurement $z[k+1]$ is distorted, the above two detection tests can not be satisfied simultaneously, unless the adding attack is a zero-power signal [11].

3.2 Real-Time Detection Algorithm

The tests (7) and (8) are built on statistical tests conducted over an infinite period of time, requiring large amounts of data. In practice, the detector is required to detect cyber attacks in real-time. It is necessary to use window detection which converts the tests to statistical tests. Consequently, detection procedure can be performed in the finite time.

To detect whether the system is attacked at the kth sampling instant in real-time, the buffer records a sequence of measurements $\{z\}$, watermark signals $\{w\}$ and controller output signals $\{u_c\}$ over a period of time. In each time window W, the detection unit processes a block of $\{z\}$, $\{w\}$ and $\{u_c\}$. At the kth sampling instant, there will be $j(j \leq k - W)$ blocks of above sequences.

The jth block of above sequences are

$$z^j = \{z[j], z[j+1], \ldots \ldots z[W+j]\}$$
$$w^j = \{w[j-2], w[j-1], \ldots \ldots w[W+j-2]\}$$
$$u_c{}^j = \{u_c[j-2], u_c[j-1], \ldots \ldots u_c[W+j-2]\}$$

At the kth sampling instant, according to the content in the jth block to check whether the measurement $z[k]$ passes detections. For convenience, define $\eta_k = z[k] - A_d z[k-1] - B_d u_c[k-1]$, and two indicator matrices R^j and L^j. Then, the detection tests (7) and (8) become

$$R^j := \frac{1}{W} \sum_{s=k-W+1}^{k} \eta_s \eta_s^T \tag{9}$$

$$L^j := \frac{1}{W} \sum_{s=k-W+1}^{k} (\eta_s - B_d w[s-2])^T (\eta_s - B_d w[s-2]) \tag{10}$$

as the expression of window detection.

In terms of real-time application, we define two indicators (α, β) on behalf of detection result. After each block is obtained, α and β can be calculated in real-time within the jth time window to check the correctness of $z[k]$, respectively, i.e.,

$$\alpha^j = tr\left(R^j\right), \beta^j = tr\left(L^j\right) \tag{11}$$

Define scalars γ_α and γ_β as threshold of α^j and β^j. If cyber attack is launched in the jth time window, $\alpha^j > \gamma_\alpha$ or $\beta^j > \gamma_\beta$ will occur. The choice of threshold is critical, if the threshold is too large, it will cause false negatives, and if the threshold is too small, it will cause false positives. The thresholds γ_α and γ_β can be obtained under normal operating condition by (11).

Finally, according to the above analysis, we can form a real-time window-based DWM detection algorithm. Algorithm 1 illustrates how the detector to detect cyber attacks in each block.

Algorithm 1. Real-time window-based DWM detection algorithm

1: $j \leftarrow 1$
2: **while** $k = 1, 2, \dots$ **do**
3:    **if** $k \geq W$ **then**
4:       obtain $z^j, w^j, u_c{}^j$
5:       compute α^j, β^j
6:       **if** $\alpha^j > \gamma_\alpha$ or $\beta^j > \gamma_\beta$ **then**
7:          trigger alarm
8:          break
9:       **else**
10:          $k \rightarrow k + 1$
11:          $j \rightarrow j + 1$
12:       **end if**
13:    **end if**
14: **end while**

4   Experimental Analysis

Two experimental analysis are operated under two cyber attacks to confirm the effectiveness, feasibility and real-time performance of the proposed detection algorithm. The detailed model parameters of the NIPSSs are: $l = 0.25\,\mathrm{m}, m = 0.109\,\mathrm{kg}, J = 10.009083\,\mathrm{kg} \cdot \mathrm{m}^2, g = 9.81\,\mathrm{m/s}^2$.

We denote some parameters, i.e., $\tau_{\max} = 0.014\,\mathrm{s}$, $h = 0.015\,\mathrm{s}$, $W = 22.5\,\mathrm{s}$, $\gamma_\alpha = 0.02165$, $\gamma_\beta = 0.0272$, $x[0] = \begin{bmatrix} 0 & 0.05 & 0 & 0 \end{bmatrix}^T$, and $w[k] \sim \mathrm{N}(0, 0.25)$. Figure 4 shows the actuator output $u[k]$ ($Kz[k-2] + w[k-1]$) and controller output $u_c[k-1]$ ($Kz[k-2]$). Although the watermark signal has certain influence, it is seen that the system is still stable.

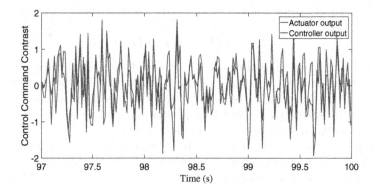

Fig. 4. Control commands.

4.1 Performance Validation Under Replay Attack

The malicious attacker records normal measurements between 75 s and 100 s and replay them from 175 s. Figure 5 shows the measurement and the actual state of cart position and inverted pendulum angle in the NIPSSs. The measurement indicates the system is still in normal operation state after attack launched. However, the inverted pendulum system has become unstable in practice. It indicates that when the replay attack occurs, it is impossible to directly identify replay attack, which may cause further damage to the system if it is not handled timely.

The proposed Algorithm 1 is applied to detect the replay attack by computing α and β based on collected data. Figure 6 presents the results, which shows that α exceeds the threshold γ_α at 177 s and β exceeds the threshold γ_β at 181 s. The results indicate that the proposed algorithm can detect the replay attack.

After the watermark signal is added in the NIPSSs, once the measurement is changed, the corresponding statistical rule of the watermark signal is also changed. Therefore, the measurements of NIPSSs are authenticated, then the malicious measurements can be detected by Algorithm 1.

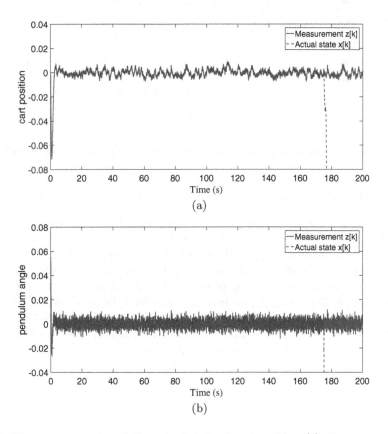

Fig. 5. The measurement and the actual state of cart position (a), the measurement and the actual state of inverted pendulum angle (b) from 0 s to 200 s under the replay attack launched at 175 s.

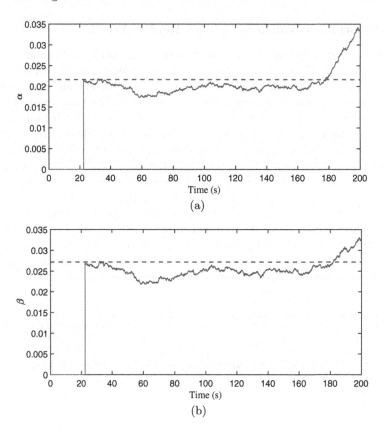

Fig. 6. The evolutions of indicator α, β in detector under the replay attack.

4.2   Performance Validation Under Noise-Injection Attack

The malicious attacker injects malicious noise signal into the measurement at 100 s. Figure 7 shows the measurement and the actual state of cart position and inverted pendulum angle in the NIPSSs. The measurements indicate the stability of the cart position state and the inverted pendulum angle is affected. However, it is hard to identify the inverted pendulum is that attacked by measurements, because the interference in the environment may lead the same results. Although the noise-injection attack does not cause great damage to the inverted pendulum, it will affect the control performance of the NIPSSs. Hence, we need to detect the attack shortly and adjust the system in time.

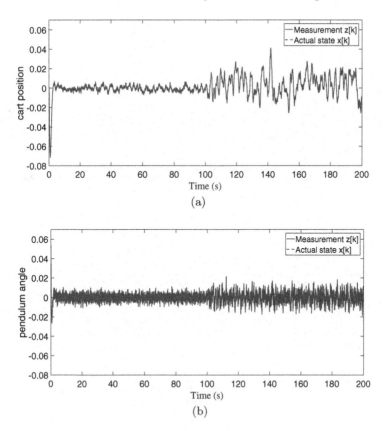

Fig. 7. The measurement and the actual state of cart position (a), the measurement and the actual state of inverted pendulum angle (b) from 0 s to 200 s under the noise-injection attack launched at 100 s.

Figure 8 demonstrates the efficacy of the proposed Algorithm 1. The results show that α exceeds the threshold γ_α at 102.9 s and β exceeds the threshold γ_β at 103.3 s. The results indicate that the proposed algorithm can detect the noise-injection attack.

The watermark signal gives an authentication to measurement to identify the process noise and malicious noise signal. Once the measurement is changed, the corresponding statistical rule of the watermark signal is also changed. Algorithm 1 can detect the malicious measurements in time.

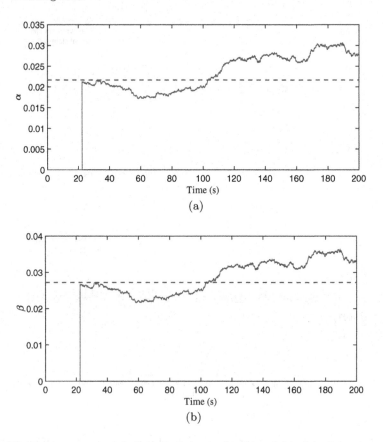

Fig. 8. The measurement and the actual state of cart position (a), the measurement and the actual state of inverted pendulum angle (b) from 0 s to 200 s under the noise-injection attack launched at 100 s.

5 Conclusion

This paper proposes a real-time window-based DWM detection method to solve the attack detection problem of NIPSSs. Firstly, the discrete model of NIPSSs is established by using detailed mathematic analysis and induction. Secondly, two statistical detection tests are constructed as an active detection mechanism to detect cyber attacks in NIPSSs. Then, Algorithm 1 is presented to conform security and real-time requirements of NIPSSs. Moreover, two indicators (α, β) are defined to identify cyber attacks that distort the measurements from the sensor fed to the NIPSSs. Finally, the effectiveness, feasibility and real-time performance of Algorithm 1 are confirmed by experimental analysis under replay attack and noise-injection attack. In the future, we will construct the optimal watermark to reduce the influence of watermark on the system which is a very interesting and significant research.

Acknowledgments. This work was supported in part by the National Science Foundation of China under Grant Nos. 61833011 and 61473182, and Project of Science and Technology Commission of Shanghai Municipality under Grant Nos.16010500300 and 19510750300, and Industrial Internet Innovation and Development Project (TC190H3WL).

References

1. Jung, S., Kim, S.S.: Control experiment of a wheel-driven mobile inverted pendulum using neural network. IEEE Trans. Control Syst. Technol. **16**(2), 297–303 (2008)
2. T. Zili, D. Pavkovi, D. Zorc.: Modeling and control of a pneumatically actuated inverted pendulum. ISA Trans. **48**(3), 327-335 (2009)
3. Du, D.: Real-time control of networked inverted pendulum visual servo systems. IEEE Trans. Cybern. (2019) https://doi.org/10.1109/TCYB.2019.2921821
4. Ge, X., Han, Q.L.: Consensus of multiagent systems subject to partially accessible and overlapping Markovian network topologies. IEEE Trans. Cybern. **47**(8), 1807–1819 (2017)
5. Hartmann, K., Steup, C.: The vulnerability of UAVs to cyber-attacks-an approach to the risk assessment. In: Proceedings of International Conference on Cyber Conflict, pp. 1–23, Tallinn, Estonia (2013)
6. Li, X., Liu, B., Zhang, C.: A real-time image encryption method of networked inverted pendulum visual servo control system. Complexity. https://doi.org/10.1155/2020/5725842(2020)
7. Ma, L., Wang, Z., Han, Q.L., Lam, H.K.: Variance constrained distributed filtering for time-varying systems with multiplicative noises and deception attacks over sensor networks. IEEE Sens. J. **17**(7), 2279–2288 (2017)
8. Zhan, G., Du, D., Wang, H.: Experimental analysis of networked visual servoing inverted pendulum system under noise attacks. In: 2018 IEEE 27th International Symposium on Industrial Electronics (ISIE), pp. 971–975, Cairns, QLD (2018)
9. Lai, S.Y., Chen, B., Yu, L.: Switching-Luenberger-observer-based redundant control under DoS attacks. Kongzhi Lilun Yu Yingyong/Control Theory Appl. **37**(4), 758–766 (2020)
10. Guo, Z., Shi, D., Johansson, K.H., Shi, L.: Optimal linear cyberattack on remote state estimation. IEEE Trans. Control Netw. Syst. **4**(1), 4–13 (2017)
11. Satchidanandan, B., Kumar, P.R.: Dynamic watermarking: active defense of networked cyber-physical systems. Proc. IEEE **105**(2), 219–240 (2017)
12. Huang, T., Satchidanandan, B., Kumar, P.R., Xie, L.: An online detection framework for cyber-attacks on automatic generation control. IEEE Trans. Power Syst. **33**(6), 6816–6827 (2018)
13. Song, Y.H., Du, D.J., Sun, Q., Zhou, H.Y., Fei, M.R.: Sliding mode variable structure control for inverted pendulum visual servo systems. IFAC-PapersOnLine **52**(11), 262–267 (2019)
14. Ahmed, C.M., Adepu, S., Mathur, A.: Limitations of state estimation based cyber attack detection schemes in industrial control systems. In: 2016 Smart City Security and Privacy Workshop (SCSP-W), pp. 6–10, Vienna, Austria (2016)
15. Mahmoud, M.S., Hamdan, M.M., Baroudi, U.A.: Modeling and control of cyber-physical systems subject to cyber attacks: a survey of recent advances and challenges. Neurocomputing **338**, 101–115 (2019)

A Fast Hybrid Image Encryption Method for Networked Inverted Pendulum Visual Servo Control Systems

Lisi Yang[1], Dajun Du[1(✉)], Changda Zhang[1], Jingfan Zhang[1], Minrui Fei[1], and Aleksandar Rakić[2]

[1] School of Mechanical Engineering and Automation,
Shanghai University, Shanghai 200444, China
lisiyangfive@163.com, ddj@i.shu.edu.cn
[2] School of Electrical Engineering, University of Belgrade,
Bulevar Kralja Aleksandra 73, 11000 Belgrade, Serbia

Abstract. This paper proposes a fast hybrid image encryption method to address the cyber security problem of image transmission in networked inverted pendulum visual servo control systems (NIPVSCSs). Firstly, the original image is encrypted by ranks crossing to resist image content leakage. Secondly, the randomly spaced watermark embedding operation are interspersed in each round of ranks cross encryption to locate the tampered image region. Furthermore, by analyzing the extracted watermark, the attack intensity can be pre-judged, which are then used to determine whether to decrypt. The tamper detection ability of digital watermark technology is combined with the security and robustness of image encryption algorithm effectively. Finally, the experimental results and analysis confirm the performance of the proposed algorithm in precisely locating regional tampering whilst maintaining high security and efficiency.

Keywords: Networked visual inverted pendulum control systems · Cyber security · Image encryption · Watermark

1 Introduction

With the rapid development of industrial camera image processing technology, visual sensor is employed to network control systems, leading to network visual servo control systems (NVSCSs) [1–4]. However, the images captured by visual sensor face cyber security risk problem under open network environment. For example, these images maybe suffer from visual information leakage, visual information tampering and etc. This will reduce data precision extracted from these images, which further decreases the control performance even causes the instability of the NVSCSs.

Aiming at the security of image information, the current research is mainly carried out from two aspects: image information leakage and tampering. Image

© Springer Nature Singapore Pte Ltd. 2020
M. Fei et al. (Eds.): LSMS 2020/ICSEE 2020 Workshops, CCIS 1303, pp. 264–278, 2020.
https://doi.org/10.1007/978-981-33-6378-6_20

information leakage refers to the illegal acquisition of an image by unauthorized party. Image encryption technology is a widely used method to prevent image information leakage. It transforms the original image data into scrambling code by some reversible algorithms, so that the illegal elements can not obtain effective information from the eavesdropped image [5–10]. Image information tampering refers to the malicious processing of the original image such as removing, adding, copying and splicing objects in the image. Image forensics technology is studied for tamper detection, which is divided into active forensics and passive forensics according to whether signature pre-extraction or information pre-embedding is carried out to identify and forensics the authenticity of the image. The active forensics technology based on digital image watermark embeds fragile watermark in the protected image in advance, so once the image is tampered, the watermark will be destroyed, and the tampering can be detected by watermark authentication [11–16]. Image passive forensics is to identify the authenticity of the image to be authenticated by feature extraction (e.g., DCT quantization coefficient, wavelet domain sub-band, PCA, SVD, etc.) and analysis [17–20].

To investigate these cyber security problems of the NVSCSs, a networked inverted pendulum visual servo control experimental platform was constructed, leading to networked visual inverted pendulum control systems (NIPVSCSs) [3,4]. However, due to the high requirements of real-time performance and control accuracy of the NIPVSCSs, the above image security research could be not effective in three defects: 1) The current research on image security focuses on the security and effectiveness of the algorithm. The speed of these algorithms is not fast enough, which is suitable for images under low frequency or low real-time requirements, not for high real-time environment. 2) If the image encryption technology is used to deal with the image security problem of the NIPVSCSs, the recovered image will still lose too much information under large-scale tampering (e.g., deletion, replication, etc.), resulting in invalid extracted state information. 3) Image forensics technology can be used to detect large-scale tampering, but the implementation of these two algorithms will lead to more serious delay problems, which aggravates the stability of the NVSCSs. Therefore, a fast hybrid algorithm of image encryption and watermarking is urgently needed. It can not only satisfy the real-time requirements of the NVSCSs, but also pre-judge the tampering before decryption, which can avoid unnecessary decryption operation and inform the NVSCSs of irreparable data loss.

This paper aims to address the real-time and security requirements of the NIPVSCSs. The main contributions of this paper are summarized as follows: 1) A ranks-based encryption algorithm is improved to satisfy the confidentiality and real-time requirements of the NIPVSCSs. 2) Digital watermark are embedded into the encrypted image to provide the encrypted image tamper detection ability, maintaining high efficiency. This algorithm embeds watermark information at random intervals in the improved fast image encryption algorithm. The attack strength of region tampering can be detected and evaluated in advance by analyzing watermark information before decryption. Once the attack intensity is greater than a certain range, the image will be considered invalid, and the decryption phase will be skipped. When the original decryption time is used in attack compensation, the delay of the control system under attack can be effectively suppressed.

The rest of this paper is organized as follows: Sect. 2 describes the image information security problem of the NIPVSCSs. Section 3 presents the hybrid algorithm and the implementation of encryption algorithm and watermark algorithm. Section 4 further confirms the security performance, tampering detection performance and real-time performance. Section 5 concludes the paper.

2   Problem Formulation

The scheme of the networked inverted pendulum visual servo control system is shown in Fig. 1, where the industrial camera is used to capture moving images of the inverted pendulum in real time and these images are sent to the remote processing unit via networks. Afterward, the remote processing unit extracts the acquired state information (cart position, pendulum angle, cart velocity and angular velocity) from the received images and calculates the control signals according to the designed control algorithm. The control signals will be transmitted to the local actuator via networks. Finally, the actuator drives the cart to achieve the stable control of the inverted pendulum.

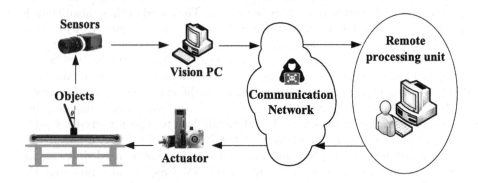

Fig. 1. Scheme of the networked inverted pendulum visual servo control system

Unfortunately, although the vision information and network bring big convenience, the following limitations emerge. On the one hand, a large amount of data of visual information are easier to be attacked during transmission. On the other hand, the introduction of visual system in the feedback loop will lead to long time delay such as image acquisition, image processing and image transmission, which has a great impact on the stability and control performance of the control system. For the image security problem of the NIPVSCSs, the image security algorithm is mainly considered from the following aspects:

1) For different image security problems, the corresponding algorithm can be used to solve, but the addition of image security algorithm will lead to the increase of image processing delay, which will affect the control performance of the system. Therefore, the designed algorithm needs to have high real-time performance.

2) Image encryption technology can be introduced to solve the problem of image information leakage, which guarantees the confidentiality of system information, but for an industrial control system, the availability of information is higher than the confidentiality. Image information tampering will destroy the integrity of information, and then affect the availability of control information. Generally, the image encryption algorithm is robust, which means that it can withstand noise and data loss. The image can be restored by decrypting after region tampering (e.g., removing, copying, etc.), which retains the integrity of the overall information to a certain extent. However, when the image information is greatly damaged (e.g., large area removing, large area copying, etc.), the information that extracted from the decrypted image can not satisfy the control accuracy requirements. In this way, the emergence of incomplete images will have a greater negative impact on the stability of the control system in the case of sacrificing part of the control performance to ensure the confidentiality.

3) Combining image forensics with image encryption technology, the encrypted image will have the function of tamper detection. However, the current research generally considers these two technologies independent processes. If the image encryption and digital watermarking are introduced by direct superposition operation into the NIPVSCSs, the time delay of the system will be greatly increased, which is detrimental more than beneficial for a long run.

It can be seen that under the risk of image information leakage and tampering, it is necessary to integrate image forensics technology into image encryption technology. Moreover, the efficiency of the image security algorithm must be improved to satisfy the control requirements of the NIPVSCSs.

3   A Hybrid Algorithm of Encryption and Watermark

The system model under the hybrid algorithm of watermark and encryption is shown in Fig. 2. In the local encryption unit, the chaotic sequence is firstly generated by using the key as the parameter of the chaotic system, and the encrypted image is obtained by using the chaotic sequence to encrypt the original image. In the process of image encryption, the ranks iterative encryption is used, where the intersections of rows and columns are considered as the watermark embedding pixels, as shown in Fig. 3. The encrypted image with watermark has strong camouflage, which is same as the general encrypted image. In the remote decryption unit, the embedded watermark can be extracted from the same position according to the key before decryption. Comparing it with the original watermark to determine whether the encrypted image has been tampered, whether the image information is credible. If the detected region tampering proportion exceeds the limit range, the image can be judged invalid and discarded. On the contrary, the decryption operation will be carried out to the image.

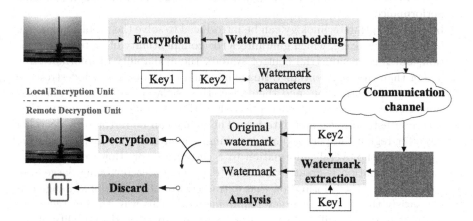

Fig. 2. Hybrid algorithm scheme of encryption and watermark

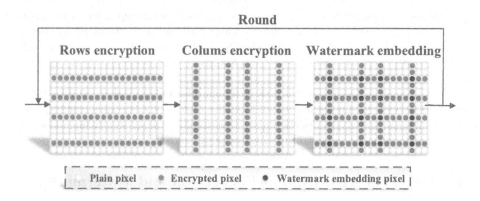

Fig. 3. Process of cross encryption and watermark embedding

Benefiting from its unpredictability and low computational complexity, the logistic map is used to generate the required random sequence. One for encryption, the other for constructing chaos watermark. The logistic map is iterated as follows:

$$x_{n+1} = \mu x_n (1 - x_n) \tag{1}$$

where $x_n \in (0,1)$, $\mu \in (0,4)$. From the existing experiments, it can be concluded that when $\mu \in (3.9,4)$, x_n can present chaotic state, and has better random performance. Therefore, $x_0 (\in (0,1))$ and $\mu (\in (3.9,4))$ are taken as key.

3.1   Encryption and Decryption

Against the image information leakage of the NIPVSCSs, an improved image encryption algorithm based on ranks interweaving is proposed. The specific cross encryption steps are as follows:

1. Key update.

$$T = T \bmod 100$$

$$x_0 = |x_0 - \frac{T}{1000}| \tag{2}$$

$$\mu_0 = 3.9 + |\mu_0 - 3.9 - \frac{T}{1000}|$$

where T denotes the sequence number of the current image collected by the control system. The sequence number T is added to the key as a disturbance, making the static key dynamic.

2. Parameters Initialization. To ensure that each row and column is encrypted only once, two recorders and the counts of unencrypted rows and columns are initialized. Recorders store the unencrypted ranks numbers in order.

$$Row[i] = i, \; i = 1, 2, \cdots, N \tag{3}$$

$$Col[j] = j, \; j = 1, 2, \cdots, M \tag{4}$$

$$count_{row} = N, \; count_{col} = M \tag{5}$$

3. Rows and columns selection. Iterate (1) once to get a new x. Using (6)–(7) a row number in row recorder is selected. Do (8) to overwrite the selected sequence number with the last unselected sequence number stored in the recorder and update $count_{row}$. After that, the unencrypted row numbers are stored in the first $count_{row}$ positions of the recorder. Repeat this step four times to select four rows. So does Columns selection, i.e.,

$$Pos_{row}[a] = \lfloor x \times count_{row} \rfloor, \; a = 0, 1, 2, 3 \tag{6}$$

$$Select_{row}[a] = Row[Pos_{row}[a]], \; a = 0, 1, 2, 3 \tag{7}$$

$$Row[Pos_{row}[a]] = Row[count_{row}], \; count_{row} = count_{row} - 1 \tag{8}$$

$$Pos_{col}[a] = \lfloor x \times count_{col} \rfloor, \; a = 0, 1, 2, 3 \tag{9}$$

$$Select_{col}[a] = Col[Pos_{col}[a]], \; a = 0, 1, 2, 3 \tag{10}$$

$$Col[Pos_{col}[a]] = Col[count_{col}], \; count_{col} = count_{col} - 1 \tag{11}$$

where $\lfloor * \rfloor$ represents a rounding down operation.

4. Encryption keys calculation.

$$k_{row,1} = mod(3 \times Select_{row}[3], 256), k_{row,2} = mod(k_{row,1} + 128, 256)$$
$$k_{row,3} = mod(5 \times Select_{row}[1], 256), k_{row,4} = mod(k_{row,3} + 128, 256) \tag{12}$$

$$k_{col,1} = mod(7 \times Select_{col}[3], 256), \; k_{col,2} = mod(k_{col,1} + 128, 256)$$
$$k_{col,3} = mod(9 \times Select_{col}[1], 256), \; k_{col,4} = mod(k_{col,3} + 128, 256) \tag{13}$$

5. Pixels Encryption.

$$P_{Select_{row}[a],w} = P_{Select_{row}[a],w} \oplus (k_{row,a} >> (2 \times w)), a = 0,1,2,3 \quad (14)$$

$$P_{v,Select_{col}[a]} = P_{v,Select_{col}[a]} \oplus (k_{col,a} >> (2 \times v)), a = 0,1,2,3 \quad (15)$$

where v and w are indexes of columns and rows ($w = 1, \cdots, M; v = 1, \cdots, N$), ">>" represents rotate right operation and "$\oplus$" represents XOR operation.

Remark 1. For the rows and columns encryption key, the new keys are generated by three times of rotate right operation of the obtained key. These four keys is called circularly to avoid the adverse effect of the large background area of the plain image. In this way, the security performance can be guaranteed without multiple rounds of encryption for the whole image, and the conditions for extracting watermark before decryption are provided.

6. Pixel value exchange. Rows and columns are swapped in pairs, rows first, then columns.

$$P_{Select_{row}[0],w} \leftrightarrow P_{Select_{row}[1],w}, \; P_{Select_{row}[2],w} \leftrightarrow P_{Select_{row}[3],w} \quad (16)$$

$$P_{v,Select_{col}[0]} \leftrightarrow P_{v,Select_{col}[1]}, \; P_{v,Select_{col}[2]} \leftrightarrow P_{v,Select_{col}[3]} \quad (17)$$

7. Repeat Step (3)–(6) until $count_{row} = 0$ and $count_{col} = 0$.

Remark 2. The traditional image encryption algorithm mostly adopts two steps of permutation and diffusion, and operates on single pixel. Although the image will have great encryption performance under such methods, it is not conducive to the algorithm efficiency. The improved algorithm can satisfy the above two performances, while the rank-based encryption reduces the complexity of the algorithm and the multiple rank crossover operation guarantees the security performance [7].

Considering the unequal length and width of the image, pre-encryption is carried out before cross encryption. After pre-encryption, the count of remaining unencrypted rows and columns is equal and multiple of four. For pre-encryption, the selection of ranks and generation of encryption key are same as those in cross encryption.

The decryption algorithm is the reverse step of the encryption algorithm. Parameter acquisition is same as encryption. In contrast to encryption, cross decryption starts with columns and then rows.

3.2   Watermark Embedding and Extraction

For image information tampering, the detection requirement can be satisfied by embedding watermark at random intervals. The parameters needed for watermark embedding are generated iteratively by the logistic map under Key_2. The specific embedding method is as follows:

1. Watermark parameter generation. Use $Key_2 = \{x_0, \mu\}$ as the initial value and parameter of the logistic map, and then iterate (1) once to get a new x. Do (18) to generate a new parameter g.

$$g = \lfloor x \times 100000 \rfloor \ mod \ 2^{16} \tag{18}$$

2. Watermark embedding. In the process of cross encryption, each round of ranks cross encryption is carried out, and the intersections of the selected ranks is used as the watermark embedding pixels. Every 4 bits of g are regarded as a watermark parameter wm. wm_L represents the lower 2 bits of wm, wm_H represents the higher 2 bits of wm. wm_L is the watermark information. wm_L determines the bits the pixel value is rotated to the right. Watermark is embedded in 2 LSBs of pixels $(Select_{row}[i], Select_{col}[j])$. Equation (1) and (18) are iterated every four embedding pixels.

$$P_{Select_{row}[i], Select_{col}[j]} = (P_{Select_{row}[i], Select_{col}[j]} \odot wm_L) >> (wm_H \times 2) \tag{19}$$

where "$\odot$" denotes the watermark embedding operation. Replace the low two bits of the pixel value P with wm_L. "$>>$" represents the rotate right operation.

Remark 3. Watermark embedding operation is interleaved in multiple ranks cross encryption. Watermark embedding points are the intersections of ranks in the multiple round cross encryption process, using the existing parameters of the encryption algorithm to save the time of embedding points selected. What's more, the selection of ranks in the proposed encryption algorithm is unique, which indicates that watermark embedding points will not be re-encrypted in the subsequent encryption rounds, so the tampering can be pre-judged before decryption.

3. Repeat Step (1)–(2) according to the round of ranks cross encryption.

Remark 4. Because of the robustness of encryption technology, image information can be recovered under small-scale attacks. Therefore, different from general digital watermark algorithms, the purpose of embedding watermark in encrypted image is not to detect and recover tampering accurately, but to detect the availability of image. Only when the attack area detected by tampered watermark is found to be too large, the encrypted image will be considered invalid. Once the image is pre-judged to be invalid, the decryption process can be directly abandoned to provide more information compensation or new image request time for the stable operation of the NIPVSCSs. For this object, the embedding watermark at random intervals is enough, which reduces the image information sacrifice.

In watermark extraction, the parameters in the encryption process are obtained according to $Key_1 = \{x_0, \mu\}$, and the watermark related parameters are generated by $Key_2 = (x_0, \mu)$. After the corresponding rotate left operation, the watermark can be extracted from the lower 2 bits of the embedding pixels. Comparing the extracted watermark with the original one, the positions of the embedding pixels which is different from the original one will be recorded. Through the location and proportion of the recorded pixels, the intensity of image attack can be analyzed.

4   Experiment and Analysis

Our experimental environment is Microsoft Visual Studio 2010 with Intel Core i5-8250U CPU@ 1.60 GHz and 8.0 GB RAM on Windows 10. 640 × 480 gray images of inverted pendulum are used to conduct the experiment. The results of the encryption and watermark embedding and decryption are shown in Fig. 4.

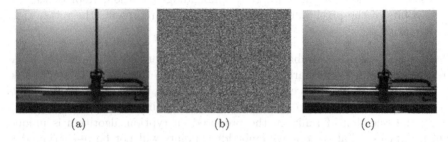

(a)   (b)   (c)

Fig. 4. Results of encryption and decryption. (a) Plain image. (b) Cipher image with watermarking. (c) Decoding image.

4.1   Security Analysis

Histogram and the correlation of two adjacent pixels can reflect the resistance of the proposed algorithm to statistical attacks. For the encryption algorithm, the more average the histogram of the image after encryption and the lower the correlation between adjacent pixels, the better the encryption effect. Histogram of inverted pendulum image before and after encryption are shown in Fig. 5. Compared with the histogram of plain image, the pixel value of the encrypted image is evenly distributed, which shows that the encryption effect is ideal. According to the statistical results of the pixel value of the embedded watermark image, the encryption performance is not affected.

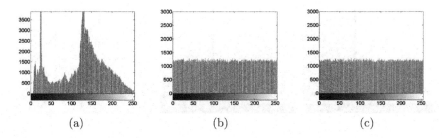

Fig. 5. Histograms of images. (a) Histogram of plain image. (b) Histogram of cipher image without watermark. (c) Histogram of cipher image with watermark.

The correlation of adjacent pixels can reflect the degree of pixel scrambling, including three directions: vertical, horizontal and diagonal. It is an important judgment basis to evaluate the scrambling effect of encryption scheme. The stronger the correlation between adjacent pixels, the closer the correlation coefficient is to 1. On the contrary, the closer the correlation coefficient is to 0. The correlation coefficient ρ_{xy} of adjacent pixels is calculated as

$$cov(x, y) = E\{(x_i - E(x))(y_i - E(y))\}$$

$$E(x) = \frac{1}{n}\sum_{i=1}^{N} x_i, \; D(x) = \frac{1}{n}\sum_{i=1}^{N}(x_i - E(x))^2 \tag{20}$$

$$\rho_{xy} = \frac{cov(x, y)}{\sqrt{D(x)D(y)}}$$

where x and y are values of two adjacent pixels; $E(x)$ and $D(x)$ represent the expectation and variance of the corresponding pixel gray value respectively; N represents the number of adjacent pixel pairs.

The correlation coefficient of plain image and cipher image are shown in Table 1. The plain image shows a high correlation, but there is almost no correlation between the adjacent pixels of the encrypted image.

Table 1. Correlation coefficients of two adjacent pixels in plain image and cipher image

Correlations	Plain image	Cipher image
Horizontal	0.999957	0.00689
Vertical	0.989676	0.005512
Diagonal	0.989738	−0.002614

To resist brute force attack, the encryption algorithm needs to have enough key space and strong key sensitivity. The keys used in the proposed encryption algorithm includes $k_{row,1} \sim k_{row,4}, k_{col,1} \sim k_{col,4}, x_0$ and μ. The key space is

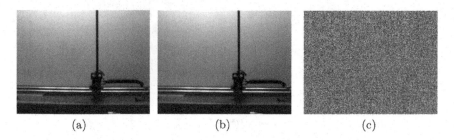

Fig. 6. Sensibility of secret keys. (a) Plain image. (b) Decoding with x_0. (c) Decoding with $x_0 - 10^{-16}$.

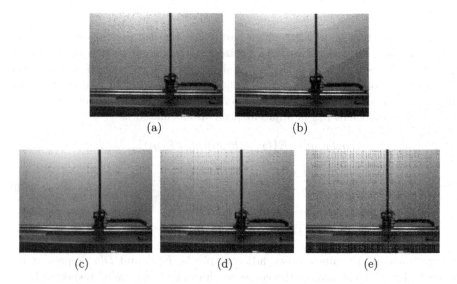

Fig. 7. Decoding images under attack. (a) Decoding image under salt and pepper noise. (b) Decoding image under Gaussian noise. (c) Decoding image under 2% cropping attack. (d) Decoding image under 5% cropping attack. (e) Decoding image under 10% cropping attack.

$(2^8)^8 \times 10^{16} \times 10^{15} \approx 2^{167} > 2^{100}$, which can ensure a good resistance to brute force attack. To test the key sensitivity of encryption algorithm, $x_0 - 10^{-16}$ are used as decryption keys. The decryption result is shown in Fig. 6. It can be seen that the encryption algorithm has good key sensitivity.

In the process of image transmission, it is a common phenomenon that image information suffers from noise attack or cropping attack. Cipher images are adding different kinds of noise and cropping different size of area to verify the robustness of the encryption algorithm. The decryption results are shown in Fig. 7, where Fig. 7(a)–(b) show that most of the effective information of the original image can be recovered after the encrypted inverted pendulum image suffers from different noises. Figure 7(c)–(e) show that after cropping attack for different areas and positions, the cropping area is scattered after decryption, and most of the effective information of the plain image can be recovered. The recovery effect is related to the cropping proportion. The smaller the area being cut, the higher the quality of the original image recovery.

4.2   Performance Analysis of Region Tamper Detection

For region tampering, the tampered region can be located by endpoint location. The tampering analysis of the embedded watermark can get the position of the tampered watermark, and record the maximum and minimum value of the row and column of the tampered position. The rectangular region with the minimum and maximum values as the vertex can be considered as the tampering region.

Although endpoint location has high detection accuracy, it is only for single region tampering. Against the multi area tampering, the morphological operation is introduced into the tamper detection. The watermark embedding points of the proposed algorithm are random and scattered, not the whole image. Therefore, the 640 × 80 image is divided into several 10 × 10 blocks for preliminary tamper detection. If a watermark embedding pixel is detected to be tampered, the block corresponding to the pixel is marked as tampered. After the watermark extraction and comparison process, the morphological closing operation are performed on the marked image to eliminate small black holes, which are present in Fig. 8. Digital watermarking is mainly used to detect image region tampering attacks such as copy-move, crop, splice and so on. The experiment takes cropping attack as an example.

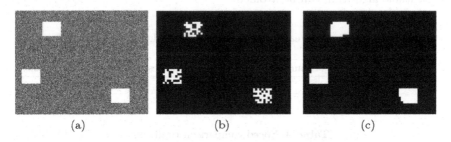

      (a)   (b)   (c)

Fig. 8. Region tamper detection. (a) Cipher images under region tampering. (b) Flag image after watermark information comparison. (c) Flag image after morphological closing operation.

The single region tampering is carried out on different size of the image region, and the error percentage between the cut ratio obtained by two methods and the real ratio is calculated. Details are shown in Table 2. The accuracy of the two methods increases with the increase of the tampering ratio. Moreover, the average of the two can achieve higher accuracy, leading to effective pre-judgment of image validity.

Table 2. Error percentage of tampering proportion calculation of two methods

Tampering proportion	Endpoint location method	Block positioning method	Average of two methods
2%	8.67%	8.12%	5.80%
5%	3.61%	5.65%	2.84%
10%	2.53%	4.04%	2.02%

4.3   Efficiency Analysis

To balance the encryption performance and efficiency, the encryption based on ranks operation is carried out. Rotate right operation is used in encryption keys, which can reduce the iteration time of key generation and avoid the poor encryption effect at the image background, so that one round of encryption can achieve the required encryption effect. To reduce the calculation of chaotic mapping, watermark embedding point selection and encryption ranks selection are combined to save the calculation time of shared data.

For analyzing the efficiency of the proposed hybrid algorithm, the computation time of the whole algorithm, watermark embedding and extraction, data preprocessing and whole algorithm without decryption are counted. The statistical results are shown in Table 3. Compared with the computing time of the whole algorithm, the computing time of the single watermarking algorithm is about 8%, and the same data preprocessing accounts for about half of it. After tampering, if the image is judged to be invalid in advance, the half time waste of the whole algorithm can be avoided.

Table 3. Speed of each part of the proposed algorithm (unit: ms)

Image size	Whole algorithm	Watermark	Data preprocessing	Without decryption
640×480	6.822	0.542	0.261	3.229

Table 4. Speed comparison (unit: ms)

Image size	Our algorithm	Ref. [5]
640×480	6.822	23.123
480×360	3.652	12.836
360×270	2.089	6.161

The efficiency of the proposed algorithm is compared with that of the algorithm [5] under different sizes of images. Details are shown in Table 4. The encryption efficiency of the proposed algorithm is three times that of algorithm [5]. Moreover, the proposed algorithm has the ability of attack detection. It can be seen that the proposed algorithm is significantly better than algorithm [5]. The proposed algorithm has the ability of tamper detection while achieving high security and efficiency.

5   Conclusion

A fast image encryption and watermarking hybrid algorithm based on chaotic mapping is proposed. To reduce the time complexity of the algorithm, some steps of encryption algorithm and watermark algorithm are overlapped. To reduce unnecessary time waste of invalid image to control system, watermark can be extracted for attack analysis before decryption. To a certain extent, it can reduce the influence of invalid image on the control system, and provide more time for the next step of information compensation. The experimental analysis of encryption security performance, attack detection performance and algorithm real-time performance shows that the proposed algorithm is feasible for high real-time networked control system.

For the NVSCSs, preventing image information leakage and detecting image attack are only a small part of the security performance. The proposed algorithm only meets the system requirements to a certain extent. There is still a long way to go in the image security research of networked vision control system.

Acknowledgments. This work was supported in part by the National Science Foundation of China under Grant Nos. 61833011 and 61473182, and Project of Science and Technology Commission of Shanghai Municipality under Grant Nos. 16010500300 and 19510750300, the 111 Project under Grant D18003 and Industrial Internet Innovation and Development Project (TC190H3WL).

References

1. Wu, H., Lou, L., Chen, C., Hirche, S., Kuhnlenz, K.: Cloud-based networked visual servo control. IEEE Trans. Ind. Electron. **60**(2), 554–566 (2013)
2. Wu, H., Lou, L., Chen, C.C., Hirche, S., Kühnlenz, K.: Performance-oriented networked visual servo control with sending rate scheduling. In: 2011 IEEE International Conference on Robotics and Automation, pp. 6180–6185, Shanghai (2011)
3. Du, D.: Real-time H8 control of networked inverted pendulum visual servo systems. IEEE Trans. Cybern. **50**(12), 5113–5126 (2020)
4. Zhan, G., Du, D., Wang, H.: Experimental analysis of networked visual servoing inverted pendulum system under noise attacks. In: 2018 IEEE 27th International Symposium on Industrial Electronics (ISIE), pp. 971–975, Cairns, QLD (2018)
5. Li, X., Liu, B., Zhang, C.: A real-time image encryption method of networked inverted pendulum visual servo control system. Complexity (2020). https://doi.org/10.1155/2020/5725842
6. Ilaga, K.R., Sari, C.A., Rachmawanto, E.H.: A high result for image security using crypto-stegano based on ECB mode and LSB encryption. J. Appl. Intell. Syst. **3**(1), 28–38 (2018)

7. Wang, X., Liu, C., Zhang, H.: An effective and fast image encryption algorithm based on Chaos and interweaving of ranks. Nonlinear Dyn. **84**(3), 1595–1607 (2016). https://doi.org/10.1007/s11071-015-2590-3

8. Wang, X., Wang, Q., Zhang, Y.: A fast image algorithm based on rows and columns switch. Nonlinear Dyn. **79**(2), 1141–C1149(2015)

9. Akkasaligar, P.T., Biradar, S.: Secure medical image encryption based on intensity level using Chao's theory and DNA cryptography. In: 2016 IEEE International Conference on Computational Intelligence and Computing Research (ICCIC), pp. 1–6, Chennai (2016)

10. Fu, X., Liu, B., Xie, Y., Li, W., Liu, Y.: Image encryption-then-transmission using DNA encryption algorithm and the double chaos. IEEE Photonics J. **10**(3), 1C15(2018)

11. Prasad, S., Pal, A.K.: A tamper detection suitable fragile watermarking scheme based on novel payload embedding strategy. Multimedia Tools Appl. **79**(3), 1673–1705 (2019). https://doi.org/10.1007/s11042-019-08144-5

12. Mathur, E., Mathuria, M.: Unbreakable digital watermarking using combination of LSB and DCT. In: 2017 International conference of Electronics, Communication and Aerospace Technology (ICECA), pp. 351–354, Coimbatore (2017)

13. Rakhmawati, L., Wirawan and Suwadi.: Image fragile watermarking with two authentication components for tamper detection and recovery. In: 2018 International Conference on Intelligent Autonomous Systems (ICoIAS), pp. 35–38, Singapore (2018)

14. Munir, R., Harlili.: A secure fragile video watermarking algorithm for content authentication based on arnold cat map. In: 2019 4th International Conference on Information Technology (InCIT), pp. 32–37, Bangkok, Thailand (2019)

15. Munir, R.: A chaos-based fragile watermarking method in spatial domain for image authentication. In: 2015 International Seminar on Intelligent Technology and Its Applications (ISITIA), pp. 227–232, Surabaya (2015)

16. Chaughule, S.S., Megherbi, D.B.: A robust, non-blind high capacity & secure digital watermarking scheme for image secret information, authentication and tampering localization and recovery via the discrete wavelet transform. In: 2019 IEEE International Symposium on Technologies for Homeland Security (HST), pp. 1–5, Woburn, MA, USA (2019)

17. Dixit, R., Naskar, R.: Review, analysis and parameterisation of techniques for copy-move forgery detection in digital images. IET Image Process. **11**(9), 746–759 (2017)

18. Gavade, J.D., Chougule, S.R.: Passive blind forensic scheme for copy-move temporal tampering detection. In: 2018 International Conference On Advances in Communication and Computing Technology (ICACCT), pp. 155–160, Sangamner (2018)

19. Wang, Y., Tian, L., Li, C.: LBP-SVD based copy move forgery detection algorithm. In: 2017 IEEE International Symposium on Multimedia (ISM), pp. 553–556, Taichung (2017)

20. Jwaid, M.F., Baraskar, T.N.: Detection of copy-move image forgery using local binary pattern with discrete wavelet transform and principle component analysis. In: 2017 International Conference on Computing, Communication, Control and Automation (ICCUBEA), pp. 1–6, Pune (2017)

A Review on Intelligent Modelling of Complex and Large Scale Systems for Unmanned Underwater Vehicles with Uncertain Disturbances

Li Liu[1,2(✉)], Qiang Tao[3], Dianli Hou[1], Fei Liu[1], Ningjun Feng[1], and Shulin Feng[1]

[1] School of Information Science and Electrical Engineering, Ludong University, Yantai 264025, China
liulildu@163.com
[2] Yantai Research Institute of New Generation Information Technology, Southwest Jiaotong University, Yantai 264025, China
[3] Shandong Commerce Vocational College, Yantai 264003, China

Abstract. With the growing measurement scale owing to engineering requirements, the network-based high-precision perception and calculation have been led to rapid develop. The dynamic networked systems of unmanned underwater vehicles (UUV) possess complex structure and large scale, due to the characteristics of uncertainties of model parameters, external disturbances for transmission and non-linear dynamics. This paper investigates the method of intelligent modelling for UUV with nonlinear dynamics, uncertain disturbance and parameters. Furthermore, the future interesting and challenging research directions are presented for the stochastic networked systems.

Keywords: Dynamic disturbances · Nonlinear dynamics · Uncertain disturbances · Transmission delays · Distributed fusion estimation

1 Introduction

The systems such as transportation, communication and energy have complex structure and large scale. They are lots of emerging problems to be solved, such as strongly non-linear dynamics and highly coupling correlation. To solve the complex mechanisms in the field of science and engineering, the intelligent computational methods are used for artificial intelligence [1], neural networks [2], genetics methods [3] and so on.

L. Liu—This work was supported by the National Natural Science Foundation of China (61903172, 61877065, 61872170, 61903174, 61873117), the Major Basic Research Project of the Natural Science Foundation of Shandong Province of China (ZR2018ZC0438) and the Key Research and Development Program of Yantai City of China (2019XDHZ085).

© Springer Nature Singapore Pte Ltd. 2020
M. Fei et al. (Eds.): LSMS 2020/ICSEE 2020 Workshops, CCIS 1303, pp. 279–293, 2020.
https://doi.org/10.1007/978-981-33-6378-6_21

The remainder section of this paper are constructed as follows. For dynamic systems, the uncertain disturbances are reviewed in Sect. 2. Some filtering approaches for the ship-environment interaction are presented in Sect. 3. The distributed estimation strategy for the stochastic networked systems is presented in Sect. 4. Then, Sect. 5 describes the system modelling. Finally, the conclusion and the research field in the future are given in Sect. 6.

2 Dynamics Disturbances

2.1 Nonlinear Dynamics

In practical applications, parameter uncertainties are widespread existence. The systems may give rise to severe performance degradation, even the system unstable. Therefore, the design of estimator is fully developed [4].

The dynamic positioning systems for UUVs are complex networked systems, which mainly owing to uncertainties of model parameters, external disturbances for transmission and the non-linear dynamics. The adaptive method is proposed to solve these problems. However, the method of the compromise between the transient system performance and the smooth control signal in UUV applications, the full potential has not been well achieved [5].

Considering the characteristics of the strongly nonlinear, stochastically time-varying of UUV dynamics as well as external disturbances, which are unpredictable, there exists the real difficulty in accurately establishing hydrodynamic system model [5]. To reduce the sensitivity for measurement noise, which cause the inaccurate measurement information owing to the difficulty of establishing the precise model for noise in practice. H_∞ filtering method is proposed as an estimation approach that is unnecessary to confirm the true noise model. Due to the considering stochastic finite signals for noise, H_∞ filtering has been applied in engineering such as state estimation and fault identification [6]. Note that the Kalman filtering and H_∞ filtering are unsuitable for the non-linear and/or non-Gaussian systems, since there are uncertainties suffer from modeling errors as well as external noise disturbances. Therefore, the aforementioned cooperative methods are only process well in the indoor, however, the performance of signal processing is unacceptable considering the effects of uncertainty such as wind and wave in UUV systems at sea [6].

2.2 Uncertain Hydrodynamic Parameters

In the underwater environment, the uncertainties contain three aspects, e.g. uncertain hydrodynamic parameters, modelling errors, as well as unknown forces owing to the currents. A command governor and disturbance observer is designed to constrain controller by the anti-disturbance technology.

First, the command governor was developed to optimize the command signals, which was constrained by the state and input information. Moreover, the disturbance observer is used to estimate the disturbances including uncertainties for parametric models, modelling errors as well as inconclusive external forces [7]. Furthermore, different from the one-step prediction, using the multi-step prediction is order to achieve the better performance for prediction [8,9].

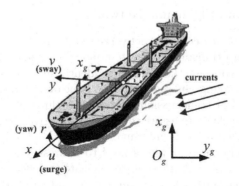

Fig. 1. Frames and states of the ship.

In the ocean, the UUV mainly suffers the uncertain disturbances such as winds, waves and currents from Fig. 1, which will cause the UUV deviating from the desired position and heading. If there are the variations in the conditions of environment and load [10], it implied that the hybrid control is able to achieve a scalable and rigorous strategy to handle some control objectives in the operational together with the changeable environmental conditions [11]. In addition, to cope with a class of complex and uncertain operations, the next generation UUV systems increase autonomy, competitiveness, security and operability when the uncertain environments [11].

For a dynamic positioning system, there are uncertain parametric model and the unbeknown time-varying disturbances, to achieve the global asymptotic stability due to the generating positioning errors, the robust adaptive control law is proposed. Simultaneously, the disturbance compensation strategy for environment was firstly converted into the issue of adaptive control by rewriting the mathematical model according to the UUV motion. Next, the external disturbances are represented as the model of multivariate linear regression. Then, the unbeknown outputs with uncertain disturbances were input and constructed an observer. Finally, the robust adaptive control law for the dynamic positioning was presented developing the theory of designed using the adaptive vectorial backstepping account of the projection method [10].

For the dynamic positioning systems, the issue of filtering and/or state estimation is valuable applications. For example, if measurement information of the velocities are unavailable, using the estimated value from the UUV to calculate and solve the actual measurement information over the state observer is a real application. Unfortunately, the measurement information from the position is disturbed from various noises such as sensor noise, winds, waves as well as currents. Note that the noise should be counteracted if it is varying slowly. That is, waves from the oscillatory motion should not be received by the feedback loop, which is named wave filtering. This approach is used to isolate from the low-frequency measurement information and wave frequency estimation [12].

2.3   Identification of Dynamic Networks

The main subjects of the complex networked systems for the UUV are consist of measurement, data transmission, system modeling, and then control systems [13]. For the prototyping of the UUV, the identification of dynamic networks from sensors is challenging. Subsequently, the issue of measurement information, signal processing, data transmission, system modeling, and controller for the complex and large scale systems are reviewed, respectively [13].

First of all, the measurement information of speed to the UUV is very important before GPS, and some problems need to address, e.g. noisy disturbances and its evolution processes. At present, to alleviate the effect of noise on the system, various filtering approaches are presented, and they are feasible [13]. For example, the traditional Kalman filtering is utilized in signal processing, state estimation and so on. Note that the position signals during the derivation do not need another sensor to perceive the measurement information. In addition, the Euler method is able to obtain a common evolutionary process. However, the precision of the calculation needs to improve [13,14].

Next, the definition of the dynamic networks is the system with the interconnected dynamic, which includes signals of measured node as well as the dynamic models. The networked structure is variable due to the dynamic link among the nodes. Suppose that the topology of the dynamic networks are known, the perceived signals with the external disturbances from the node signals is spatially correlated from the dynamic structure and the spectral density is singular [15]. The developed methods of the identifying dynamic networks are mainly consist of three aspects. The first aspect is described as follows. If the dynamic network is known, the interconnection structure as the condition, the considered identification problem is a signal module. The second aspect, for the full networked dynamics, the networks needs to know as the given topology. Finally, the last aspect, the topology and dynamic from the complex networks are able to be identified [15]. Importantly, because of the influence of external environment observation, the positioning errors would inevitably emerge. To improve the filtering effect, the conditional constrained filtering from the positioning value is introduced. That is, the extended Kalman filtering was proposed, the results are robustness and smoothness [16].

3   Filtering for the Ship-Environment Interaction

For the discrete systems, Sato et al. proposed an adaptive gain tuning algorithm using Kalman filtering, as well as switching Kalman filtering and H_∞ filtering. Note that the method of gain tuning and switching is dependent on the square mean for innovations [17]. Thus, the solutions for filtering relied on the force feedforward control was presented. For instance, taking into account of the crane force for the controller of the dynamic positioning systems, the low-frequency component is introduced. Moreover, the accurate estimation is necessary to achieve the feedforward solution. Unfortunately, in the ship-environment interaction, obtaining the accurate estimation is difficult owing to the time-varying disturbances and non-linear dynamics. Therefore, the low-frequency estimation for the dynamic systems need to operate in complex environment conditions [18]. The two conditions for assumption are listed as follows.

Assumption 1. Simplify the system modeling according to the waves and currents. They are designed by white noise.

Assumption 2. The reasonable offset scheme for the ship in surge, sway as well as yaw, since the smaller size compared to that one [18].

To improve the ships motion control performance, while reduce the disturbances over communication. Some relevant better results are proposed which are effective over analyzing the ships motion law. Especially, model the characteristics of the external disturbances. On the one hand, the dynamic characteristics of the non-linear and uncertain are studied. A backstepping controller is devised for the fuzzy neural network, the methods of estimation and compensation are investigated [19]. On the other hand, the ships motion is interfered over winds and waves during navigation, and the estimation approach is essential for the control of ships navigation [19].

Based on the time series which are auto-covariance, the classes of random signals are validated by a designed estimator. The proposed estimator is used to quantify the mean from the random uncertainties. Note that the time series are stationary. Obviously, the classes of time series are representative, which are vilified from the stability testing using the time series perceived [20]. Establishing the dynamic model for the UUV is the increasing interesting in engineering application fields. For the significant and important technique, the dynamic modeling should be sufficient equilibration for models considering the complex structure and the accurate calculation. Based on the above analysis, Zhu et al. [21] investigated an approach of a robust ship dynamic modeling from a conventional system modeling.

4   Distributed Estimation Strategies

In recent decades, the issue of dynamic positioning needs to the high-precision measurement, to solve the space dynamic positioning, the relevant measurement techniques have been researched quickly. With the rapid development of the complex networked structure, it contains the complexity for the communication links, and the limitation for the calculating power. All nodes use the strategy sharing signal processing workload. At the same time, coordinate the signal processing and communication task configuration structure. If the signals are dealt with all the nodes in the network, the system performance will be attenuated with the increasing of the scale for the networked dynamic systems. As a result, the distributed configuration structure is applied to improve the capacity of collaborative computing as well as reduce the communication burden [22].

In the train of the distributed estimation method is widely applied, such as target dynamic positioning as well as failure diagnosis. To further improve the state estimation accuracy, the significant distributed filtering and fusion methods have been proposed, for example, distributed Kalman filtering, distributed H_∞ filtering as well as distributed particle filtering [23]. Importantly, the distributed filtering approaches are used to achieve the information coordination and fuse the error estimation for the complex and large scale systems [24].

On the one hand, for the networked system, the unpredictable disturbances mainly stem from the restricted communication facilities. To overcome the difficulties of the traditional Kalman filtering is unable to deal with the colored noise, the robust Kalman filtering is investigated by integrating into the various noise disturbances. For instances, the state-dependent noise is used to describe the uncertainties of the system [25]. Considering the correlation of noise, the distributed fusion Kalman filtering [26,27] is represented considering the cross-correlation of noise for system modeling with measurement as well as process noise. On the other hand, the received valid information from sensors may be asynchronous due to the uncertain transmission delays over network. Proposing the measurement transformation approach used the Kalman filtering of restructuring the measurement and innovation sequences, respectively. More importantly, the time-delayed system is equivalent to the time-free dynamic system to alleviate the computational burden [28].

5   System Description and Modelling

5.1   Dynamic Network Definition

For dynamic network, which is consist of L internal nodes denoted as ω_i satisfying $i = 1, \cdots, L$. Note that introducing the external variables r_k, $k = 1, \cdots, K$, where K is a scalar. Therefore, the internal variable ω_i is determined as follows [15]:

$$\omega_i(t) = \sum_{\substack{l=1 \\ l \neq i}}^{L} G_{il}^0(q)\,\omega_l(t) + \sum_{k=1}^{K} R_{ik}^0(q)\,r_k(t) + v_i(t) , \tag{1}$$

where q^{-1} signifies the time-delay. For one-step delay of the i-th node, i.e. $q^{-1}\omega_i(t) = \omega_i(t-1)$.

Some symbols are expressed as follows:

(i) G_{il}^0 are one of the transfer functions, which are strictly proper rational. Its worth noting that G_{il}^0 equivalent to the single networked modules.

(ii) The external variables r_k are designed, which are determined according to the user. The other transfer functions R_{ik}^0 are the proper rational.

(iii) For the process noise v_i, define the process vector as $v = [v_1 \cdots v_i]^T$, which is established the model with the aid of the rational spectral density based on the stationary stochastic process. For example, if $p = 1, 2, \cdots$ and $p \leq L$, there is a p-dimensional white noise, which is $e \overset{\Delta}{=} [e_1 \cdots e_p]^T$, and the error covariance matrix is denoted as $P^0 > 0$. Therefore, the process noise is expressed as $v(t) = H^0(q)\,e(t)$. Note that $H^0(q)$ plays the same role with R_{ik}^0.

5.2 Mathematical Model of Ship Motion

For ship motion environment, the state space model Nomoto handles the input and output for dynamic systems, the rudder angle and the heading angle are input and output, respectively. From the input and out signals, the response model is established with the aid of the first-order input-output. Next, the corresponding second-order response model is also modelled from the linear state space [19]. That is,

$$G\dot{\gamma} + \gamma = K\sigma\,, \tag{2}$$

in which, the course angular velocity is denoted as γ, σ represents the rudder angle. In addition, symbols G and K are the indices of the ship maneuvering character.

Owing to the nonlinearity in the course of the ship navigation, establish the model as follows:

$$G\dot{\gamma} + \gamma + \alpha\gamma^3 = K\sigma\,, \tag{3}$$

where α denotes the coefficient of the nonlinearity.

Based on Nomoto model, the first-order response model for the rudder angle is constructed, which displays the manipulation according to the ship course change. That is

$$\dot{\sigma} = K_E(\sigma_\gamma - \sigma)/G_E\,. \tag{4}$$

Note that for the rudder manipulation, σ_γ expresses the command angle, the rudder gain is represented as K_E, and G_E is used to record the time constant.

Based on the above analysis, for the ship motion, the disturbances as well as the state space model from the rudder characteristic are able to establish as follows:

$$\begin{cases} \dot{\psi} = \gamma \\ \dot{\gamma} = f(\gamma) + w(t) + \beta\sigma \\ f(\gamma) = -\frac{1}{G}\gamma - \frac{\alpha}{G}\gamma^3 \\ \dot{\sigma} = K_E(\sigma_\gamma - \sigma)/G_E \end{cases}\,, \tag{5}$$

here $f(\gamma)$ represent the internal uncertainties. On the contrary, the external disturbances are denoted as $w(t)$. Meanwhile, β is control gain, as well as ψ is used to describe the ship course.

5.3 Two Frames Modelling for AUV

This section will introduce two reference frames for the autonomous underwater vehicle (AUV) model [7], the one is the earth-fixed reference frame, while the other is the body-fixed reference frame.

First, according to Cartesian position, define $[x, y, z]^T \in \Re^3$ as the earth-fixed reference frame. At the same time, the Euler angles are described as

$[\phi, \theta, \psi]^T \in S^3$, and S^3 is the denotes three-dimension sphere. Second, the body-fixed reference frame is used. Adopting $[u, v, r]^T \in \Re^3$ express the linear velocities including the surge, sway and heave, respectively. Corresponding to that, $[p, q, r]^T \in \Re^3$ denote the angular velocities, such as roll, pitch as well as yaw. As a result, there is the kinetic model for the AUV, which is six degrees-of-freedom. Therefore, it is shown as follows:

$$M\dot{v} + C(v)v + D(v)v + \Delta(v, \eta) + g(\eta) = \tau + \tau_w(t). \tag{6}$$

It should be noted that $\eta = [x, y, z, \phi, \theta, \psi]^T \in \Re^3 \times S^3$ remarks a vector, which is denoted as the position and altitude, respectively. Similarly, $v = [u, v, w, p, q, r]^T \in \Re^6$ denotes a vector of linear as well as angular velocities. $M = M^T > 0 \in \Re^{6 \times 6}$ is signified as the inertial matrix. The matrix $C(v) \in \Re^{6 \times 6}$ is the terms of the centripetal and Coriolis. The vector $\tau = [\tau_u, \tau_v, \tau_w, \tau_p, \tau_q, \tau_r]^T \in \Re^6$ is used to describe the control forces as well as moments. Then, vector τ_w is determined as $\tau_w = [\tau_{wu}, \tau_{wv}, \tau_{ww}, \tau_{wp}, \tau_{wq}, \tau_{wr}]^T \in \Re^6$, they are the environmental forces. Symbol $\Delta(v, \eta) \in \Re^3$ is used for representing the un-modeled hydrodynamics as well as modeling errors. In spite of the under-actuated AUVs are frequent, many fully-actuated AUVs are wide application rapidly. Therefore, lots of control strategies are researched.

The upper boundaries for the linear and angular velocities are limitations. Thus, define the velocity vector as the angular velocity, which satisfies the following boundary, i.e.

$$v_{\min} \leq v \leq v_{\max}. \tag{7}$$

Obviously, $v_{\min}, v_{\max} \in \Re^6$ are the six-degree-of freedom vectors, and the constant values are able to be achieved from the experiments of the AUV.

Then, the control input τ is constrained between the minimum constant vector $\tau_{\min}$ and the maximum constant vector $\tau_{\max}$, that is,

$$\tau_{\min} \leq \tau \leq \tau_{\max}. \tag{8}$$

where $\tau_{\min}, \tau_{\max} \in \Re^6$ are constant vectors.

From the above analysis, the purpose of the control is to design the control low, which breaks through the limitation of disturbances. For tracking the constrained signal $v_r \in \Re^6$, which is a desired signal, the constrained signal could not violate the constrains of the state in (7) as well as input in (8). On the other hand, the internal and external disturbances should be compensated for the six-degree-of-freedom kinetics of the AUV.

5.4   Identification Algorithm of the Two-Stage Stepwise

The two-stage stepwise identification algorithm (TSIA) is used to optimize the model structure. Subsequently, estimate the model parameters [29]. In general, the TSIA can able to select the most valuable input terms. In addition, for the neural network, the optimal number for the hidden layer neurons is also determined by the forward selection step in the TSIA.

There are two steps in the TSIA, which are the forward selection as well as the second refinement. Based on the selection scheme, matrices M_k and R_k are involved from a recursive matrix M_k and a residual matrix R_k, respectively. The definitions are

$$M_k = P_k^T P_k \, , \ k = 1, \ \cdots \, , \ n \, ,$$
$$R_k = I - P_k M_k^{-1} P_k^T \, , \ R_0 = I \, . \tag{9}$$

From (9), the symmetrical matrix $P_k = [p_1, \ p_2, \ \cdots \, , \ p_k]$ is consist of the first k selected terms which are the full regression matrix. For the matrix R_k, the properties are attracted as follows:

$$R_k^T = R_k \, , \quad R_k \cdot R_k = R_k \, , \tag{10}$$

$$R_k R_i = R_i R_k = R_k \, , \quad \text{for all } k \geq i \, , \tag{11}$$

$$R_{k+1} = R_k - \frac{R_k p_{k+1} p_{k+1}^T R_k^T}{p_{k+1}^T R_k p_{k+1}} \, , \ k = 0, \, 1, \, \cdots \, , \, n-1, \tag{12}$$

$$R_{k, \, \cdots \, , \, p, \, \cdots \, , \, \cdots \, , \, q, \, \cdots \, , \, k} = R_{1, \, \cdots \, , \, q, \, \cdots \, , \, \cdots \, , \, p, \, \cdots \, , \, k}, \ p, q \leq k \, . \tag{13}$$

From (13), the selection order is able to change for the repressor terms $p_1, \ p_2, \cdots \, , \ p_k$ are similarly to the residual matrix R_k. For the second step of refinement, the selection order can able to alleviate the computational burden.

Secondly, to eliminate the invalid input, it is selected from the forward constraint and the input selection terms.

It is assumed that a selected item p_k is transferred into the n-th position in P_k. The two adjacent terms are inter-exchanged, which are repeated as

$$p_q^* = p_{q+1} \, ,$$
$$p_{q+1}^* = p_q \, ,$$
$$q = k, \ \cdots \, , \ n-1 \, . \tag{14}$$

In which, symbol $*$ represents the updated value. Using the relationship, the residual matrix R_q will be refreshed as following

$$R_q^* = R_{q-1} - \frac{R_{q-1} p_q^* (p_q^*)^T R_{q-1}^T}{(p_q^*)^T R_{q-1} p_q^*} \, . \tag{15}$$

Then, similarly to p_q^* and R_q^*, the following terms such as A, P_n^* and R_k^* will be refreshed at one time.

The elements of A in the q-th and $(q+1)$-th columns from the rows 1 to $q-1$ should be updated as

$$a_{i,q}^* = a_{i, \, q+1} \, ,$$
$$a_{i,q+1}^* = a_{i,q} \, , \ i = 1, \ \cdots \, , \ q-1 \, . \tag{16}$$

Next, the corresponding row, such as the elements in the q-th row from column q to column n are refreshed from the following definition

$$a^*_{q,i} = \begin{cases} \dfrac{a_{q+1,\,q+1} + a^2_{q,\,q+1}}{a_{q,q}}, & i = q, \\ a_{q,\,q+1}, & i = q+1, \\ \dfrac{a_{q+1,\,i} + a_{q,q+1}a_{q,i}}{a_{q,q}}, & i \geq q+2. \end{cases} \tag{17}$$

And then, the next row, i.e. the $(q+1)$-th row $a_{q+1,\,i}$ satisfying $i = q+1, \cdots, n$ is also updated as follows:

$$a^*_{q+1,i} = \begin{cases} \dfrac{a_{q,\,q} + a^2_{q,\,q+1}}{a^*_{q,q}}, & i = q+1, \\ \dfrac{a_{q,\,i} - a_{q,q+1}a^*_{q,i}}{a^*_{q,q}}, & i \geq q+2. \end{cases} \tag{18}$$

Finally, the q-th and the $(q+1)$-th from the vector b will be only updated from

$$b^*_q = \frac{b_{q+1} + a_{q,q+1}b_q}{a_{q,q}}, \tag{19}$$
$$b^*_{q+1} = \frac{b_q + a_{q,q+1}b^*_q}{a^*_{q,q}}.$$

Here, b^*_q and b^*_{q+1} are the modified values.

It is worth notation that the updated process is repeated and continue, until the k-th terms of the values are set to the n-th position. Therefore, the updated matrices, such as the new regression and the residue matrix, are defined as follows:

$$P^*_n = [p_1, \cdots, p_{k-1}, p_{k+1}, \cdots, p_n, p_k], \tag{20}$$
$$R^*_k = [R_1, \cdots, R_{k-1}, R^*_k, \cdots, R^*_n].$$

If the terms have been moved to the n-th position given in P_k, contributions of the moved term will be reviewed from (11). The whole process of moving and comparing is repeated and continue, until throughout the selected terms are more significant than those remaining from the candidate term pool.

5.5   Filtering Algorithms

The generated discrete time-invariant system model is established as follows, which is linear process [17]

$$x_{i+1} = Ax_i + Bw_i, \tag{21}$$

$$y_i = Cx_i + v_i, \tag{22}$$

where the state vector is denoted as $x_i \in \Re^n$, the measurement is $y_i \in \Re^q$, $w_i \in \Re^p$ represents the process noise and $v_i \in \Re^q$ expresses the measurement noise at time i. The coefficient matrices are defined as $A \in \Re^{n \times n}$ and $B \in \Re^{n \times p}$ as well as $C \in \Re^{q \times n}$. There is the estimated state-function $z_i \in \Re^l$, whose quantity is defined as follows:

$$z_i = Lx_i \,, \tag{23}$$

where $L \in \Re^{l \times n}$ is used to express the coefficient matrix.

This section will introduce a novel approach, including an adaptive gain tuning Kalman filtering, H_∞ filtering as well as selectively switch filtering, which are used to estimation.

Adaptive Gain Tuning Kalman Filtering. Based on the statistic signals whose gaps contains measurements and estimations, $v_i = y_i - \hat{y}_i$ is named innovation with white noise for an adaptive gain tuning Kalman filtering.

First of all, the purpose of the conventional Kalman filtering methods is able to reduce the influence from Gaussian noise, that is

$$\hat{x}_{i+1|i} = A\hat{x}_{i|i} \,, \tag{24}$$

$$\hat{x}_{i|i} = \hat{x}_{i|i-1} + K_i \left(y_i - C\hat{x}_{i|i-1} \right) \,, \tag{25}$$

$$\hat{x}_{0|i} = \bar{x}_0 \,, \tag{26}$$

$$\hat{z}_i = L\hat{x}_{i|i} \,. \tag{27}$$

Note that $\hat{x}_{i|j}$ is the estimated value of x_i from the measurements at time j. The initial state x_0 with the mean $\bar{x}_0$. Therefore, the filtering method is shown with filter gain K_i as follows:

$$K_i = P_i C^T \left(C P_i C^T + R \right)^{-1} \,, \tag{28}$$

$$P_{i+1} = A \left(I - K_i C \right) P_i A^T + BQB^T \,, \tag{29}$$

$$R = E \left[v_i v_i^T \right] \,, \tag{30}$$

$$Q = E \left[w_i w_i^T \right] \,, \tag{31}$$

$$\bar{x}_i = x_i - \hat{x}_{i|i-1} \,, \tag{32}$$

$$P_i = E \left[\tilde{x}_i \tilde{x}_i^T \right] \,, \tag{33}$$

where $E[\bullet]$ is the expectation operation, and $P_i \in \Re^{n \times n}$ represents the covariance, which is satisfied the Riccati difference equation (RDE) from (29). Similarly, the initial state x_0 has the initial condition P_0. The covariance matrices from process and measurement noise are represented as $Q \in \Re^{p \times p}$, $Q \geq 0$ and $R \in \Re^{q \times q}$, $R > 0$, respectively. It is worth noting that noise such as $\{w_i\}$ and $\{v_i\}$ are supposed to be uncorrelated with Gaussian processes. Furthermore, P_i satisfies the Gaussian process, which denotes the state-estimation error given in (33). The identity matrix is represented as I.

For the sake of simplicity, define matrices X and Y from the function $\psi(X,Y)$ satisfying the RDE

$$\psi(X,Y) \overset{\Delta}{=} A(I - YC)X(I - YC)^T A^T + AYRY^T A^T, \tag{34}$$

with

$$P_{i+1} = \psi(P_i, K_i) + BQB^T. \tag{35}$$

Here, the adaptive Kalman filtering is investigated according to the innovations $\{v_i\}$, i.e.,

$$v_i = y_i - C\hat{x}_{i|i-1}. \tag{36}$$

Therefore, the covariance matrix from v_i is represented as follows:

$$E\left[v_i v_i^T\right] = CP_i C^T + R. \tag{37}$$

Suppose that the innovation $\{v_i\}$ is gently varying, the covariance $E\left[v_i v_i^T\right]$ is approximately computing $\overline{v_i v_i^T}$ as

$$\overline{v_i v_i^T} = \frac{1}{m} \sum_{j=i-m+1}^{i} v_j v_j^T, \tag{38}$$

in which, scalar m is used to record the size of the window from data. If $1 \le i < m$, m is substituted with i.

Next, the feedback information from $\{v_i\}$ in (38), P_i is updated from (35) is periodically tuning at the preset interval ΔT, that is

$$P_{i+1} = \frac{Tr\left[\overline{v_i v_i^T} - R\right]}{Tr\left[E\left[v_i v_i^T\right] - R\right]} \left\{\psi(P_i, K_i) + BQB^T\right\}. \tag{39}$$

Note that the trace is denoted as $Tr[\bullet]$. The stability analysis is validated from the following two cases to P_{i+1}:

$$Tr\left[\overline{v_i v_i^T} - R\right] > 0, \tag{40}$$

and

$$P_{i+1} - \psi(P_{i+1}, K_{i+1}) - BQB^T > 0. \tag{41}$$

The above conditions are used for ensuring the asymptotic stability from the closed-loop filtering during the tuning process. It is worth notation that the proposed adaptive gain tuning Kalman filtering from (39) is supported the stability analysis, and it processes the similarity to the conventional method.

H_∞ **Filtering.** For the state estimation, the conventional H_∞ filtering is presented, which is effectively for the deterministic noise.

The conventional H_∞ filtering is solved according to the filter gain H_i, which satisfies as follows in systems (21) and (22):

$$H_i = \bar{P}_i C^T \left(C \bar{P}_i C^T + R\right)^{-1}, \tag{42}$$

$$\bar{P}_{i+1} = A \bar{P}_i V_i^{-1} A^T + B Q B^T, \tag{43}$$

$$V_i = I + \left(C^T R^{-1} C - \gamma^{-2} L^T L\right) \bar{P}_i, \tag{44}$$

different from Kalman filtering, the scalar $\gamma > 0$ is non-unique. It satisfied the following existence condition based on the finite-horizon H_∞ filtering

$$\sup_{x_0,v,w} \frac{\sum_i \|z_i - \hat{z}_i\|^2}{\tilde{x}_0^T \bar{P}_0^{-1} \tilde{x}_0 + \sum_i \left(v_i^T R^{-1} v_i + w_i^T Q^{-1} w_i\right)} < \gamma^2. \tag{45}$$

In which, $\bar{P}_i > 0$ and $V_i \bar{P}_i^{-1} > 0$ are satisfied. If $\gamma \to \infty$, H_∞ filtering is coincided with Kalman filtering. However, the difficulty is how to appropriately design the scalar parameter γ.

Switching Filtering. The switching filtering method is investigated from the adaptive Kalman filtering and the H_∞ filtering, which has been designed to exploit the advantage from each filtering. To achieve the switching filtering from the innovations computation in (38) is used to the square means.

Finally, the criterion of the switching filtering with the power ratio is defined as

$$v_{ratio} \triangleq \frac{Tr\left[v_{K_i} v_{K_i}^T\right]}{Tr\left[v_{H_i} v_{H_i}^T\right]} < r_v, \tag{46}$$

where parameters v_K and v_H are the innovations from the adaptive Kalman filtering and H_∞ filtering, respectively. Parameter r_v denotes the preset threshold. The switching filtering is used to select the appropriate filtering. If (46) is satisfied, the adaptive gain tuning Kalman filtering is utilized. On the contrary, the H_∞ filtering is selected.

6   Conclusions and Future Topics

This paper has reviewed the issue of state estimation for the complex and large scale systems in UUV with uncertain disturbances, nonlinear dynamics, uncertain hydrodynamic parameters, identifications of dynamics networks and filtering for the ship-environment interaction. Finally, the distributed estimation strategy for high-precision measurement and positioning has been reviewed to improve the spatial positioning accuracy. Based on the previous review, lots of relevant topics for further research are able to be listed as follows:

(i) Establishing a unified framework model under various communication mechanisms is continue to be an interesting research issue, which constrained by multiple network-induced phenomena, for instance, signals with random transmission delays, packet loss and disorder packet and so on.

(ii) Furthermore, for the networked nonlinear and complex systems with network-induced phenomena, the fusion strategy for distributed estimation probably is a challenging research direction.

(iii) Moreover, the issue of the distributed estimation for the non-linear networked systems with time-varying environment is significance in the engineering application. Therefore, the research direction is quite meaningful.

(iv) Finally, the design of the event-triggered mechanism may be a future potential research trend for non-linear complex and large scale systems, which is tackled the issue of the estimation and filtering.

References

1. Aydin, A.: Applications of artificial intelligence techniques to enhance sustainability of industry 4.0: design of an artificial neural network model as dynamic behavior optimizer of robotic arms. Complexity **2020**(8564140), 1–10 (2020)

2. Zhang, Z., Dong, Y.: Temperature forecasting via convolutional recurrent neural networks based on time-series data. Complexity **2020**(3536572), 1–8 (2020)

3. Ge, M.-F., Liang, C.-D., Zhan, X.-S., Chen, C.-Y., Xu, G., Chen, J.: Multiple time-varying formation of networked heterogeneous robotic systems via estimator-based hierarchical cooperative algorithms. Complexity **2020**(8357428), 1–18 (2020)

4. Zheng, Z., Huang, Y., Xie, L., Zhu, B.: Adaptive trajectory tracking control of a fully actuated surface vessel with asymmetrically constrained input and output. IEEE Trans. Control Syst. Technol. **26**(5), 1851–1859 (2018)

5. Makavita, C.D., Jayasinghe, S.G., Nguyen, H.D., Ranmuthugala, D.: Experimental study of command governor adaptive control for unmanned underwater vehicles. IEEE Trans. Control Syst. Technol. **27**(1), 332–345 (2019)

6. Wu, H., Mei, X., Chen, X., Li, J., Wang, J., Mohapatra, P.: A novel cooperative localization algorithm using enhanced particle filter technique in maritime search and rescue wireless sensor network. ISA Trans. **78**, 39–46 (2018)

7. Peng, Z., Wang, J., Wang, J.: Constrained control of autonomous underwater vehicles based on command optimization and disturbance estimation. IEEE Trans. Industr. Electron. **66**(5), 3627–3635 (2019)

8. Liu, L., Yang, A., Zhou, W., Tu, X., Wang, G., Wang, H.: Event-based finite horizon state estimation for stochastic systems with network-induced phenomena. Trans. Inst. Measurement Control **41**(6), 1580–1589 (2019)

9. Liu, L., Yang, A., Zhou, W., Naeem, W., Wang, G., Wang, H.: Modelling and estimation for uncertain systems with transmission delays, packet dropouts, and out-of-order packets. Complexity **2018**, 1–15 (2018)

10. Du, J., Hu, X., Krstić, M., Sun, Y.: Dynamic positioning of ships with unknown parameters and disturbances. Control Eng. Practice **76**, 22–30 (2018)

11. Brodtkorb, A.H., Værnø, S.A., Teel, A.R., Sørensen, A.J., Skjetne, R.: Hybrid controller concept for dynamic positioning of marine vessels with experimental results. Automatica **93**, 489–497 (2018)

12. Fossen, T.I., Strand, J.P.: Passive nonlinear observer design for ships using lyapunov methods: full-scale experiments with a supply vessel. Automatica **35**(1), 3–16 (1999)

13. Melek, E., Gokhan, T.T., Philip, A.W., Seniz, E.: Marine measurement and real-time control systems with case studies. Ocean Eng. **159**, 457–469 (2018)

14. Zhang, G., Huang, C., Zhang, X., Zhang, W.: Practical constrained dynamic positioning control for uncertain ship through the minimal learning parameter technique. Digit. Signal Proc. **12**(18), 2526–2533 (2018)
15. Weerts, H.H.M., Hof, P.M.J.V.D., Dankers, A.G.: Prediction error identification of linear dynamic networks with rank-reduced noise. Automatica **98**, 256–268 (2018)
16. Xu, C., Xu, C., Wu, C., Qu, D., Liu, J., Wang, Y., Shao, G.: A novel self adapting filter based navigation algorithm for autonomous underwater vehicles. Ocean Eng. **187**, 106146 (2019)
17. Sato, M., Toda, M.: Adaptive algorithms of tuning and switching Kalman and H_∞ filters and their application to estimation of ship oscillation with time-varying frequencies. IEEE Trans. Industr. Electron. **67**(1), 501–511 (2020)
18. Ye, J., Godjevac, M., Baldi, S., Hopman, H.: Joint estimation of vessel position and mooring stiffness during offshore crane operations. Automat. Constr. **101**, 218–226 (2019)
19. Chen, Z., Qin, B., Sun, M., Sun, Q.: Q-learning-based parameters adaptive algorithm for active disturbance rejection control and its application to ship course control. Neurocomputing (2019)
20. Brouwer, J., Tukker, J., Klinkenberg, Y., Rijsbergen, M.V.: Random uncertainty of statistical moments in testing: Mean. Ocean Engineering **182**, 563–576 (2019)
21. Zhu, M., Sun, W., Hahn, A., Wen, Y., Xiao, C., Tao, W.: Adaptive modeling of maritime autonomous surface ships with uncertainty using a weighted LS-SVR robust to outliers. Ocean Eng. **200**, 107053 (2020)
22. Ge, X., Han, Q.-L.: Distributed event-triggered H_∞ filtering over sensor networks with communication delays. Inf. Sci. **291**, 128–142 (2015)
23. Zhong, X., Mohammadi, A., Premkumar, A., Asif, A.: A distributed particle filtering approach for multiple acoustic source tracking using an acoustic vector sensor network. Sig. Process. **108**, 589–603 (2015)
24. Li, D., Kar, S., Moura, J.M., Poor, H.V., Cui, S.: Distributed Kalman filtering over massive data sets: analysis through large deviations of random Riccati equations. IEEE Trans. Inf. Theory **61**(3), 1351–1372 (2015)
25. Keshavarz-Mohammadiyan, A., Khaloozadeh, H.: Consensus-based distributed unscented target tracking in wireless sensor networks with state-dependent noise. Sig. Process. **144**, 283–295 (2018)
26. Feng, J., Wang, Z., Zeng, M.: Distributed weighted robust Kalman filter fusion for uncertain systems with autocorrelated and cross-correlated noises. Inf. Fusion **14**(1), 78–86 (2013)
27. Yan, L., Li, X.R., Xia, Y., Fu, M.: Optimal sequential and distributed fusion for state estimation in cross-correlated noise. Automatica **49**(12), 3607–3612 (2013)
28. Liu, L., Yang, A., Tu, X., Fei, M., Naeem, W.: Distributed weighted fusion estimation for uncertain networked systems with transmission time-delay and cross-correlated noises. Neurocomputing **270**, 54–65 (2017)
29. Liu, K., Li, K., Peng, Q., Guo, Y., Zhang, L.: Data-driven hybrid internal temperature estimation approach for battery thermal management. Complexity **2018**, 9642892 (2018)

Research on Connectivity Restoration of Multiple Failures in UAV Cooperative Reconnaissance System

Shengfeng Zhang$^{(\boxtimes)}$ and Yeteng Zhu

The 723 Institute of CSIC, Yangzhou 225001, China
zhang_shengfeng@126.com

Abstract. Aiming at the problem of multiple failures in UAV cooperative reconnaissance system, a connectivity restoration algorithm VG-PTRA is proposed based on the topology reconstruction. By dividing the topology into grids, this algorithm transforms the global repair issue into inter-grids and intra-grid problem. In addition, two optimization mechanisms named "combining failure area" and "decreasing weight of inter-connected nodes" are also proposed to reduce the calculation overhead of VG-PTRA. Finally, simulation results confirm the effectiveness of this algorithm.

Keywords: Connectivity restoration · UAV cooperative reconnaissance system · Topology reconstruction

1 Introduction

In recent years, as one of the core platforms in information warfare, UAVs have been widely used in many fields such as reconnaissance and battlefield surveillance [1]. However, due to the complexity and change of reconnaissance tasks and the requirements for reconnaissance efficiency, single UAV may not meet the mission requirements [2]. Using multiple UAVs for coordinated reconnaissance, the dynamic information of targets could be obtained in real time. Moreover, through the information fusion processing technology, auxiliary decision support is provided for troops to implement accurate strikes and battlefield assessments [3].

The research of UAV cooperative reconnaissance system can be divided into reconnaissance planning, reconnaissance area coverage, reconnaissance path planning, asymptotic trajectory tracking and many other aspects. Among them, the coverage problem of reconnaissance area usually manifests as "point to area" collaborative reconnaissance, that is, a control method is designed for a specific large-area mission, so that multiple UAVs can quickly and efficiently cover the entire known environment with minimal cost [4]. However, considering the harsh environment of battlefield deployment, the studies of failure should be conducted after UAV nodes are disturbed or destroyed. Moreover, since the data transmission between reconnaissance nodes is the primary

© Springer Nature Singapore Pte Ltd. 2020
M. Fei et al. (Eds.): LSMS 2020/ICSEE 2020 Workshops, CCIS 1303, pp. 294–308, 2020.
https://doi.org/10.1007/978-981-33-6378-6_22

condition for collaboration, it is particularly important to ensure the connectivity of the reconnaissance UAV system, especially after the failure of multiple UAV nodes.

At present, when conducting research on deployment problems of multi-agents cooperative reconnaissance, the agents are usually abstracted into simple points in two-dimensional plane [5, 6]. Similarly, in this paper, we also assuming that all UAVs for reconnaissance are in the same plane, they are all isomorphic and the single UAV could be approximated as a mobile node with certain reconnaissance and communication radius which are denote as r and R in Fig. 1. Therefore, the connectivity restoration problem of multiple UAV failures in a cooperative reconnaissance system could be approximated as multipoint failures connectivity repair problem in an undirected graph. For the convenience of presentation, the following content will name the UAV as node for short.

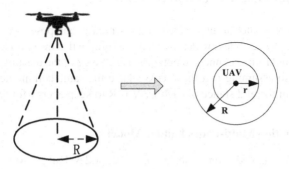

Fig. 1. Approximate model of single UAV.

The problem of connectivity restoration for multi-nodes failure is different from single node failure, because there may be some association between these failed nodes. Research on connectivity repair of multi-nodes failure in connected networks usually converts this problem into connectivity restoration of multi-connected branches or simply of regional failures. For instance, LPI utilizes the integer programming method, which is a centralized multi-nodes failure connectivity repair strategy [7]. However, this strategy is only designed for the failure of cut-vertexes in the network, it is not applicable to the failure of multiple ordinary nodes and cannot handle the normal failure situation in the network. Subsequently, researchers also proposed a multi-actor nodes failure repair algorithm SFRA for wireless sensor-actor networks (WSAN) [8]. SFRA is a typical hybrid distributed algorithm, it can comprehensively analyze the related parameters such as node energy consumption and node degree, and select the best repair nodes among the children of the failed ones. Similarly, reference [9] proposes a repair strategy capable of handling large-scale nodes failures in the network by performing tree-like planning on the nodes and starting from the sink node to record the hierarchical paths of all nodes. Due to the construction of tree topology and the damage caused by non-associative multi-nodes failure is inestimable, similar algorithms cannot restore the connectivity of non-associated multi-nodes failure effectively.

As a consequence, aiming at the problem of multiple failures in UAV cooperative reconnaissance system, this paper proposes a connectivity restoration strategy named as

Virtual Grids-based Partial Topology Reconstruction Algorithm (VG-PTRA). Topological rasterization can divide the network into virtual grids, and it is a typical distributed method, which has been applied to related topological control problems [10]. VG-PTRA is mainly divided into two aspects of intra-grid and inter-grid repair. The strategy of "centripetal shrinkage" is adopted in the restoration of intra-grid, and the inter-grid repair needs to ensure that all the corresponding inter-points are keeping workable. In addition, by analyzing the connectivity characteristics of the nodes between grids, this paper also proposes optimization strategies to reduce the calculation of algorithm execution. Finally, simulation results show that the algorithm can repair the connectivity of multiple failures effectively.

2   System Model and Problem Statement

In this paper, we consider an undirected graph which is composed by reconnaissance UAV nodes. Once these nodes are deployed, it's assuming that every node could connect with each other and form a connected network using some of the existing strategies such as C2AP. We also assuming that these UAVs have the same communication range R, and the maximum travel distance of each UAV is R to keep energy for reconnaissance.

2.1   Non-association Multi-nodes Failure Model

Firstly, the model named Non-Association Multi-nodes Failure Model (NMFM) is proposed.

$$\begin{cases} Component(V - V^*) \geq 2 \\ Component(V^*) \geq 2 \end{cases} \tag{1}$$

Where V means the set of initial nodes, V* means the set of disabled nodes, and Component ($\blacksquare$) is a function could calculate the number of connected components. Intuitively, it's indicating that the non-association multi-nodes failure could destroy the initial topology into at least two connected components, and the failed non-association multi-nodes are composed of at least two connected components. For instance, set V* consists of all hollow cycles in Fig. 2. It's clearly that Component (V*) equals to 4, and the four components are presented as dotted areas from A to D.

2.2   Virtual-Grid Model

Assuming that the network has been deployed in a two-dimensional area with horizontal and vertical lengths no more than X and Y, respectively. And hence, this target region could be covered by a rectangle whose abscissa is X and ordinate is Y.

We also assume that the deployed region of reconnaissance system is A in Fig. 3(a), after virtual grids are partitioned, the coordinate of the grid who contains an arbitrary node i (x_i, y_i) is expressed as follows:

$$G_x^i = x_i/g, G_y^i = y_i/g \tag{2}$$

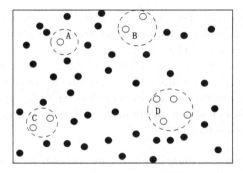

Fig. 2. Legend of NMFM.

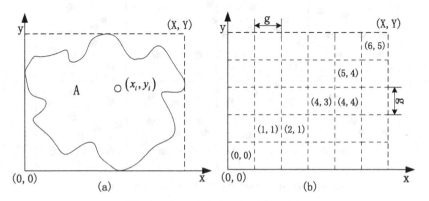

Fig. 3. Legend of virtual-grid model.

Where g represents the step size of grid partition, x_i and y_i is abscissa and ordinate of the grid which node i is located. It's clearly that, each node has an independent grid coordinate, and the number of grids is depended on the value of g. And then, the connectivity restoration of multi-nodes failure could be subdivided into each grid to complete.

3 Virtual Grids-Based Partial Topology Reconfiguration Algorithm

Firstly, the repair strategy should be designed based on the local reconfiguration of topology, which can effectively avoid the huge hardware consumption and uncertainty caused by large scale node deployment. Secondly, the repair algorithm usually cannot quickly and accurately determine the nature of multiple failed nodes, a more reasonable strategy should not analyze the failure type of a single node or the impact of its failure. It can avoid the tedious decision procedure and the redundant computation cost during the execution of the algorithm, which could effectively avoid the misjudgment caused by the existence of "vertex- cut set". Finally, this repair strategy should adopt a distributed architecture to improve execution efficiency.

VG-PTRA can be divided into two components: intra-grids and inter-grids restoration. Through the completion of intra-grid and inter-grids repair, connectivity of the entire network restoration could be completed.

In this paper, the information transfer between UAV nodes is used to determine their working status. That is, if a node fails to receive the information from its 1-hop neighbor in two consecutive communication cycles, then this neighbor is treated as failed node. In order to analyze the connectivity restoration problem in the grid more clearly, three possible scenarios are listed below:

1. Multi-interrelated nodes failure in the grid, like nodes B and C in Fig. 4.

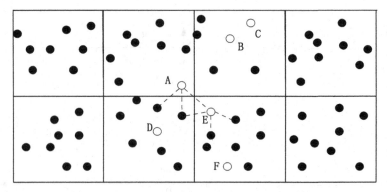

Fig. 4. Legend of gridded network with multiple failures.

2. Multi non-correlated nodes failure in the grid, like nodes E and F in Fig. 4.
3. Single node failure in the grid, like nodes A and D in Fig. 4.

According to the composition of sensor nodes in each grid after partition, the following two definitions are given for the relative position of sensor nodes in the grid:

Definition 1 Inter-Connected Node: Refers to those nodes whose 1-hop neighbors in different grids. That is, if node i is an inter-connected node, there must be node j in its one-hop neighbors to meet conditions $G_x^i \neq G_x^j or G_y^i \neq G_y^j$, like nodes A and E in Fig. 4.

Definition 2 Intra-Connected Node: Refers to those nodes whose one-hop neighbors in the same grid with them. That is, if node i is an intra-connected node, there must be node j in its 1 hop neighbors to meet conditions $G_x^i = G_x^j, G_y^i = G_y^j (d_{i,j} \leq R)$, like nodes B, C, D and F in Fig. 4.

If a single failed node in grid is intra-connected, according to the definition 2, this failed node is not directly connected with other nodes in other grids, and all 1-hop neighbors of the failed node are in the same grid. Therefore, the failure effect of this node on network connectivity is limited to the corresponding grid. And so on, if multi-nodes failed in a grid are all intra-connected, their effect is also limited.

Through the above analysis, we could find that all the grids can be divided into two types after the gird partition of initial connected network. In detail, internal nodes are fully and non-fully connected. If multi-nodes failed in fully connected grids, we should restore connection between remaining nodes. However, if multi-nodes failed in non-fully connected grids, more complex discussions are needed. Figure 5(a) presents the topology of initial network, 5(b) presents the topology after gird partition, in which I is non-fully connected grid, II is fully connected grid.

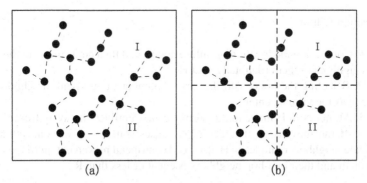

Fig. 5. Legend of grid partition.

3.1   Restore Strategy of Intra-grid Failure

When carrying out restoration in a grid, the idea of "centripetal contraction" is adopted. That is, regard the failed node as center, dispatching the remaining surviving nodes move to the original location of the failed node. It is necessary to point out that, the different distributions (fully connected or non-fully connected) of the nodes in the initial grid need to be discussed separately. The specific scheduling steps of the repair nodes are discussed as follows:

1. Nodes in the grid are fully connected

    • Single failure

    Step 1: Confirming the location of failure node and its 1-hop neighbors.
    Step 2: Dispatching 1-hop neighbors move to the position R/2 far away from the failure node, respectively.

Step 3: If this movement cause some other disconnection, those disconnected neighbors also need to move, until the distance between them and their one-hop neighbors is equal or less than R.

• Multiple failure

Multi-nodes failures in grids also need to be divided and discussed separately into two categories: association and non-association.

– Association failure

Step 1: Confirming the location of failure nodes, and then confirming all the normal one-hop neighbors of them (denoted as set H).
Step 2: Selecting the node with the largest number of surviving neighbors in the failure node as the virtual center.
Step 3: All nodes in H move to this virtual center until R/2 far away from it.
Step 4: If the movement of nodes in H causes some other disconnection, those disconnected neighbors of nodes in H also need centripetal movement, until the distance between them and their one-hop neighbors is equal or less than R.

– Non-Association failure

Since nodes in the initial grid are fully connected, and the movement of 1-hop neighbors of failure node may cause cascaded movement of other nodes, all nodes in the grid will involve once the first moving starts. Therefore, executing steps similar to association failure, intra-grid restoration of non-association failure could be accomplished.

In the case of multi-nodes failure in a grid, all 1-hop neighbors of the failure node need to move until R/2 away from the virtual center. Therefore, choosing the node with the largest number of surviving neighbors as the virtual center may reduce the moving distance of all neighbor nodes in the one-hop of the failure node. It's important to note that, if there are multiple failure nodes with the same number of surviving 1-hop neighbors, we should select the one with the nearest distance between the largest connected components as the virtual center. The following figure explains how to restore the failure of case 1.

The four sub-graphs in Fig. 6 represent the legends of single node failure, restoration of single node failure, multiple associated nodes failure and restoration of multiple associated nodes failure, respectively. If node A fails as expressed in the sub-graph (a), and its one-hop neighbor nodes are B, C, D and E. When performing the repair strategy, they will move toward the location of node A and move to the center R/2 away from it. The repaired internal structure of this grid is shown in the sub-graph (b). On the other hand, if nodes A, B, C, D, E are all disabled in the sub-graph (c), the one-hop neighbors of all failed UAVs are F and G, but node G is in the existing connected branch, so it is selected as the virtual center. It can be seen from the above four sub-graphs that the mobile scheduling method adopted by the failure repair strategy has not changed, but the movement range is limited to the mobile branch where the failed nodes were originally located.

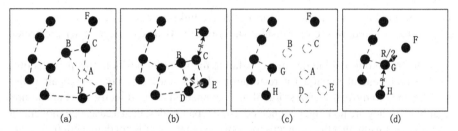

Fig. 6. Restoration of full connectivity of initial nodes in grids.

2. Nodes in the grid are partially connected

For the case where nodes in the grid are not fully connected at the initial moment, the problem is handled in a similar way as case 1, but there are differences between them. The main manifestation is that when the nodes in a grid are fully connected, all nodes are in the same connected branch; when the nodes are not fully connected, there are multiple connected branches in the same grid, and the failed nodes may also exist in multiple connected branches. When the failed nodes are repaired in a non-fully connected grid, the centripetal movement of the failed nodes or the block of failed nodes and the subsequent nodes to maintain the connection with the failed nodes' one-hop neighbor, the adjustment range is limited to their respective connection subset. The repair process of nodes failure in case 2 is described with reference to the following figures.

Similar to Fig. 6, sub-graphs in Fig. 7 represent the legends of single node failure, restoration of single node failure, multiple associated nodes failure and restoration of multiple associated nodes failure, respectively.

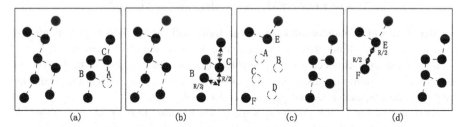

Fig. 7. Restoration of non-full connectivity of initial nodes in grids.

3.2 Restore Strategy of Inter-grids Failure

The previous part of this paper has discussed the case where the failed nodes are all intra-connected node, if the failure of inter-connected nodes occurs, the connectivity between the grid and its neighboring grids may also be damaged. In addition, the centripetal contraction strategy used in the restoration may also cause the movement of some inter-connected nodes. Considering that the network is whole connected before the grid division, so if there are nodes in the initial grid, there must be inter-connected nodes. The inter-connected nodes in each grid form effective links between grids, and the connectivity restoration requires re-establishing the connectivity links between inter-connected nodes.

If the inter-connected nodes move or fail at theirs initial position, this paper adopts a direct replacement method for restoration, that is, the corresponding repair nodes are scheduled to move to the position of inter-connected nodes directly. For the one-hop neighbors, if they are no longer connected to the repair nodes, then a cascaded movement is required to ensure that the connectivity between other nodes is not damaged. In fact, the repair strategy of the inter-connected nodes can be summarized as follows: completing the construction of the connection links between the corresponding connected subset to an inter-connected node or multiple inter-connected nodes. The specific scheduling steps for repairing nodes can be described as follows:

Step 1: Confirming the inter-connected node (e.g. node A) which needs to be repaired, select the node closest to it as the repair node (e.g. node B), and move node B to the location of node A.

Step 2: Assuming that after node B moves, it is no longer connected to its neighbors. Select a node among all the one-hop neighbors of B who has the maximize value of Eq. 1, where num_i represents the degree of the candidate node, d_{iA} represents the distance to the failed node A, and α is an arbitrary constant between 0–1 (sets to 0.5), p is the candidate index of the node. If node C has the largest candidate index, then it needs to move centripetal to the position where the distance from node B is R to ensure the connectivity between them. Furthermore, if other remaining one-hop neighbors of node B cannot guarantee the connectivity with the moved position of B and C, then they need to select the node of connected branch which is closer and move it to a location with a distance of R.

$$p = [\alpha \bullet num_i + (1 - \alpha)/d_{iA}]/\sqrt{num_i^2 + (1/d_{iA})^2} \tag{3}$$

Step 3: If node C cannot maintain connectivity with its one-hop neighbors after it moves, it needs to continue to select and schedule subsequent nodes by referring to Eq. 1.

If multiple inter-connected nodes disabled, the calculation needs to be performed sequentially. In addition, considering the possibility of node movement conflicts during the repair process, each node only can move just once.

As shown in Fig. 8, sub-graph (a) is the initial structure in the grid, and sub-graph (b) is the topology of the grid after restoration. Assume that the inter-connected nodes D and F fail, where F has only one neighbor C, and D has two neighbors A and E, respectively. According to the repair strategy, node C moves to the position of F, and node E is selected to complete the restoration of D according to Eq. 1. In addition, since the movement of node C will destroy the connectivity with B, node B needs to move to a position that is R away from node C, which will in turn cause the similar movement of node G. Moreover, considering the analysis in step 2, after node B moves, node A will no longer be connected with it, and it will choose a closer node G to build a connected link, and move ahead to the position R away from it.

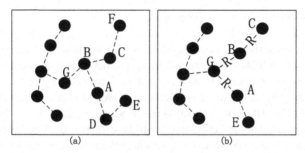

Fig. 8. Restoration of inter-connected nodes in a grid.

3.3 Optimization Strategy

Through the implementation the above repair strategy, basic connectivity restoration of multiple failures in the network could be accomplished. When analyzing the topology structure of grid division, it is obvious that the repair strategy still has some aspects could be optimized, and they are specifically manifested as "Combining failure area" and "Decreasing weight of inter-connected nodes".

● Combining failure area

If non-associated multi-nodes failure occurs in the grid, and there are multiple non-associated failed nodes in the one-hop neighbor of some surviving nodes, some non-associated failures can be converted into associated failures through the local movement of the surviving node, and this process is named as "Combining failure area".

As shown in Fig. 9, sub-graph (a) is the intra-grid topology with initial failure, and two non-associated failed nodes I and B are the one-hop neighbor of node G. By scheduling node G to move to the position of node I, the problem of non-associated failures is transformed into associated failures, and the merged grid structure of the failure area is shown in sub-graph (b).

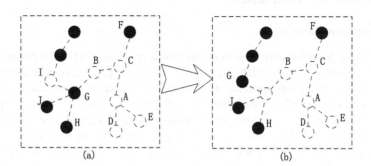

Fig. 9. Legend of combining failure area.

• Decreasing weight of inter-connected nodes

Since the inter-connected nodes always appear in pairs, by retaining the inter-connected nodes with highest weight, and treating other inter-connected nodes as intra-connected nodes, the amount of calculation during connectivity restoration could be reduced, and this process is named as "Decreasing weight of inter-connected nodes".

As shown in Fig. 10, sub-graph (a) corresponds to the case where there are many pairs of inter-connected nodes between same grids, such as node A and B, node C and D. While sub-graph (b) corresponds to the case where there is one to many inter-connected nodes between the grids, such as node A and B, C, D.

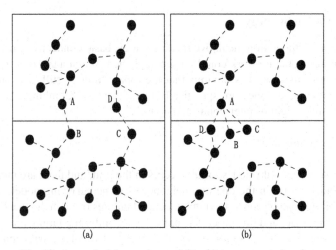

      (a)   (b)

Fig. 10. Legend of decreasing weight of inter-connected nodes.

4   Performance Evaluation

4.1   Simulation Parameters

This section conducts a simulation analysis of the performance of the algorithm, and the simulation is based on the matlab 2010b platform. On the premise of different network scales and node communication radius, the multi-nodes failure repair results after the execution of the VG-PTRA algorithm are compared and analyzed. The comparison parameters used in the simulation are the number of reconstructed nodes required for repair and the total distance that the nodes need to move. Using these parameters for comparison can reflect the topology reconstruction scale and overhead required by the algorithm.

In the case of multi-nodes failure, the simulation is designed in two groups, the first one is the simulation of total number of repair nodes that the VG-PTRA algorithm needs to move with the failure scale, and the second is the simulation of total moving distance of the repair node with the failure scale. The simulation parameters are shown in Table 1.

Table 1. Simulation parameters.

Parameters	Value
Target area	50 km * 50 km
Number of UAVs	200–400
Number of grids	9–25
Scale of failure	10%–30%
Communication range	1 km

4.2 Total Number of Repair UAVs

This section sets the different parameters to simulate the total number of nodes required for VG-PTRA. The number of deployed nodes is set from 200 to 400. Considering that the repair results are also affected by the number of grids, we also discuss the simulation results of different virtual grid numbers. The simulation results are shown as follows.

It is clearly that when the scale of deployed UAVs increases gradually, the number of UAVs required for repair will increase accordingly, but the number of repair UAVs is inversely correlated to the number of partioned grids. This is because when the scale of nodes increases, the number of one-hop neighbors of failed ones will gradually increase, and the number of repair UAVs that VG-PTRA requires will increase too. Furthermore, when the number of grids increases, the distribution of failed UAVs may be more evenly, and the number of moving UAVs will be constrained by the range of grid. Therefore, the number of repair UAVs will decrease. According to the above simulation results, it can be seen that when the scale of failed nodes becomes larger, the number of UAVs that VG-PTRA needs to move will gradually increase, but the increasing trend will be slower, this is because if the multiple failed nodes are partially associated, the number of corresponding one-hop neighbors will not cause a large increase (Fig. 11).

4.3 Total Moving Distance of Repair UAVs

In this section, different parameters are utilized to simulate the total distance moved by the VG-PTRA algorithm, and the simulation results are shown in Fig. 12.

As shown in sub-graph (a), under the condition that the number of girds is fixed, the total distance that repair nodes need to move will increase with the incensement of the initial deployment nodes. This is because the increase of network scale will lead to an increase in the number of nodes required for repair. However, the slight fluctuations in the simulation results may be due to the relatively dense deployment of each node when the network size increases, and the initial distance between nodes will gradually decrease. In addition, in the case of a constant network scale, increasing the number of divided grids, the total distance required to repair the nodes will show a similar decreasing trend as the total number of nodes required to move. This value may also fluctuate, because when the number of grids increases, the possibility of the blank grid increases, the number of remaining nodes in the grid will become smaller and smaller, and the relative distance between nodes will increase, which will lead to the increase of the total moving distance

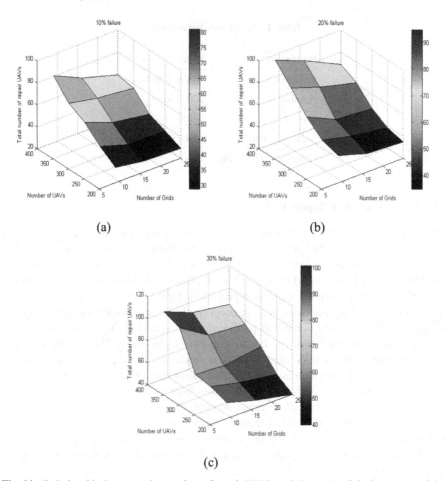

Fig. 11. Relationship between the number of repair UAVs and the scale of deployment and the number of grids.

of repair nodes. Sub-graph (b) and (c) are only different in the size of failed nodes, the change of total moving distance of repair nodes is similar to sub-graph (a).

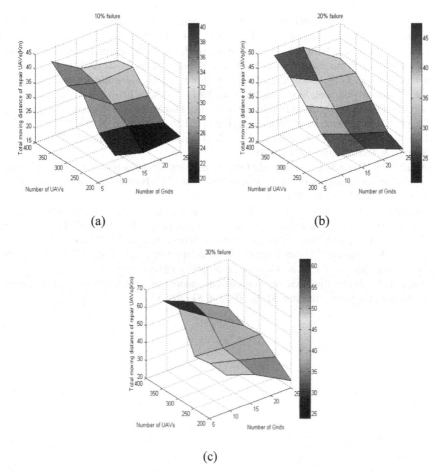

Fig. 12. Relationship between the moving distance of repair UAVs and the scale of deployment and the number of grids.

5 Conclusion

This paper investigates the connectivity restoration problem of multi-nodes failure in the UAV cooperative reconnaissance system, and proposes an algorithm named as VG-PTRA which is based on the topology reconstruction. This algorithm avoids the judgment of the properties of each failed node, and could reduce the computational overhead required for the connectivity restoration. In addition, this study simplifies the problem model to a two-dimensional plane, and the research of three-dimensional model will be studied in our future work.

References

1. Liu, H.X., et al.: Research on technologies for multi-UAV cooperative reconnaissance. Fire Control Command Control **42**(12), 1–4 (2017)

2. Pang, Q.W., et al.: Research on multi-UAV cooperative reconnaissance mission planning methods: an overview. Telecommun. Eng. **59**(6), 741–748 (2019)
3. Chu, Z., Guo, F.L., Nian, S.L.: Programming model of UAV cooperative reconnaissance mission based on multi-agent. Command Control Simul. **38**(5), 21–27 (2016)
4. Fu, X.W., Wei, G.W., Gao, X.G.: Cooperative area search algorithm for multi-UAVs in uncertainty environment. Syst. Eng. Electron. **38**(4), 821–827 (2016)
5. Yin, J.J.: Design of Distributed Optimization Protocol for Multi-agent Systems over Directed Graphs. Dalian University of Technology, China (2019)
6. Luo, K.: Dynamic Coverage Control of Multi-Agent Systems. Huazhong University of Science & Technology, China (2019)
7. Alfadhly, A., Baroudi, U., Younis, M.: Optimal node repositioning for tolerating node failure in wireless sensor actor network. In: Proceedings of the 25th Biennial Symposium on Comm, Kingston, Canada, pp. 67–71 (2010)
8. Alfadhly, A., Baroudi, U., Younis, M.: An effective approach for tolerating simultaneous failures in wireless sensor and actor networks. In: Proceedings of the First ACM International Workshop on Mission-Oriented Wireless Sensor Networking, pp. 21–26 (2012)
9. Akkaya, K., Senturk, I.F., Vemulapalli, S.: Handling large-scale node failures in mobile sensor/robot networks. J. Netw. Comput. Appl. **36**(1), 195–210 (2013)
10. Wang, L.L., et.al.: Research on even-driven underwater sensor networks deployment. Transducer Microsyst. Technol. **32**(9), 50–52 + 57 (2013)

Improved Bundle Adjustment
Based on ADMM

Ruijie Ma[1], Xin Jiang[1], Jingjing Liu[1(✉)], Mingyu Wang[1(✉)], Wenhong Li[1(✉)],
and Ting You[2]

[1] State Key Laboratory of ASIC and System, School of Microelectronics,
Fudan University, Shanghai, People's Republic of China
{19212020118,19212020102,liujingjing,mywang,wenhongli}@fudan.edu.cn
[2] Key Laboratory of Air-driven Equipment Technology of Zhejiang Province,
Quzhou University, Quzhou, People's Republic of China
josonjodys@163.com

Abstract. Bundle adjustment plays an significant role in SLAM (simultaneous localization and mapping), which is utilized in both front-end visual odometry and global back-end optimization. In many SLAM systems, bundle adjustment is employed to estimate the location of 3D landmarks and 6 DOF camera pose. However, as the dimension of the optimization variable increases, bundle adjustment consumes more and more time in SLAM, mainly in the iterative process. In this paper, an improved algorithm for this large scale bundle adjustment problem has been proposed. Firstly, according to the pose consensus, the classic algorithm ADMM (alternating direction method of multipliers) is introduced into the bundle adjustment problem. Secondly, for the non-convex optimization problem, the sub problem optimization method is introduced, and convergence and stopping criteria of the algorithm are discussed. Finally, the semi-dense direct method visual odometry for verification is implemented, and the experiments prove that the improved bundle adjustment algorithm has a speed advantage and can be applied to BAL (bundle adjustment in the large) problem.

Keywords: Bundle adjustment · Simultaneous localization and mapping · Alternating direction method of multipliers · Visual odometry · Pose consensus

1 Introduction

In the past decades, with the development of robots and autonomous driving, SLAM (simultaneous localization and mapping) have become increasingly important [3,7]. In order to locate and map more accurately, people use higher quality and higher resolution images to extract more feature points. However, this caused the bundle adjustment process in SLAM to spend more time and computing resources, and some even required GPUs to speed up. Bundle adjustment is the problem that optimizes the visual reconstruction to obtain the best 3D

© Springer Nature Singapore Pte Ltd. 2020
M. Fei et al. (Eds.): LSMS 2020/ICSEE 2020 Workshops, CCIS 1303, pp. 309–322, 2020.
https://doi.org/10.1007/978-981-33-6378-6_23

structure and estimation of viewing angle parameters (camera calibration and pose) [22] in the SLAM. Bundle adjustment adjusts to obtain optimal camera parameters and world landmarks coordinates by utilizing the camera's pose and the 3D coordinates of the measurement points as unknown parameters, and using the coordinate of the feature points observed on the camera at the front intersection as observation data. The optimized parameters include the coordinates of the world landmarks and the pose of the camera. Both of these determine the projection bundle during the camera projection process, so we are adjusting the bundle, which is also the origin name of the bundle adjustment.

Bundle adjustment is a non-linear optimization problem, which can be optimized through iteration using the least squares method. The Gauss-Newton and Levenberg-Marquadt methods [13] are the most popular methods in optimizing the bundle adjustment formula. The time consumed by the bundle adjustment process is related to the dimension of the optimization variable. As the dimension of the optimization variable increases, the time of the bundle adjustment will also greatly increase. In some dense and semi-dense construction maps, there are tens of thousands of optimized world coordinate points, which makes the matrix dimension of the least squares solution very large, resulting in a slow speed. Even in recent years, people have realized the sparseness of the bundle adjustment problem [6,19], but it still takes a lot of time for large-scale reconstruction.

In order to increase the speed in bundle adjustment optimization, a related algorithm is proposed in this paper. The proposed algorithm decomposes the large Hessian matrix into two small Hessian matrices, where the dimensions of the small matrix are about half that of the large matrix, and the speed of inverting the two small matrices is much greater than that of the large matrix. The main contributions of this paper are as follows: (1) The algorithm proposes that the large matrix in the bundle adjustment process can be decomposed into two mutually constrained small matrices. The two small blocks have different world points, but their poses are the same. In this way, the ADMM [5] algorithm can be applied to constrain it for optimization. (2) The projection function is non-convex, so the convergence and termination conditions of the algorithm need to be analyzed. We analyze that projection function has local Lipschitz-continuous, and thus the augmented Lagrangian function in ADMM is convex and it can converge.

The structure of the paper is organized as follows. Section 1 reviews the time-consuming problems in large-scale bundle adjustment and proposed an improved method. In Sect. 2, we discuss the related work of incremental bundle adjustment, explain their advantages and disadvantages, and lead to our research content. Section 3 introduces the bundle adjustment and the ADMM algorithm, and discusses how to apply the ADMM algorithm in bundle adjustment. In the Sect. 4, we discussed the convergence and termination conditions of the algorithm. In Sect. 5, the experimental results using our proposed algorithm has been shown. At last, our conclusion will be showed in Sect. 6.

2   Related Work

When the robot moves in a long time and space, there are many methods using the sliding window [8] method to discard some historical data due to the many optimized variables. Or according to the practice of pose graph [17], abandon the optimization of the landmark points, and only keep the edges between poses. There are also some attempts to solve Structure-from-Motion in cities and even the earth, such as the work in [11]. However, none of these tasks has been globally optimized because the cost of global bundle adjustment optimization is very large and cannot be completed with limited time and resources.

Some approaches have been proposed for incremental bundle adjustment. The iSAM [15] transforms the graph optimization problem in bundle adjustment into a Bayesian tree establishment, update, and inference problem. The whole architecture of this method is based on probabilistic reasoning, and it uses QR matrix factorization to optimize on sparsity. But only a small part of the decomposition result has been updated in each iteration instead of the whole graph. SLAM++ [20] recover estimates and variances, and update Schur complement space incrementally in bundle adjustment. However, the above algorithm is only appropriate for handling the sparse camera problem (most of the key points are only visible in few frames), which is consistent with the large-scale SfM problem. But in the SLAM problem, most of the frames in the local sliding window share a large part of the key points, so that the above incremental algorithm evolves into a conventional BA solver, and the positioning accuracy cannot be better than other latest algorithm. In visual inertial fusion of SLAM, $ICE - BA$ [18] efficiently uses the previously optimized intermediate results to avoid new redundant calculations. This algorithm significantly improves the solution speed and can be applied to most VI-SLAM based on sliding window method, but this method only uses the repeatability of intermediate calculation results. At the same time, the good engineering in Ceres [1] and g2o [14] implements BA and is used in various SLAM systems. However, there is an obvious disadvantage in these methods: the complexity increases twice with the number of image frames. In order to achieve real-time pose estimation, SLAM systems based on these solvers can only utilize very limited measurements.

In order to deal with above problems, referring to ADMM algorithm [5], we propose an improved bundle adjustment algorithm to optimize the world points and pose. Main contributions of this paper are as follows: (1) Our algorithm proposes that the large Hessian matrix in the BA process can be decomposed into two mutually constrained small Hessian matrices. The two small blocks have different world points, but their poses are the same. In this way, the ADMM algorithm can be used to constrain it for optimization. (2) The projection function is non-convex, so the convergence and termination conditions of the algorithm need to be analyzed. We analyze that projection function has local Lipschitz-continuous, and thus the augmented Lagrangian function in ADMM is convex.

3   Proposed Method

3.1   Bundle Adjustment

Bundle adjustment is a reprojection process. With the camera parameters, the measured world landmarks are reprojected on the pixel plane and then compared with the points observed by the camera. Camera parameters include external and internal parameters. The external parameters include the camera's translation vector and rotation matrix. The internal parameters include the distortion coefficient and focal length. At different times, the external camera parameters are not the same, but internal parameters are generally the same. Suppose we have M observed 3D points, the external parameter set of N cameras is recorded as $\Xi = \{\xi_i \in R^C \mid i = 1, ..., N\}$, and the internal parameter set is recorded as $D = \{di_i \in R^I \mid i = 1, ..., N\}$, record the observed 3D point set $P = \{p_j \in R^3 \mid j = 1, ..., M\}$ and the observation as $Z = \{z_{ij} \in R^2 \mid i = 1, ..., N; j = 1, ..., M\}$, z_{ij} indicates that the i-th camera observes the j-th point. The cost function can be expressed as follows

$$f(\Xi, D, P) = \frac{1}{2} \sum_{i=1}^{N} \sum_{j=1}^{M} \|z_{ij} - h(\xi_i, d_i, p_j)\|_2^2 \tag{1}$$

where $h(\xi_i, d_i, p_j)$ is the reprojection equation of the j-th point on the i-th camera, is nonlinear and non-convex. The pose here includes two parameters, namely translation vector and rotation matrix. Solving this nonlinear least squares problem is equivalent to adjusting camera parameters and landmark points at the same time, and this process is called bundle adjustment.

3.2   ADMM

The proposal of ADMM algorithm [4] is to handle the following format problems:

$$\min \quad f(x) + g(z) \quad \text{subject to} \quad Ax + Bz = c \tag{2}$$

According to the augmented Lagrangian multipliers method, we get

$$L(x, z, y) = f(x) + g(z) + y^T(Ax + Bz - c) + \frac{p}{2}\|Ax + Bz - c\|_2^2 \tag{3}$$

The ADMM algorithm consists of iterations:

$$x^{k+1} = \underset{x}{argmin}\, L_p(x, z^k, y^k) \tag{4}$$

$$z^{k+1} = \underset{z}{argmin}\, L_p(x^{k+1}, z, y^k) \tag{5}$$

$$y^{k+1} = y^k + p(Ax^{(k+1)} + Bz^{k+1} - c) \tag{6}$$

The updates of parameters are implemented by alternating iterations.

And we want to use the ADMM algorithm to solve the general form of the consistency optimization problem:

$$\min \sum_{i=1}^{N} f(x_i) \quad \text{subject to} \quad \hat{x}_i = \hat{z}, i = 1, ..., N \tag{7}$$

Here, because the parameter space is divided into blocks, the parameter dimensions of each sub-objective function $f_i(x_i)$ are different, which are called local variables. Local variables no longer correspond to global variables but are part of global variables. Map a part of x_i to a part of the global variable z as $\hat{z}_i$. Using ADMM, the expressions capable of the deriving iteration are

$$x_i^{k+1} = \underset{x_i}{argmin}(f_i(x_i) + y_i^k(\hat{x}_i - \hat{z}) + \frac{p}{2}\|\hat{x}_i - \hat{z}\|_2^2) \tag{8}$$

$$\hat{z}^{k+1} = \frac{1}{n}\sum_{i=1}^{N}\hat{x}_i^{k+1} \tag{9}$$

$$y_i^{k+1} = y_i^k + p(\hat{x}_i^{k+1} - \hat{z}^{k+1}) \tag{10}$$

The augmented Lagrangian constant $p > 0$, and Eq. (9) is the average of the optimization results of different nodes.

3.3   Pose Consensus

According to the algorithm given above, for a large bundle adjustment problem, we can decompose it into multiple small blocks for processing. In fact, dividing into two small blocks is the most common and can achieve better speed and accuracy [10].

   In the bundle adjustment problem, a camera will observe many points at one location, and then optimize the feature points and locations together. In the optimization, the characteristics of these feature points are the same, and the constraints on the camera pose are also the same. This inspired us to divide them into two separate optimizations. When we are processing bundle adjustment, the observation set Z is divided into $Z_1 \in Z$ and $Z_2 \in Z$, that is z_{ij} is randomly divided into two parts. At the same time, we record the external and internal parameters of each camera as T as a whole and define T_1 and T_2 as the external parameters of the N cameras in the observation sets Z_1 and Z_2, respectively. The relationship between these blocks is shown in Fig. 1.

   Then bundle adjustment problem in the (1) can be modified as follows:

$$\min \sum_{i=1}^{2} f(T_i, P_i) \quad \text{subject to} \quad A_i T_i = T, i = 1, 2 \tag{11}$$

The top left corner of A_i is an identity matrix, and all other elements are 0. The size of the identity matrix is related to the pose that needs to be optimized.

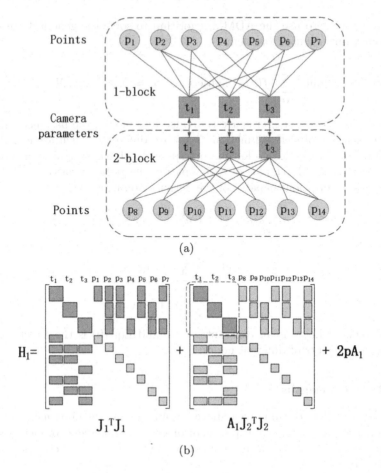

Fig. 1. Camera parameters are represented by squares, and circles represent world coordinate points. The blue line indicates that a world coordinate point is observed by a camera. Red double arrows indicate that these two poses are local poses and are constrained by the global pose. Another figure shows calculation of the first iteration of the Hessian matrix. $A_1 J_2^T J_2$ means to take out the position-constrained part in 2-block. (Color figure online)

According to the augmented Lagrangian multiplier method, the cost function obtained is as follows:

$$L_p = \sum_{i=1}^{2} (f(T_i, P_i) + y_i^T (T_i - T) + \frac{p}{2}\|T_i - T\|^2) \tag{12}$$

$$(T_1, P_1)^{k+1} = \underset{T_1, P_1}{argmin}\, L_p(T_1, P_1, T_2^k, P_2^k, y_1^k, y_2^k) \tag{13}$$

$$(T_2, P_2)^{k+1} = \underset{T_2, P_2}{argmin}\, L_p(T_1^{k+1}, P_1^{k+1}, T_2, P_2, y_1^k, y_2^k) \tag{14}$$

$$T = \frac{1}{2}(T_1 + T_2) \tag{15}$$

$$y_1^{k+1} = y_1^k + p(A_1 * T_1 - T) \tag{16}$$

$$y_2^{k+1} = y_2^k + p(A_2 * T_2 - T) \tag{17}$$

Taking the first block as an example, it can be finally derived that the first-order Jacobian matrix in the derivation of poses and landmark points is as follows.

$$\frac{\partial L_p}{\partial T_1} = \frac{\partial f(T_1, P_1)}{\partial T_1} + y_1 + p(T_1 - T)$$
$$+ A_1(\frac{\partial f(T_2, P_2)}{\partial T_2} + y_2 + p(T_2 - T)), \tag{18}$$
$$\frac{\partial L_p}{\partial P_1} = \frac{\partial f(T_1, P_1)}{\partial P_1}$$

From above analysis, we can also see that the constraints between the two small blocks are only poses and no landmark points. The calculation method for the first block's Hessian matrix is shown in Fig. 1. So we calculate the Jacobian matrix independently and constrain it before iteration.

4 Convergence and Stopping Criterion

4.1 Convergence

The reprojection function f(x) in Eq. (7) should be a convex function in ADMM algorithm. But in fact, the objective function of bundle adjustment in Eq. (1) is non-convex. Some work, such as [12,16], the proximal splitting method applied to non-convex problems has been analyzed. Work [9] proposed a convergence state and consistency of coordinate points, and [23] suggested that the camera consistency problem was convergent. Referring to the analysis above, the convergence of the improved bundle adjustment algorithm in this paper will be analyzed.

The proof of the ADMM method's convergence on the convex function is shown in [5]. In Eq. (12), the convexity of f(x) determines the convexity of L_p. If f(x) is a convex function, so is L_p. But the reprojection function is not convex. With the theory in [16], f(x) should be local Lipschitz-continuous and have Lipschitz constant p_{min}, which ensures that when $p > p_{min}$, L_p is a convex function. The proximity operator should handle on all landmarks in Eq. (12), because Lipschitz-continuous is defined on all variables. Based on the requirement of Lipschitz-continuous, we discuss the objective function in bundle adjustment.

Because the internal parameter matrix does not change with changes in camera motion, we can ignore the internal parameters when analyzing convergence. Therefore, for the reprojection function of Eq. (1), we consider $R_i^{(1:2)}$ as the first two rows of R_i, and $R_i^{(3)}$ is the third row of R_i (similarly for $t_i^{(1:2)}$ and $t_i^{(3)}$).

Then the projection function of the j-th point on the i-th camera can be written as follow:

$$f(P_j|R_i, t_i) := \frac{R_i^{(1:2)} P_j + t_i^{(1:2)}}{R_i^{(3)} P_j + t_i^{(3)}} \tag{19}$$

t and R are the translation vector and rotation matrix in the external parameters, respectively. For this formula, the denominator is the only part of the f(x) gradient that may disrupt Lipschitz-continuous. To avoid infinitely small denominators, we must assume that the denominator $R_i^{(3)} P_j + t_i^{(3)} > d_{min} > 0$. This is that the depth of any point observed by any camera in bundle adjustment must be greater than $d_m in$, and this is a reasonable assumption in the SLAM problem.

Then we consider the Lipschitz-continuous of the rotation parameter of the objective function in Eq. (1). We can utilize quaternions and Euler angles to represent the rotation, and the rotation matrix R is denoted as r_R. The work [25] proposed that to warrant the Lipschitz-continuous of the cost function, the Lipschitz-continuous of $\partial R/\partial r_R$ must be guaranteed. The mapping of a quaternion to a rotation matrix is designed to normalize the quaternion, so when the quaternion is close to zero, the gradient is not limited. Although the gradient $R^3- > SO(3)$ of the exponential mapping of the angular axis to the rotation matrix also has a part of $\|\theta v\|^{-a}, a > 0$, it can be proved that when $\|\theta v\|- > 0$, the Jacobian tensor of the exponential mapping is a constant. Therefore, the gradient of the exponential map Lipschitz-continuous can be proved. Based on the above analysis, we can see that the augmented Lagrangian function L_p is convergent.

4.2   Stopping Criterion

For the problem in Eq. (12), when the ADMM method finally reaches the optimal solution, the following two conditions will be satisfied:

$$A_1 T_1 - T = 0, A_2 T_2 - T = 0 \tag{20}$$

$$0 \in \partial f(T_1^*) + A_1^T y_1^*, 0 \in \partial f(T_2^*) + A_2^T y_2^* \tag{21}$$

We note the dual and original residuals as $r^k = A_1 T_1 - T + A_2 T_2 - T$ and $s^k = pA_1^T A_2(T_2^k - T_2^{k-1})$, respectively. According to these two formulas, it can be derived as follows:

$$
\begin{aligned}
0 \in \partial f(T_1^{k+1} &+ A_1^T y_1^k + pA_1^T(A_1 T_1 - T + A_2 T_2 - T) \\
&= \partial f(T_1^{k+1} + A_1^T(y_1^k + pr^k + pA_2(T_2^k - T_2^{k+1})) \\
&= \partial f(T_1^{k+1} + A_1^T y^{k+1} + pA_1^T A_2(T_2^k - T_2^{k+1})
\end{aligned}
\tag{22}
$$

That is

$$pA_1^T A_2(T_2^{k+1} - T_2^k) \in \partial f(T_1^{k+1} + A_1^T y^{k+1}) \tag{23}$$

In this algorithm, when the dual residual term approaches 0, the above conditions can be satisfied. We assume that the dual residuals and the initial residuals are reduced to a certain error range and then stop iterating. Generally set to

$$\|s^k\|_2^2 < \epsilon^{dual}, \|r^k\|_2^2 < \epsilon^{pri} \tag{24}$$

From the above introduction and analysis of the algorithm, we can get the improved bundle adjustment algorithm process. Based on pose consensus, an improved bundle adjustment algorithm is summarized in Algorithm 1.

Algorithm 1 - Improved Bundle Adjustment based on the ADMM

1: **input** Camera parameters Ξ,D and landmarks P
2: **Initialize** $\xi_i = 0, i = 1, ..., N$, get D and P from measurements
3: **repeat**
4:     **update x1** with $Eq.(13)$
5:     **update x2** with $Eq.(14)$
6:     **update T** with $Eq.(15)$
7:     **update** y_1^{k+1} with $Eq.(16)$
8:     **update** y_2^{k+1} with $Eq.(17)$
9: **until** the criterion in $Eq.(24)$ is satisfied

---

5 Experiments

5.1 Semi-dense Visual Odometry

In the direct method, there is no correspondence between feature points because the descriptors are not calculated and feature matching is not performed. We don't know which point p_2 on the second photo corresponds to point p_1 on the first photo. Therefore, the idea of the direct method is to provide an initial value of pose, and to look for position of p_2 from the current estimate. If the pose estimates given are poor, the appearance (brightness) of p_2 will be significantly different from p_1. To reduce this difference, we constantly adjust the camera pose to find p_2, which is more similar to p_1. The object of minimization is the photometric error, and the premise of the direct method is the photometric constant assumption.

The semi-dense direct method is to calculate pixels with gradients in the image and then track these pixels. The advantage of the direct method is that it uses more observation information, which can be used in situations where features are missing, and to build a semi-dense map. At the same time, the shortcomings are also very clear, and the movement must be small to ensure convexity, and the photometric constant is a strong assumption.

In this paper, a semi-dense direct method of visual odometry has been implemented, which has three main characteristics. First, pick some points randomly

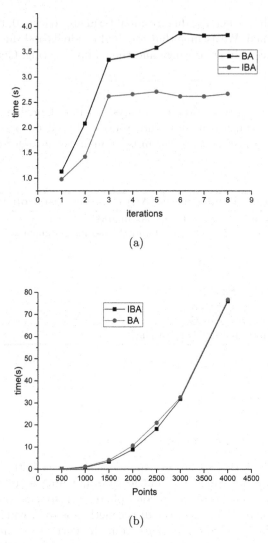

Fig. 2. We use the rgbd_dataset_freiburg1_xyz for testing to find the average time required for each frame and the iteration time changes with the number of feature points. IBA and BA represent the results of using improved and unimproved bundle adjustment, respectively.

when taking pixels, which can speed up the speed. This also utilizes the feature that the direct method does not require feature matching. Second, a four-layer image pyramid is used when estimating poses, which can improve accuracy. Third, the improved bundle adjustment algorithm proposed above is used when estimating the pose and landmark points.

TUM [21] is a large data-set, which includes ground truth data and RGB-D images, and we use it for experiments. The convergence of the proposed algorithm is tested first. According to stopping criterion, we set to stop iteration when the error is less than a certain range. Because we do not know whether the algorithm converges, we need to set the maximum number of iterations, so that even if the algorithm does not converge, it will not enter the iteration loop. If the algorithm is convergent, the time required by the algorithm will tend to stabilize with the number of iterations increases. We assume each frame takes 1000 points, and then set a different maximum iterations. Simultaneously, we analyze influence of the number of extracted feature points on algorithm time by changing the number of feature points and compare iteration time. The above experiment result is shown in the Fig. 2.

It can be obtained from the result, for a single-frame picture, when the maximum number of iterations of the bundle adjustment process is greater than 3, the time required is basically unchanged. The reason for this is that when the number of iterations has not arrived the maximum iterations, iteration process has been stopped according to the stopping criterion. This also reflects the convergence of the algorithm, because if the algorithm does not converge, the time will increase as the number of iterations increases. And it can be seen from the figure that as the number of extracted feature points increases, time required by the algorithm also increases dramatically. This is because the complexity of the matrix decomposition algorithm used in the bundle adjustment process is $O(n_3)$, where n is the side length. It can also be seen from the curve that our improved bundle adjustment algorithm is superior to the previous one.

After getting our algorithm is convergent, we set up experiments to compare time and error. We tested four sets of data-sets, the time required for testing each set of data-sets and output estimated trajectory. Then calibrate with the real trajectory to get the RPE (relative pose error) and ATE (absolute trajectory error). At the same time, calculate average time consumed by each frame. We set 1000 points for each frame, and the results are shown in Table 1.

Table 1. Results of different data-sets under different optimization methods

Data-set	Images	Algorithm	Total time	Average time	ATE	RPE
f1_xyz	1353	BA	1876.64	1.39	0.18	0.24
		IBA	1546.37	1.14	0.23	0.39
		BAP	9.20	0.01	0.55	0.82
f1_desk	574	BA	965.61	1.68	0.87	1.04
		IBA	740.21	1.29	0.69	1.20
		BAP	25.94	0.05	2.65	3.76

The above table is the test results of four different data-sets. Among them, BAP means that only the pose is optimized, and the landmark points are not optimized, BA and IBA are the same as the previous experiment. The four sets of experiments only optimize the pose and the speed is fast. This is because most of the time in BA is used in the calculation of matrix inversion. The matrix that only optimizes the pose is small, so the speed is fast, but the error is also large. It can be obviously obtained from the experiments in the first and second groups that the algorithm we proposed is more efficient than the previous, and the errors of the two algorithm are relatively similar. Overall, our algorithm is faster than the previous one when the error is not large, and it is more effective when it is more translation and less rotation.

5.2 BAL Problem

Bundle adjustment in the large [2] (BAL) is a geometric reconstruction problem. BAL provides a data-set that includes observation points, camera parameters, and world coordinates. We need to optimize these parameters and coordinate

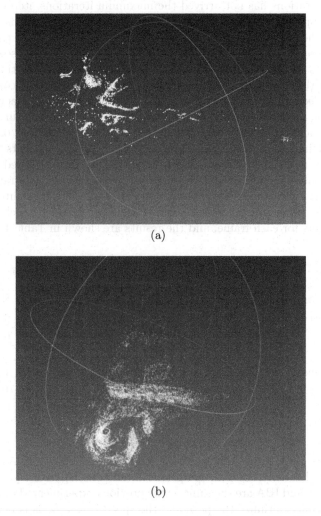

(a)

(b)

Fig. 3. The result of applying the proposed algorithm to the BAL problem.

points according to the bundle adjustment. In our experiments, the proposed algorithm has been applied to this large-scale bundle adjustment, and some of the results obtained are shown in the following Fig. 3.

The above experiment selected a small data-set in Trafalgar Square and Dubrovnik scenes, respectively. From the reconstruction results, the reconstruction outline can be clearly seen. At the same time, the reconstruction process takes less time than the previous algorithm.

6   Conclusions and Future Work

In this paper, an improved bundle adjustment algorithm is proposed. The main contribution of this work consists of the following two parts: (1) The ADMM algorithm is introduced in the bundle adjustment process based on pose consensus, so that bundle adjustment can be processed more efficiently. (2) The convergence of proposed algorithm is analyzed in detail, and the stopping criterion is derived according to the convergence. The experiment results of semi-dense visual odometry and BAL problem show that the proposed algorithm has a faster speed when testing the data-sets. In future, the improved algorithms in global bundle adjustment in SLAM and SFM will be explored, simultaneously we will explore more efficient algorithms for different optimization problems.

Acknowledgments. This work was partially supported by the National Natural Science Foundation of China with Grant Nos. (61525401, 61234002), the Natural Science Foundation of Shanghai with Grant No. (19ZR1420800).

References

1. Agarwal, S., Mierle, K., et al.: Ceres solver. http://ceres-solver.org
2. Agarwal, S., Snavely, N., Seitz, S.M., Szeliski, R.: Bundle adjustment in the large. In: Daniilidis, K., Maragos, P., Paragios, N. (eds.) ECCV 2010. LNCS, vol. 6312, pp. 29–42. Springer, Heidelberg (2010). https://doi.org/10.1007/978-3-642-15552-9_3
3. Bailey, T., Durrant-Whyte, H.: Simultaneous localization and mapping (SLAM): Part ii. IEEE Robot. Autom. Mag. **13**(3), 108–117 (2006)
4. Bertsekas, D.P., Tsitsiklis, J.N.: Parallel and Distributed Computation: Numerical Methods, vol. 23. Prentice hall, Englewood Cliffs (1989)
5. Boyd, S., Parikh, N., Chu, E., Peleato, B., Eckstein, J., et al.: Distributed optimization and statistical learning via the alternating direction method of multipliers. Foundations Trends® Mach. Learn. **3**(1), 1–122 (2011)
6. Cadena, C., et al.: Past, present, and future of simultaneous localization and mapping: Toward the robust-perception age. IEEE Trans. Rob. **32**(6), 1309–1332 (2016)
7. Durrant-Whyte, H., Bailey, T.: Simultaneous localization and mapping: part i. IEEE Robot. Autom. Mag. **13**(2), 99–110 (2006)
8. Engel, J., Koltun, V., Cremers, D.: Direct sparse odometry. IEEE Trans. Pattern Anal. Mach. Intell. **40**(3), 611–625 (2017)

9. Eriksson, A., Bastian, J., Chin, T.J., Isaksson, M.: A consensus-based framework for distributed bundle adjustment. In: Proceedings of the IEEE Conference on Computer Vision and Pattern Recognition, pp. 1754–1762 (2016)
10. Fortin, M., Glowinski, R.: Augmented Lagrangian Methods: Applications to the Numerical Solution of Boundary-Value Problems. Elsevier, Amsterdam (2000)
11. Frahm, J.-M., et al.: Building Rome on a cloudless day. In: Daniilidis, K., Maragos, P., Paragios, N. (eds.) ECCV 2010. LNCS, vol. 6314, pp. 368–381. Springer, Heidelberg (2010). https://doi.org/10.1007/978-3-642-15561-1_27
12. Fukushima, M., Mine, H.: A generalized proximal point algorithm for certain nonconvex minimization problems. Int. J. Syst. Sci. **12**(8), 989–1000 (1981)
13. Gao, X., Zhang, T., Liu, Y., Yan, Q.: 14 Lectures on Visual SLAM: From Theory to Practice. Publishing House of Electronics Industry, Beijing (2017)
14. Grisetti, G., Kümmerle, R., Strasdat, H., Konolige, K.: g2o: a general framework for (hyper) graph optimization. In: Proceedings of the IEEE International Conference on Robotics and Automation (ICRA), Shanghai, China, pp. 9–13 (2011)
15. Kaess, M., Johannsson, H., Roberts, R., Ila, V., Leonard, J.J., Dellaert, F.: iSAM2: incremental smoothing and mapping using the Bayes tree. Int. J. Robot. Res. **31**(2), 216–235 (2012)
16. Kaplan, A., Tichatschke, R.: Proximal point methods and nonconvex optimization. J. Global Optim. **13**(4), 389–406 (1998). https://doi.org/10.1023/A:1008321423879
17. Lee, D., Myung, H.: Solution to the slam problem in low dynamic environments using a pose graph and an RGB-D sensor. Sensors **14**(7), 12467–12496 (2014)
18. Liu, H., Chen, M., Zhang, G., Bao, H., Bao, Y.: Ice-ba: incremental, consistent and efficient bundle adjustment for visual-inertial slam. In: Proceedings of the IEEE Conference on Computer Vision and Pattern Recognition, pp. 1974–1982 (2018)
19. Lourakis, M.I., Argyros, A.A.: SBA: a software package for generic sparse bundle adjustment. ACM Trans. Math. Softw. (TOMS) **36**(1), 1–30 (2009)
20. Salas-Moreno, R.F., Newcombe, R.A., Strasdat, H., Kelly, P.H., Davison, A.J.: Slam++: simultaneous localisation and mapping at the level of objects. In: Proceedings of the IEEE Conference on Computer Vision and Pattern Recognition, pp. 1352–1359 (2013)
21. Sturm, J., Engelhard, N., Endres, F., Burgard, W., Cremers, D.: A benchmark for the evaluation of RGB-D SLAM systems. In: Proceedings of the International Conference on Intelligent Robot Systems, IROS, October 2012
22. Triggs, B., McLauchlan, P.F., Hartley, R.I., Fitzgibbon, A.W.: Bundle adjustment — a modern synthesis. In: Triggs, B., Zisserman, A., Szeliski, R. (eds.) IWVA 1999. LNCS, vol. 1883, pp. 298–372. Springer, Heidelberg (2000). https://doi.org/10.1007/3-540-44480-7_21
23. Zhang, R., Zhu, S., Fang, T., Quan, L.: Distributed very large scale bundle adjustment by global camera consensus. In: Proceedings of the IEEE International Conference on Computer Vision, pp. 29–38 (2017)

Design and Implementation of High Performance Elliptic Curve Coprocessor Based on Dual Finite Field

Guang Han[1], Guanghua Chen[1(✉)], Yupeng Wu[3], Weimin Zeng[2(✉)], and Jingjing Liu[4,5(✉)]

[1] School of Mechatronic Engineering and Automation,
Shanghai University, Shanghai 200072, China
hg1682qd@163.com, chghua@shu.edu.cn
[2] Shandong Hua Yi Micro-Electronics Technology Co., Ltd., Jinan 250101, China
gavinsh2003@163.com
[3] School of Microelectronics, Shandong University, Jinan 250101, China
1473487086@qq.com
[4] State Key Laboratory of ASIC and System, School of Microelectronics, Fudan University,
Shanghai 201203, China
liujingjing@fudan.edu.cn
[5] Key Laboratory of Air-Driven Equipment Technology of Zhejiang Province,
Quzhou University, Quzhou 324000, China

Abstract. In order to realize the scalar multiplication algorithm in GF (p) and GF (2^m), and reducing the module implementation area, a high performance elliptic curve cryptographic coprocessor supporting dual finite field is designed and implemented. By multiplexing the shift register, the operation efficiency of the modular multiplication algorithm is greatly improved. On the other hand, combining multiplication and inverse operations, which reduces the extra modular multiplication operation after inverse operation and reduces the complexity of implementation. After verification by comparing SYSTEMC model with RTL coding, the function of the improved algorithm is correct. The result of hardware simulation shows that the equivalent unit gate is 27092 by 0.13 um SMIC standard cell process library. The modified modular multiplication requires 37.2 us, and the modular inverse requires only 96.1 us under 50 MHz clock constraint, average of 233 bit operation in GF (2^m) and 256bit operation in GF (p). Compared with previous works, it has a large advantage in terms of area, and the speed has reached the expected goal.

Keywords: Elliptic curve crypto coprocessor · Modular multiplication · Modular inverse · Dual-field

1 Introduction

With the frequent occurrence of network intrusions and data breaches, network security has attracted more and more attention. Traditional methods such as firewalls, intrusion

© Springer Nature Singapore Pte Ltd. 2020
M. Fei et al. (Eds.): LSMS 2020/ICSEE 2020 Workshops, CCIS 1303, pp. 323–336, 2020.
https://doi.org/10.1007/978-981-33-6378-6_24

detection, and patching vulnerabilities have been unable to fundamentally solve information security problems. As the core root for the underlying hardware platform of trusted computing, TPM (Trusted Platform Module) [1] can effectively avoid the occurrence of such security events. It adopts ECC (Elliptic Curve Cryptosystem) [2] with high encryption strength, short key length, small bandwidth and low hardware implementation complexity to ensure information security. However, TPM uses the ECC algorithm for data encryption and decryption, which faces the problems of slow operation speed and too large area of the hardware circuit, and the Dual-Field operations are not supported. Therefore, there is an urgent need to design a high performance ECC coprocessor in VLSI (Very Large Scale Integration).

To improve the scalar multiplication means that the ECC coprocessor can get high performance. Many schemes and designs have been proposed in [4–9]. In the literature [4–6], the authors modified the multiplication unit to achieve a faster modular multiplication speed. On the other hand, paper [7, 8] proposed a new architecture based on the extended Euclidean algorithm. Literature [9] successfully reduced power consumption in dual finite fields by fusing modular multiplication and modular inverse algorithms. However, these documents are more or less faced with the following three problems: (1) unable to support dual finite field operations, (2) too large to achieve the area, (3) operation speed encountered a bottleneck.

There are two ways to solve the above problems in terms of area, reducing the use of registers and using serial multipliers. For another problem, it is solved by fusing different finite field algorithms. In the process of implementation, the bottom-up hardware design method is adopted. First, the basic underlying computing module is researched and implemented, and then the higher-level point multiplication design is performed. In this paper, the improved and unified modular multiplication and modular inverse algorithms are implemented in two finite fields: GF (p) and GF (2 m). At the same time, in order to expand the operation of different lengths, the algorithm design uses a variable length method.

The rest of this paper is described as follows. Section 2 overview the theoretical work. Section 3 analyze and improve the modular multiplication algorithm and modular inversion algorithm based on double finite fields. The combination of the multiplexing of the shift register and the reduction process makes the area further reduced. In Sect. 4, the hardware circuit according to the improved algorithm is designed. Statistics and analysis of experimental results in Sect. 5. Finally, this paper summarized in Sect. 6.

2 Theoretical Overview

2.1 Elliptic Curve Cryptography

An elliptic curve is not an ellipse in the usual sense. In cryptography, it refers to a non-super-singular elliptic curve defined on the field F by the simplified WEIERSTRASS equation. The point P based on the equation and a special point O at infinity constitutes the elliptic curve in the domain F [3]. The calculation in the field F can usually be performed in the rational number field, the complex number field and the finite field. But for the purpose of accurately calculating the information in the application of cryptography, it

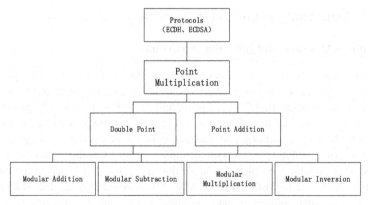

Fig. 1. Elliptic curve cryptographic operation structure

is often used as GF (p) and GF (2^m). The elliptic curve equation on the prime finite field GF (p) and the binary finite field GF (2^m) can be described as follows,

$$y^2 = x^3 + ax + b \tag{1}$$

$$y^2 + xy = x^3 + ax^2 + b \tag{2}$$

where a and b are the parameters of the equation. Because the two curve equations are different, the calculation method of the elliptic curve encryption algorithm is also different, which requires that we need to design separately in hardware implementation.

2.2 Elliptic Curve Cryptography Architecture

The security of Elliptic Curve public key cryptosystem based on the difficulty of the elliptic curve discrete logarithm problem, that is, given a finite field Fq, $E(Fq)$ represents an elliptic curve defined on the finite field Fq, for a point P on a known ellipse, it is easy to find $Q = kP$, whereas it is very difficult to find k when P and Q are known. Therefore, the essence of the elliptic curve public key cryptosystem is the kP operation.

According to the different levels of operation structure, the elliptic curve cryptosystem is implemented from top to bottom in order of protocol layer, scalar multiplication operation, point addition-double point operation, and finite field operation [10]. The scalar multiplication operation is the core of the whole system, which determines the encryption and decryption operation speed of the entire elliptic curve cryptographic design scheme. The bottom layer is the finite field operation, including basic algorithms such as modular addition, modular subtraction, modular multiplication, and modular inversion (Fig. 1). From the perspective of hardware implementation, compared with operations such as modular addition and modular subtraction, the algorithm of modular multiplication and modular inverse operation is the most complex and the most difficult to implement. As the core underlying algorithm, they are the top priority of this paper. Therefore, how to reasonably plan the k of the scalar multiplication operation and optimize the modular inverse and modular multiplication operation, and improve it in terms of area and speed, has become the bottleneck of the cryptographic coprocessor.

3   Algorithm Analysis and Improvement

3.1   Modified Modular Multiplication Algorithm

The modular multiplication operation on the finite field is different in GF (p) and GF (2 m), but ultimately it must be converted into basic addition, subtraction and XOR operations. Usually in the finite field of prime numbers, it is necessary to pay attention to the carry and borrow of the addition and subtraction operations, but compared with the carry of GF (p), the addition and subtraction of GF (2 m) will eventually perform a mod 2 operation, which requires only a simple bitwise XOR during implementation. Therefore, the modular multiplication algorithm has a big difference in operation on the two finite fields and is difficult to be unified. On the other hand, the usual method of calculating modular multiplication is to calculate the multiplication firstly, and then perform the modular operation on the reduced polynomial or modulus. However, this calculation method has an intermediate value of at most 2 k bits, which is not allowed for a cryptographic coprocessor with very strict area requirements. Therefore, how to eliminate the problem of excessive intermediate values becomes the key.

In order to solve the above problems, we will improve the modular multiplication algorithm in the binary domain and the prime domain. By designing the intermediate variable c, the intermediate value multiplexing of the two finite fields is unified. And at the same time, the intermediate value is shifted in the process of calculating addition and subtraction. The improved dual finite field modular multiplication algorithm is as follows (Table 1):

The principle of the improved algorithm is to judge and shift the intermediate value of the multiplication of two numbers in the high-order priority multiplication algorithm [10]. The reduction operation is completed during the shift process. The improved algorithm has the following advantages: (1) In the process of calculating multiplication, the reduced polynomial or modulus p is reduced in each iteration. In this way, we can ensure that the intermediate value is always kept within k bits, and the intermediate value of the multiplication calculation can be used for register multiplexing, which greatly reduces the area of the coprocessor. (2) In each iteration, the highest bit of the intermediate result is judged, to avoid the subtraction operation in each iteration, and to improve the calculation speed.

3.2   Modified Modular Inverse Algorithm

For the speed improvement, area improvement and fusion of the two finite fields of the modular inverse algorithm, it has always been a research hotspot of universities and research institutions at home and abroad. To sum up, there are mainly three efficient algorithms for solving modular inverse operations: the Fermat algorithm based on Fermat's theorem, the Montgomery inversion algorithm and the extended Euclidean algorithm. The operation of the modular inverse algorithm based on Fermat's theorem is relatively complicated, and it cannot effectively support calculations in the prime finite field and the binary finite field. Montgomery inversion algorithm needs to be transformed into a specific domain for calculation, and it is suitable for RSA and other modular exponentiation about large number. Although there are relevant papers [11, 12]

Table 1. Improved dual finite field modular multiplication algorithm

---

Input: The k-bit binary numbers a, b, p and the field judgment *field*, where p is
a prime number or a binary reduced polynomial $f(x)$.
Output: $c = a \cdot b \bmod p$
1.$c = 0$
2.*for* $i = 0$ *to* $k-1$
field $= GF(2^m)$:
    2.1 *if* $(c[k-1] = 0)$
        2.1.1 $c \ll 1$;
        2.1.2 $c = c + a \cdot b[i]$;
    2.2 *else*
        2.2.1 $c \ll 1$;
        2.2.2 $c = c + a \cdot b[i] + p$;
    2.3 *go to step 3*;
field $= GF(p)$:
    2.4 $c \ll 1$;
    2.5 $c = c + a[k-1-i] \cdot b$
    2.6 *if* $(c >= p)$ $c = c - p$;
    2.7 *if* $(c >= p)$ $c = c - p$;
    2.8 *go to step 3*;
3.*return c*;

---

that use the Montgomery algorithm in the inversion algorithm, but considering the multiplexing of registers and domain transformation and other factors, this paper chooses to use the extended Euclidean algorithm to implement the modular inverse operation after comparison and balance.

Among the ECC implementations that support dual finite fields, the extended Euclidean algorithm is the most widely used, because the algorithm can complete the modular inversion in the process of shifting and judgment, and there is no need for pre-calculation and domain conversion operations. And after fusion improvement, this paper unifies the inversion algorithm in GF (p) and GF (2 m), and the operations of shift and XOR are also more suitable for hardware implementation. The improved modular inverse algorithm is to find the Greatest Common Divisor (GCD) of two polynomials to perform the modular inverse operation. Although addition and subtraction have different representations in the prime and binary domains, we can combine the two very well by adding domain judgment variables (Table 2).

3.3 Scalar Multiplication Algorithm

As the most important operation step of the entire ECC cryptographic coprocessor, the choice and implementation of the scalar multiplication algorithm determines the performance of the entire coprocessor. After careful consideration and analysis, in order to balance the area and speed of hardware implementation, scalar multiplication operation

Table 2. Improved dual finite field modular inverse algorithm

Input: k-bit binary numbers y, x, p and the field judgment variable *field*, where p is a prime number or binary reduced polynomial, and $0 <x, y <p$.
Output: $y * x^{-1} mod p$
1.$a = x, b = p, u = y, v = 0$;
2.$while(a \neq 1)$
  2.1 $if(a_0 = 0)$
      2.1.1 $a>>1$;
      2.1.2 $if(u_0 = 0) u >> 1$;
      2.1.3 $else\ u = (u + p) >> 1$;
  2.2 $if(b_0 = 0)$
      2.2.1 $b>>1$;
      2.2.2 $if(v_0 = 0) v >> 1$;
      2.2.3 $else\ v = (v + p) >> 1$;
  2.3 $if(a > b)a = a - b, u = u - v$;
      2.3.1 $else\ b = b - a, v = v - u$;
3.$if(a = 1) return\ v$;

adopts Left-to-Right binary method for point multiplication. In the process of calculation, double point and point addition can be calculated in parallel, which greatly improves the speed of scalar multiplication operation. This method requires l times of double point operations totally, and an average of l/2 times of double point operations, and only needs to store two points of P and Q, which has a large area and speed advantage. The algorithm is as follows: (Table 3)

Table 3. Scalar multiplication algorithm

Input: point P on the elliptic curve, l-bit binary number k
Output: kP
  1. Initialization: $Q = O$, O is the point of infinity.
  2.$for\ i = l - 1\ down\ to\ 0$
    2.1 $Q = 2Q$;
    2.2 $if(k_i \neq 0)$
       2.2.1 $Q = Q + P$;
  3.$return\ Q$;

Based on the above algorithm and analysis, it can be concluded that the goal of the improved algorithm always focuses on the area reduction, and takes into account the increase in speed. Modular multiplication algorithm has been improved through fusion, greatly reducing the width of the intermediate value, while completing the modular operation in the process of multiplication operation, in addition, the multiplexing of the intermediate value register further saves area. During the improvement of the modular inverse algorithm, according to the characteristics of the scalar multiplication algorithm, the multiplication and the modular inverse operation are performed simultaneously, which saves area and greatly improves the speed of implementation. The rest of the article will give the structure, storage unit and simulation results after the above improved algorithm is implemented as a cryptographic coprocessor.

4 Hardware Design and Implementation

4.1 The Structure of Modified Modular Multiplier

As shown in Fig. 2, A, B, P is loaded into the corresponding register before the calculation starts, and the intermediate variable c is input through the *MUX*. In the calculation process, the base $w = 32$ is selected, and $c[max]$ and $b[i]$ are judged after passing through the *2–4 Decoder*. Corresponding to addition and subtraction in GF (p), and simple *XOR* operation in GF (2 m), the final result is saved in register RAM_C.

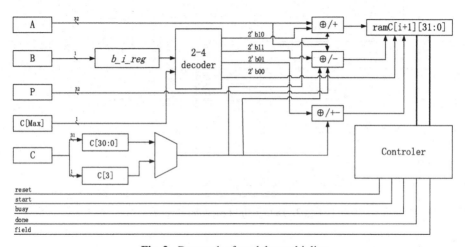

Fig. 2. Data path of modular multiplier

In terms of hardware implementation, the improved algorithm reads 32-bit data every cycle, and outputs the result according to the combination of b and c bits as 00, 01, 10, and 11. The modified modular multiplier requires only one MUX, one 2–4 Decoder, an Adder, an XOR, and a subtraction unit, without pre-calculation. As the core computing module, the design of the modular multiplier determines the speed and area of top-level computing. The complexity of the multiplier structure is low, and the delay of the critical path is only related to the unit XOR gate delay, which reduces the execution time during the hardware implementation.

4.2 The Structure of Modified Modular Inverse

As shown in Fig. 3, first initialize x, p, and y to intermediate variables a, b, and u, and then judge a_0 and b_0 respectively, if it is 0, the right shift operation is performed. Next, determine the magnitude of a or b, if a is greater than b, then $a = a - b$, $u = u - v$, otherwise $b = b - a$, $v = v - u$. Finally, when $a = 1$, the operation is ended, and the result is output through the domain judgment variable.

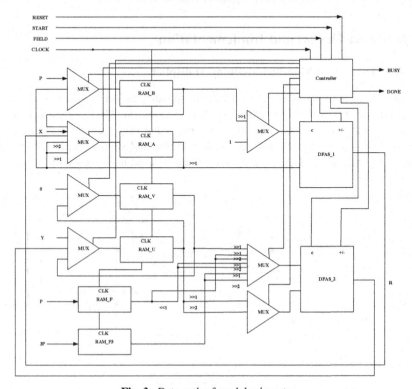

Fig. 3. Data path of modular inverter

In the improved modular inverse algorithm, we take advantage of the feature that the addition and subtraction in the binary domain operation are both XOR operations, which is well combined with the operation in GF (p), so as to realize the modular inverse operation in the dual finite field. On the other hand, in this paper, y is directly brought into the modular inverse operation during the calculation, and the multiplication operation is fused in each iteration of the inverse operation. With this method, the operation result is $y * x^{-1} \bmod p$, which eliminates the multiplication operation after the inversion is completed. This is another highlight of the improved algorithm in this paper.

4.3 ECC Coprocessor Structure

The system architecture of the ECC cryptographic coprocessor is shown in Fig. 4, which mainly includes several parts: API interface unit, ECC core computing unit, control unit, and RAM (Random Access Memory) storage unit. Among them, the API interface unit represents the peripheral interface circuit module, which is mainly responsible for the communication between the coprocessor and the MCU. The ECC core computing unit is mainly divided into two parts. The function of the internal control state machine is to sequentially control and coordinate the work of the basic computing module. Another ALU algorithm implementation unit, includes arithmetic units such as modular multiplication, modular inversion, and XOR. The control unit is mainly responsible for controlling the data storage sequence of RAM and ECC Core. At the same time, the RAM storage unit can be expanded, such a design is beneficial to expand the design. If more functions are added, it is only necessary to reuse the underlying algorithm unit and change the RAM Controller.

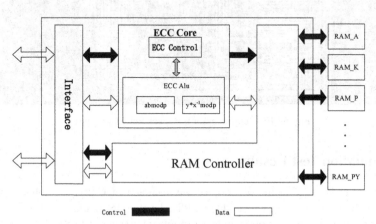

Fig. 4. ECC coprocessor system architecture

ECC Core is the core computing unit of the entire coprocessor, responsible for all data processing and calculation. In this paper, the cryptographic coprocessor architecture has the characteristics of clear hierarchy, clear structure and strong scalability.

The hardware circuit is implemented in Verilog language. In order to transform the designed RTL code into the gate-level netlist and optimize the area and timing of the circuit, the Design Compiler of Synopsys is adopted for synthesis, and SMIC's 0.13 um process library was selected. As shown in Fig. 5, the top-level netlist after being synthesized, the leftmost side is the API interface of the ECC cryptographic coprocessor, the middle is the RAM storage unit, and the right side is the ECC core arithmetic unit. The division of labor in each part is clear, the arrangement is reasonable, and the connection is clear.

In the calculation process, the pre-stored large numbers are first split into 32-bit words and stored in off-chip RAM, while the intermediate values in the calculation process are stored using on-chip 32-bit registers. Under the control of the RAM Controller, the 32-bit data in the corresponding RAM unit is read at each clock cycle for operation in the on-chip register, and stores the result in the corresponding RAM area. Factors such as the area and speed of the circuit need to be considered comprehensively during design, and finally design the required RAM unit. A part of the storage unit is set aside for expansion, and when this part of the storage unit is expanded, the function of the coprocessor can be greatly increased.

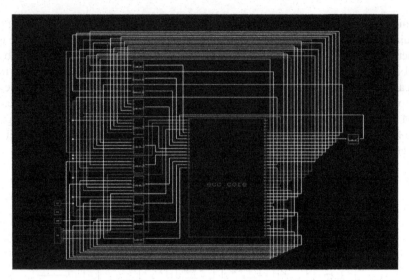

Fig. 5. ECC coprocessor top-level Netlist structure

5 Simulation Test Results

In order to verify whether the function of the designed ECC cryptographic coprocessor meets the initial setting, we conducted RTL-level code simulation. The public key parameters required for encryption and decryption are first generated by the software and saved in a file in advance, and then Testbench will read this parameter into SRAM during simulation. In addition, the private key k for encryption and decryption is randomly generated each time. Using the automatic verification function of the self-designed cryptographic coprocessor automatic simulation system, we have performed up to a million levels of verification, and all results are correct, ensuring the correctness of the designed circuit.

5.1 Implementation Results of ECC Core Module

In order to view the verification results more intuitively, this paper extracts the process of calculating scalar multiplication, as shown in Fig. 5. After the start signal is valid, the coprocessor starts data reading and calculation. After the calculation is completed, the done signal is set to high level, and the calculation result is compared with the software calculation result. In this simulation, kPx and kPy calculated by the software are shown in Table 4. It can be seen that the result of kP calculated by SystemC is consistent with the result obtained by the coprocessor, indicating that the function of the coprocessor conforms to the initial setting and the verification is correct (Fig. 6).

On the other hand, in order to verify the correctness of the calculation in the two finite fields, we performed the modular inverse operation in GF(P) and GF(2 ^{m}). As shown in Fig. 7, before the calculation starts, a, b, u, v, p will be initialized and written to the corresponding SRAM storage unit. The circuit starts working when the START signal set to 1, and ends at the position which DONE signal set to 1. The result of the

Table 4. ECC kP operation data

K = 0x000000da_2daf81ad_f4a941d5_4b04e80e_8225c5dc_0b7951b2_58e9c762_245c219d
kP$_x$ = 0x00000121_5a90274c_676ad5ee_9606b152_d0a15899_1d86b28d_89be3091_dfd9c560
kP$_y$ = 0x000001d4_a5b90033_f22dcc99_cc9222b8_5af8d8af__b49ffa08_dfa4ad04_6916d190

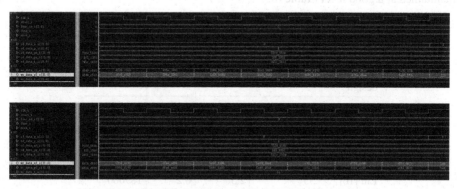

Fig. 6. kP simulation calculate results of coprocessor

modular inverse operation has been saved in the storage unit represented by u. SystemC simulation results are shown in Table 2, which is completely consistent with the results in Fig. 7, indicating that the modular inverse operation we designed is completely correct under the two finite fields (Table 5).

Table 5. ECC Modular inversion operation data

U_GF(p) = 0x0000009e_14c7fadc_1e1b60e5_e82547df_2855ac10_fd8b8f8d_f8913840_4fcf6f50
U_GF(2 m) = 0x000001ce_8e1580b7_7339a6e3_838187e2_a3797ee4_7af8ce57_627d63bf_29921801

5.2 Performance Analysis

In the design process of the cryptographic coprocessor, the main goal of this paper is to reduce the area while taking into account the speed increase. The hardware circuit is implemented in Verilog HDL. The functional simulation of each module uses NC-Sim of Cadence Company, and uses the Design Compiler of Synopsys Company for synthesis. The results after DC synthesis show that, based on SMIC 0.13 um process library, the area has obvious advantages compared with other similar designs, and the implementation speed can also meet the design requirements, details as follows:

By comparing the results of ASIC implementation, Wang's [13] implementation is based on high clock frequency, and the speed of modulus inversion after equivalent conversion is lower than that of this paper, and the speed of modulus inversion of [14] is also higher than that of this paper. Although literature [15] modular multiplication and modular inverse operations are fast, they are weaknesses in terms of area. In comparison, this paper is much better than other literatures in terms of area, and the speed can be maintained at a good level (Table 6).

Table 6. ECC circuit performance comparison

Reference design	Clock/MHz	MUL/us	INV/us	Area/Gate
Wang [13]	167	–	30.72	54628
Literature [14]	–	–	324	43521
Li [15]	487	–	–	73096
Liu [16]	55	2.3	6.2	189000
Our design	**50**	**37.2**	**96.1**	**27092**

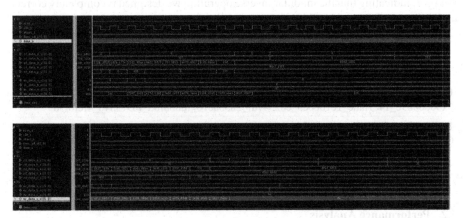

Fig. 7. Modular inversion calculate results of coprocessor

6 Conclusion

Aiming at the pain points of the TPM security module, this paper improves and implements a variable-length dual-field ECC cryptosystem based on the mathematical background of elliptic curve cryptography.

1. The underlying algorithm supports extended 160–512 bit ECC calculation.
2. The application of double point-point addition algorithm effectively reduces the implementation area of kP operation, and can also meet the speed requirements.
3. The improved modular multiplication and modular inverse operations greatly reduce the area of the hardware circuit and increase the running speed through multiplexing registers and dual-domain fusion.
4. The use of off-chip dual-port RAM speeds up the data access speed. By reserving RAM space, the scalability of the system is enhanced.
5. Using the automatic simulation system combining SystemC and RTL for verification and simulation, the circuit can be verified multiple times without the manual labor, which greatly improves the efficiency of chip design.

The experimental results show that the ECC cryptographic coprocessor designed in this paper achieves a large area saving, and the calculation rate has a great advantage compared with similar designs, which meets the design requirements for the area of the cryptographic coprocessor in TPM.

Acknowledgements. This work was supported by the Natural Science Foundation of Shanghai(19ZR1420800).

References

1. TPM Main Specification v1.2 (2008). https://www.trustedcomputinggroup.org/specs/TPM/mainP1DPrev103.zip
2. Koblitz, N.: Elliptic curve cryptosystems. Math. Comput. **48**(177), 203–209 (1987)
3. Christof, P., Jan, P.: Understanding cryptography (2011)
4. Rezai, A., Keshavarzi, P.: High-Throughput modular multiplication and exponentiation algorithms using multibit-scan–multibit-shift technique. IEEE Trans. Very Large Scale Integr. Syst. **23**(9), 1710–1719 (2015)
5. Aris, A., Ors, B., Saldamli, D.: Architectures for fast modular multiplication. In: Euromicro Conference on Digital System Design. IEEE (2011)
6. Lin, W.C., Ye, J.H., Shieh, M.D.: Scalable montgomery modular multiplication architecture with low-latency and low-memory bandwidth requirement. IEEE Comput. Soc. (2014)
7. Naofumi, T., Jun-ichi, Y., Kazuyoshi, T.: A fast algorithm for multiplicative inversion in GF(2 m) using normal basis. IEEE Trans. Comput. **50**(5), 394–398 (2001)
8. Ibrahim, H., Fayez, G., Atef, I.: High speed and low area complexity extended euclidean inversion over binary fields. IEEE Trans. Comput. Electron. **65**(3), 408–417 (2019)
9. Andrew, D., Owen, D.E., Israel, F.L., Gratz, P.V.: Energy-efficient implementations of GF(p) and GF(2 m) elliptic curve cryptography. In: 2015 33rd IEEE International Conference on Computer Design (ICCD). IEEE (2015)

10. Hankerson, D., Menezes, A., Vanstone, S.: Guide to Elliptic Curve Cryptography, pp. 25–147. Springer Verlag New York Inc, New York (2004)

11. Savaš, E., Tenca, A.F., Koç, C.K.: A scalable and unified multiplier architecture for finite fields $GF(p)$ and $GF(2^m)$. In: Koç, C.K., Paar, C. (eds.) CHES 2000. LNCS, vol. 1965, pp. 277–292. Springer, Heidelberg (2000). https://doi.org/10.1007/3-540-44499-8_22

12. Gutub, A.A., Tenca, A.F.: Efficient scalable VLSI architecture for montgomery inversion in GF(p). Integr. VLSI J. **37**(2), 103–120 (2004)

13. Wang, J., Jiang, A., Sheng, S.: A dual field modular inversion algorithm and hardware implementation. Acia Scientiarum Naturalium Universitatis Pekinensis **43**(1), 138–143 (2006)

14. Wolkerstorfer, J.: Dual-field arithmetic unit for $GF(p)$ and $GF(2^m)$. In: Kaliski, B.S., Koç, C.K., Paar, C. (eds.) CHES 2002. LNCS, vol. 2523, pp. 500–514. Springer, Heidelberg (2003). https://doi.org/10.1007/3-540-36400-5_36

15. Li, J., Dai, Z.: Research and design of add-based length-scalable dual-field modular multiplication-addition-subtraction. In: IEEE 2nd International Conference on Integrated Circuits and Microsystems (2017)

16. Liu, Z., Liu, D., Zou, X.: An efficient and flexible hardware implementation of the dual-field elliptic curve cryptographic processor. IEEE Trans. Ind. Electron. (2017)

Research on UAV Signal Recognition Technology Based on GRU Neural Network Model

Peng Ding, Yuancheng Zhu, Zhipeng Lei[(⊠)], Wei Zheng, Jian Wu, and Qian Gao

State Grid Liaoning Electric Power Co., Ltd., Yingkou Power Supply Company,
No. 40, East Bohai Street, Yingkou, Zhanqian District, China
1036227088@qq.com

Abstract. With the rapid development of the drone market, the drone black flight incident has become the norm. In order to be able to take countermeasures against black flying drones in a more targeted manner and accurately identify the models of black flying drones, the existing identification technology cannot identify the types of unmanned aerial vehicles in real time and accurately. This paper proposes the UAV signal recognition technology based on the GRU neural network model. Taking the rotor UAV as the research object, the GRU neural network model is used to extract the characteristics and model design of the UAV communication signal according to the characteristics of the UAV communication signal., Model practice and training, and analyze the test results of the UAV recognition technology designed in this article. Experimental results show that the technology can identify common UAV models more accurately.

Keywords: UAV identification · GRU neural network · Spectrum characteristics

1 Introduction

Due to the rapid development of the drone industry, drones have entered the public field of vision. A variety of remote-controlled aircraft, helicopters, and multi-rotor aircraft have been discovered and used more and more. The elements use and participate in combat operations. In order to respond to the threat posed by the increasing use of drones (especially small drones) by the enemy, the development and equipment of effective anti-drones (C-UAV) systems have become the current focus of attention of governments and the military of various countries. In recent years, the domestic market for small civilian drones has grown rapidly, reaching an annual growth rate of more than 50%. It is expected that the total domestic drone market will reach 75 billion by 2025 [1]. With the rapid development of the drone market, the drone black flight incident has become the norm. How to deal with various safety problems caused by drone black flight and the potential danger of drone black flight have also caused all aspects of the whole society of high concern [2].

This word is supported by Science and Technology Project of State Grid Corporation: Research on UAV Defence System and Key Technology Based on Cracking of Flight Control Protocol (2020YF-46).

© Springer Nature Singapore Pte Ltd. 2020
M. Fei et al. (Eds.): LSMS 2020/ICSEE 2020 Workshops, CCIS 1303, pp. 337–348, 2020.
https://doi.org/10.1007/978-981-33-6378-6_25

With the proposal of the strong smart grid plan, the continuous improvement of the substation's intelligence and the continuous application of unattended substations, strengthening the substation's safety technology and prevention level, and improving the substation's intelligence have become the key development goals of the strong smart grid [3]. Patrol inspection has also become the norm of the power system. Figure 1 is the scene photos taken by the drone during the inspection of a substation in Shenyang. At present, there are frequent incidents of unmanned aerial vehicles affecting safety, and some criminals intend to use drones to transport all kinds of communication equipment, monitor candid camera equipment, interference equipment, ignition devices (such as gasoline bottles), explosives, blades and saw blades. Endangering the safety of substations or power lines [4]. Therefore, it is necessary to accurately identify the enemy aircraft type on the basis of distinguishing between enemy and enemy drones, and use different countermeasures based on the recognition results, so as to achieve the purpose of accurate countermeasures. At present, the most commonly used radar detection type and spectrum detection type have the problem that they can't accurately identify the model of the UAV. Based on the analysis of the UAV identification technology at home and abroad, this article uses the rotor UAV as the identification object Using the GRU neural network model, the training process and training results of the network are given. When the drone enters the monitoring area, it can obtain the drone signal and identify the drone model.

Fig. 1. The scene photos taken by the drone during the inspection of a substation in Shenyang

Literature [5] summarizes the methods of radar detection, laser detection, photoelectric detection, acoustic detection, metal detection, and radio detection, compares the performance differences between the principles, and proposes the development directions of mobile detection, fusion detection, and tracking detection. Literature [6] uses radio detection, optical detection and radar detection to identify the UAV in the surveillance area. These methods are greatly affected by the external environment. When the external environment is relatively harsh or the UAV performs special camouflage can't be accurately identified. Literature [7] proposed a UAV monitoring algorithm based on radio signal feature recognition, which can not only improve the accuracy of radio signal detection, but also extract more monitored UAV information. This paper uses the GRU neural network model to the unmanned aerial vehicle model can be accurately identified, so that the corresponding countermeasures can be accurately taken according to the characteristics of the model. Literature [8] proposed a pigeon-inspired optimization (PIO) as an optimized search strategy, and proposed a hybrid model of SVR and pigeon-inspired optimization (SVRPIO) to achieve the target recognition of the UAV with target rotation and proportional change.

2   Paper Preparation UAV Model Identification Based on GRU Neural Network Model

In this paper, four common and representative UAVs are selected, and the software platform is used to collect and process UAV signals. At present, mainstream drones on the market mainly use two frequency bands of 2.4 GHz and 5.8 GHz. The drones selected in this article are Mavie 2, Spark, Mavie Air 2, Phantom 4, Phantom 4 pro, Mavie Air. Figure 2 shows the drones selected in this article. The signal collected in this article is the time domain information of the UAV communication signal. According to the sampling theorem, when performing signal processing, it is necessary to ensure that all the information carried in the signal is collected, which will lead to a large amount of sampled data. It is suitable to directly use the GRU neural network model for processing, so it is necessary to perform feature extraction on the collected time-domain data, so that it is convenient for neural network processing.

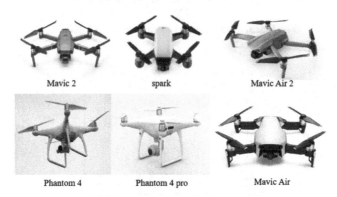

Mavic 2   spark   Mavic Air 2

Phantom 4   Phantom 4 pro   Mavic Air

Fig. 2.   Selected drones in this article

2.1   Feature Extraction

Since the collected time-domain signal data is not suitable for direct processing by neural network model, firstly, the feature extraction of the collected UAV time-domain signal is performed. Therefore, the training data input in the UAV model recognition using the GRU neural network model is the feature vector of the UAV communication signal after feature extraction. In fact, the feature is a statistical result, which is only related to the drone model, and the amount of data is not particularly large. In communication signals, the commonly used features are general features and high-order accumulation features. In this paper, the frequency spectrum of the UAV communication signal is used as the input signal of the neural network model. Therefore, this paper uses the method of general feature extraction to convert the time domain data to frequency Domain data, that is DFT:

$$X(k) = \sum_{n=0}^{N-1} x(n)\, e^{-j2\pi\pi/\cdot kn} \quad (k = 0,1,2\dots, N-1) \tag{1}$$

x(n) represents the sampled analog signal, and X(k) represents the data after DFT conversion.

Figure 3, 4, 5 and 6 are the spectrum diagrams of the communication signals of Phantom 4 pro 2.4G, Spark 5.8G, Phantom 4 2.4G, Phantom 4 pro 5.8G.

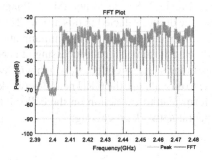

Fig. 3. Spectrogram of Phantom 4 pro 2.4G

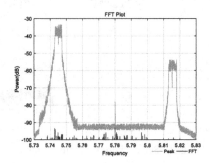

Fig. 4. Spectrogram of Spark 5.8G

After determining the method of feature extraction, the software platform is used to target many different types of UAV equipment, each model of equipment selects multiple UAVs, and 50 sets of training data and 10 sets of tests are collected for each UAV. Data, and label calibration.

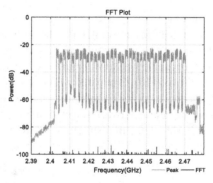

Fig. 5. Spectrogram of Phantom 4 2.4G

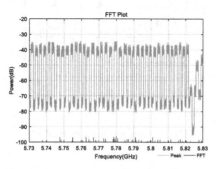

Fig. 6. Spectrogram of Phantom 4 pro 5.8G

2.2 Model Design

Network selection: In the modeling process of traditional neural networks, general neural networks, such as convolutional neural networks, usually accept a fixed-length vector as an input. When a variable-length vector is input, the original input can be converted into a fixed-length vector representation can capture some local features of the input, but the long-distance dependency between the input vector and the front is still difficult to learn. This makes this neural network seem to be a bit weak for sequence data with high correlation between before and after. For example, although CNN can extract local features, but lacking the ability to learn sequence-related information, it is easy to cause the lack of input information before and after, and destroy the sequence structure [9].

Recurrent neural networks can handle input sequences that become long and orderly. Looping means that this type of neural network will record the current output and input it into the model together with the subsequent data to calculate. It simulates the sequence of people reading an article, reads every word in the article from front to back, and encodes the useful information read before into the state variable, so that it has a certain memory ability and can better understand the text. The application in this paper is to identify the characteristics of each curve in the spectrogram from front to back, and encode the previous spectrum feature information into the state variable, so as to have a certain memory capacity and better extract the features of the entire spectrum. Commonly used recurrent neural networks include RNN, LSTM and GRU. RNN is suitable for serializing information but cannot extract features in parallel. For the problem of a single model, by reasonably mixing depth models, the advantages of the model can be synthesized, the bottleneck of the single model can be weakened, and the effect of spectrum recognition and classification can be improved [10]. Compared with RNN and LSTM, GRU is easier to learn the long-term dependence between sequences, and the structure is simpler and easier to calculate, because the neural network selected in this paper is GRU [11].

Network structure: Fig. 7 shown the unit structure of GRU:

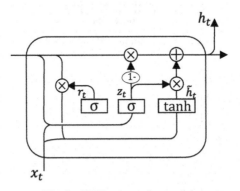

Fig. 7. GRU propagation model.

The repeating network module of GRU realizes the calculation of two gates, namely update gate and reset gate, corresponding to z_t and r_t in the figure below. The design of these gates is to remove or enhance the input information into the nerve cell unit. The update gate is used to control the degree to which the state information at the previous moment is substituted into the current state. A larger value of the update gate indicates that more state information is substituted at the previous moment. The reset gate is used to control the degree of ignoring the state information at the previous moment. The smaller the reset gate is, the more it is ignored. The update gate z and the reset gate r jointly control the information source of the current hidden layer state. If the update gate z is 0 and the reset gate r is 1, then the GRU unit degenerates into the same structure as the simplest RNN loop unit. The connections between GRU neurons are expanded as shown in the time dimension Fig. 8:

In the GRU network, the activation h_t of the hidden layer node at time t is a linear interpolation between the activation h_{t-1} at the previous moment and the candidate activation $\tilde{h}_t$:

$$h_t = (1 - z_t) \cdot h_{t-1} + z_t * \tilde{h}_t \tag{2}$$

During the calculation, two steps of forgetting and remembering were carried out at the same time. Here, the update gating z is used. The closer the gating signal of z is to 1, the more data is memorized, and the closer to 0 is, the more forgotten. The calculation formula of z_t is as follows:

$$z_t = \sigma\left(W_z \cdot [h_{t-1}, x_t]\right) \tag{3}$$

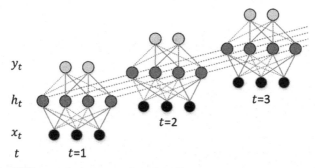

y_t

h_t $t=3$

x_t $t=2$

t $t=1$

Fig. 8. GRU expands the connection diagram of neurons in the time dimension.

Where σ is the sigmoid function.

A linear summation process is used between the existing state and the newly calculated state, so $\tilde{h}$ is calculated as:

$$\tilde{h} = \tanh \cdot \left(W_{\tilde{h}} \times \left[r_t \times h_{t-1}, x_t \right] \right) \tag{4}$$

In the formula, tanh represents the tanh activation function, r_t represents a set of reset gates, and the calculation of r_t is:

$$r_t = \sigma \times \left(W_r \times \left[h_{t-1}, x_t \right] \right) \tag{5}$$

According to the activation h_t of the obtained GRU at time t, the output of the model can be further obtained:

$$y_t = \sigma \times \left(W_o \times h_t \right) \tag{6}$$

Among them, W_r, W_z, W_h, W_o are the weight parameters that need to be learned. The weight coefficients are the parameters that the GRU recurrent neural network model needs to learn. X_t is the current input, corresponding to the length of each input sequence in this article Considering the selected time step length, h_t is the output of the hidden layer value at the current moment. W_o is the element in the weight matrix from the hidden layer to the output layer; W_z is the element in the connection matrix from the input layer to the update gate z; W_r is the element in the connection matrix from the input layer to the reset gate r; W is the input layer to be activated Elements in the connection matrix of state h.

Select activation function: Next, choose the activation function. In the neural network, the activation function can add some nonlinear factors to the neural model, which makes the neural network's ability to solve nonlinear problems have been greatly improved. According to the background of this article, select the appropriate activation function [12]. The activation function selected in this article is the Leaky ReLu activation function. Its advantage is that when the input is less than 0, there is no problem of gradient disappearance, the learning speed is faster, and the Dead ReLu Problem can be solved Some neurons may never be activated, causing the corresponding parameters to never be updated. Figure 9 is the Leaky ReLu function diagram.

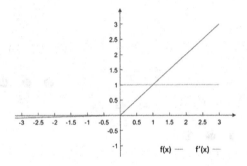

Fig. 9. Leaky ReLu function diagram.

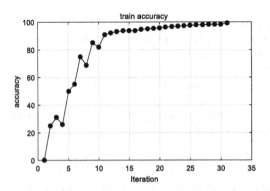

Fig. 10. The neural network training results.

The loss function is used to measure the error between the output of the network and the true value during training. Because the UAV signal recognition and classification aimed at in this paper is actually a multi-classification problem, the loss function is selected as the cross-entropy loss function. In this paper, the training data output samples for different classifications, the sample data is uneven, due to the different market share of different UAV models, and the difference in the number of samples of different UAVs during training. For this type of unbalanced data set, this paper uses the focal loss function Focal Loss. Focal Loss has been proposed for dense object detection tasks. It can be used to train high-precision dense object detectors. It can solve the imbalance of the proportion of class samples in target detection, and can make one-stage detector achieve the accuracy of two-stage detect. And does not affect the original speed. detect accuracy and does not affect the original speed.

The Loss of ordinary cross entropy can be defined as:

$$CE(p_t) = -\log(p_t) \tag{7}$$

p_t represents the probability that the samples belong to the true class. One-stage detector has a certain gap between the number of positive and negative samples during training. The Loss defined by this formula is less effective when the sample ratio is unbalanced, and it is impossible to control easy and difficult classification The weight of the sample.

Therefore, the improved Focal Loss is defined as:

$$FL(p_t) = -(1 - p_t)^\gamma \log(p_t) \tag{8}$$

For the sample ratio imbalance problem, a common practice is to add weights to positive and negative samples. If the frequency of negative samples appears more, the weight of negative samples is reduced. If the number of positive samples is small, the weight of positive samples is relatively increased, which is $(1-p_t)$. The function played by the multiplicative factor. γ here is called focusing parameter, $\gamma \geq 0$, $(1-p_t)^\gamma$ is called modulation factor, the purpose is to reduce the weight of easily classified samples, so that the model is more focused on difficult classification during training Samples. When $\gamma = 0$, Focal Loss is the traditional cross-entropy Loss. When γ increases, the modulation coefficient also increases. The concentration parameter γ smoothly adjusts the proportion of the lower weights of the easy-to-divide samples. The core of Focal Loss is to use a suitable function to measure the contribution of difficult and easy-to-classify samples to the total Loss. Combined with the actual data of this article and the characteristics of the problem, select $\gamma = 0.23$.

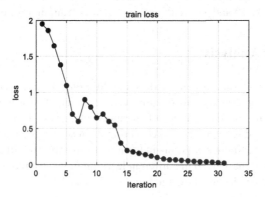

Fig. 11. The neural network training results.

In addition, combined with the characteristics of the actual problems in this paper and the advantages of the existing methods, batch gradient descent (BGD) is selected as the optimization method. If all the data is input into the network, the memory may overflow. Using batch, random input of different data each time can make the network more robust. When using the batch gradient descent algorithm, first divide the training data set into batches, and then optimize the loss function of only one batch at a time during optimization.

The RMSprop algorithm is used to improve the loss function algorithm. Since the steeper gradient is larger, the learning rate will decay faster, which is conducive to the parameters moving in the direction closer to the slope bottom, thereby accelerating convergence, and can solve the learning problem of too fast rate decay prevents the training from stopping prematurely. The specific formula is as follows:

$$s \leftarrow \gamma \times s + (1 - \gamma) \times \nabla_\theta J(\theta) \odot \nabla_\theta J(\theta) \tag{9}$$

$$\theta \leftarrow \theta - \eta / \sqrt{s + \varepsilon} \odot \nabla_\theta J(\theta) \tag{10}$$

θ is the parameter, $J(\theta)$ is the objective function, $\nabla_\theta J(\theta)$ is the gradient of the parameter, s can be considered to only accumulate the gradient closer to time, where γ generally takes the value of 0.9, in fact, this is an index The decayed mean term reduces the occurrence of explosions, thus helping to avoid the problem of a rapid decline in learning rate.

2.3 Model Implementation and Training

First extract feature vectors from the collected data, then build the model, and finally train the model and test. According to the above model audit, model training and testing are performed. The picture shows the test results on the test data set after feature extraction and training through the GRU neural network model. It can be seen that with the increase of the number of iterations, the model converges relatively quickly. When the iteration reaches the 10th round, the recognition rate has reached more than 90%. Although the loss value will fluctuate slightly at the beginning of training, as the number of iterations increases during the training process, the recognition rate of the final model stabilizes above 99.2%. Figure 10 and 11 are the neural network training results.

3   Test Result Analysis

The 6 types of UAVs with different models and different frequency bands participating in the experiment can quickly and steadily learn and recognize when collecting stable time-domain signals. As the number of iterations increases, their recognition ability becomes stronger and stronger. However, during the test, it was found that noise and if there are interference frequency points in the monitoring area, it will affect the learning effect to a certain extent. Due to the different learning times, the recognition results will be biased. The test results are shown in Figs. 11, 12 and 13.

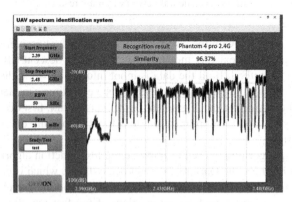

Fig. 12. Phantom 4 pro 2.4G recognition results

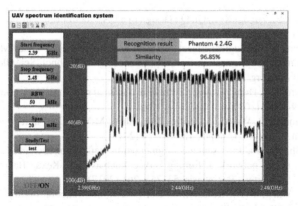

Fig. 13. Phantom 4 2.4G recognition results

4 Conclusion

In this paper, the rotor drone is taken as the research object, and the UAV signal recognition technology based on the GRU neural network model is proposed, and the technology is tested. Multiple test results show that the UAV based on the GRU neural network model proposed in this paper the signal recognition technology can realize the accurate identification of the model of the drone, and can quickly and accurately identify the model of the illegal drone, which lays the foundation for the subsequent countermeasures against the illegal drone. However, in actual application scenarios, there will be many kinds of interference. If the interference frequency in the monitoring area coincides with the communication frequency of the drone, then before performing feature extraction, it is necessary to cooperate with a certain signal detection method to convert the drone signal stripped from many complicated input signals, and then the UAV identification. Therefore, in the future, under the condition of more noise and interference, the communication signal of the UAV can be processed and further studied, so that the UAV recognition technology proposed in this paper can be better applied in practice.

References

1. Zhang, J., Zhang, K., Wang, J.Y., Wang, H.B.L.V.P.: Current status and development trend of low-altitude anti-drone technology. J. Aeronautical Eng. Progress, **9**, 1–8(2018)
2. Fang, L.: Analysis on the development of anti-micro and small UAV electronic equipment abroad. J. Civil and Civil Technol. Products, **08**, 22–26 (2019)
3. Tang, M.W., Dai, L.H., Lin, Z.H., Wang, F.D., Song, F.G.: The application of drones in power line patrols. J. China Electric Power **46**, 35–38 (2013)
4. Yu, H.B., Tang, W., Du, Z.X., Gong, W.: Design of patrol flight control system based on UAV substation power system. J. Electronic Des. Eng. **27**, 49–52 (2019)
5. Qu, X.T., Zhuang, D.G., Xie, H.B.: Overview of "low, slow, and small" UAV detection methods. J. Command Control and Simul. 1–8 (2020)
6. Wang, T.J., Huang, W.H., Fan, Z.X., Wu, Y.B.: Research on UAV detection technology method. J. China Radio **11**, 57–58 (2017)

7. He, X.Y., Han, B., Zhang, X.Y., Qi, Q.: Design of a UAV monitoring algorithm based on radio signal feature recognition. J. China Radio **11**, 72–74 (2019)

8. Xin, L., Xian, N.: Biological object recognition approach using space variant resolution and pigeon-inspired optimization for UAV. Science China Technol. Sci. **60**(10), 1577–1584 (2017). https://doi.org/10.1007/s11431-016-0641-4

9. Zhou, F.V., Jin, L.P., Dong, J.: A review of convolutional neural network research. J. Comput. Sci. **40**, 1229–1251 (2017)

10. Guo, B., Zhou, J.X., Wu, C.L., Zhang, Z.H., Zhou, M.M.: Gated unit for recurrent neural networks. Int. J. Autom. Comput. **13**, 226–34 (2016)

11. Wang, B., Shi, X., Su, J.: A sentence segmentation method for ancient chinese texts based on recurrent neural network. J. Beijing Daxue Xuebao Ziran Kexue Ban/acta Scientiarum Naturalium Universitatis Pekinensis, **53**, 255–261 (2017)

12. Jiang, A.B., Wang, W.W.: Research on ReLU activation function optimization. J. Sensors Microsyst. **37**, 50–52 (2018)

Comprehensive Analysis for the Effect of Thermal Barrier Coating Porosity on Ultrasonic Longitudinal Wave Velocity

Shuxiao Zhang[1,3,4,5], Zhiyuan Ma[2,5], Li Lin[2,5], and Wei Feng[1,3,4,5(✉)]

[1] Shenzhen Key Laboratory of Smart Sensing and Intelligent Systems, Shenzhen Institutes of Advanced Technology, Chinese Academy of Sciences, Shenzhen 518055, China
{sx.zhang,wei.feng}@siat.ac.cn
[2] NDT & E Laboratory, Dalian University of Technology, Dalian 116000, China
[3] Guangdong Ultrasonic Nondestructive Testing Engineering Technology Research Center, Shenzhen Institutes of Advanced Technology, Chinese Academy of Sciences, Shenzhen 518055, China
[4] Shenzhen College of Advanced Technology, University of Chinese Academy of Sciences, Shenzhen, China
[5] Guangdong Provincial Key Lab of Robotics and Intelligent System, Shenzhen Institutes of Advanced Technology, Chinese Academy of Sciences, Shenzhen 518055, China

Abstract. According to SEM photos of ZrO_2 coatings prepared by plasma spraying, 24 physical models with different porosities were constructed, and ultrasonic testing was simulated by finite-difference time-domain method, and the change of longitudinal wave velocity under porosities was analyzed according to the ultrasonic reflection coefficient amplitude spectroscopy. The simulation shows that the longitudinal wave velocity and porosity of the coating show a linear correlation given that the pore morphology and distribution state are unchanged. The simulation is verified by experiments, and it is found that the experiments and the simulation are in good agreement. This research work has important value for the determination of longitudinal wave velocity of thermal barrier coating in ultrasonic testing.

Keywords: Thermal barrier coatings (TBCs) · Porosity · Change of longitudinal wave velocity

1 Introduction

Thermal barrier coatings (TBCs) are widely used in the aerospace industry and play an important role in protecting the engines [1]. Typically, 250 μm thick ceramic coatings deposited on turbine blades can reduce the metal average temperature by up to 80 °C and the hot-spot temperature by 170 °C or more [2]. The thermal conductivity, insulation, and durability performance of TBCs are highly dependent upon its porosity. Porosity is an integral part of thermal barrier coatings (TBCs) and is required to provide thermal insulation and to accommodate operational thermal stresses [3]. Therefore, it is

© Springer Nature Singapore Pte Ltd. 2020
M. Fei et al. (Eds.): LSMS 2020/ICSEE 2020 Workshops, CCIS 1303, pp. 349–357, 2020.
https://doi.org/10.1007/978-981-33-6378-6_26

of great significance to characterize porosity effectively. Kulkarni et al. [3] studied the correlation between thermal conductivity and coating porosity. The results show that the increase of porosity can effectively hinder the heat transfer to the coating. At the same time, high porosity will aggravate the corrosion of coating, which is mainly because pores will lead to the penetration of air and other combustible gases. Portinha et al. [3] showed that the residual stress of TBCs gradually increased with porosity increasing. Therefore, the porosity, pore size and distribution are intentionally introduced in the coating spraying process. Hence, how to effectively characterize the microstructure is a mandatory requirement for optimizing TBCs performance, monitoring service life, and developing new TBCs. Ultrasonic techniques have been previously demonstrated useful for characterization of porosity [6–10].

However, TBCs prepared by plasma spraying or electron beam physical vapor deposition suffer from (1) the pores being randomly distributed, (2) the micro-cracks being random and complex, and (3) the large number of pores leading to obvious dispersion in ultrasonic testing of coatings, fluctuations in sound velocity and attenuation coefficient, distortion of waveforms, etc. Accurate characterization of the TBCs porosity is difficult due to the complex pore morphology and ultra-thin coating thickness [3]. P-wave sound velocity is an important parameter to characterize the microstructure and properties of materials. There is an urgent need for TBCs ultrasonic testing to establish the relationship between porosity and P-wave sound velocity. At the present, research shows that the longitudinal wave velocity of TBCs decreases monotonously with porosity [11], but these studies are still in the qualitative and semi-quantitative stage. Therefore, it is of great theoretical and practical significance to further explore the influence of thermal barrier coating porosity on ultrasonic longitudinal wave velocity.

Built on previous work, this paper continues to promote the work on the influence of porosity on sound velocity in thermal barrier coatings (TBCs). Due to the randomness of the diameters and the locations of the pores in TBCs, it is difficult to prepare samples that meet the experimental requirements. Therefore, this paper intends to build a physical model of TBCs that matches the actual pore morphology of ZrO2 coating by SEM photo in-situ modeling method, and to combine with ultrasonic numerical simulation technology to study the influence of TBCs porosity on longitudinal wave velocity. Finally, the reliability of numerical simulation is verified by ultrasonic testing experiment. This paper establishes the influence of porosity in TBCs on the sound velocity of longitudinal waves and reveals the propagation mechanism of ultrasonic waves in heterogeneous coatings.

2 Principle

2.1 Principle of the Effect of Pores on Longitudinal Wave Velocity

The longitudinal wave velocity of a material depends on the young modulus and density of the material, and its expression is:

$$V \propto \sqrt{\frac{E}{\rho}} \tag{1}$$

In which E is the young modulus of the material and ρ is the density of the material. The existence of porosity in the coating will inevitably reduce the material density, and then affect the sound velocity. In addition, the influence of porosity on young modulus of coatings has been studied by predecessors, and the relationship between porosity and young modulus of coatings can be expressed by the formula (2) [12]:

$$E = E_0 e^{-bp} \tag{2}$$

In which E_0 is the young modulus of dense material, p is the overall porosity of coating, and b is a constant. The relationship between young modulus and porosity of coating can be expressed as [13]:

$$E = E_0\left(1 - 1.9p + 0.9p^2\right) \tag{3}$$

These results show that the existence of pores in the coating will affect the longitudinal wave velocity of the coating.

2.2 SEM in-Situ Modeling Theory

Even if the porosity of the coating is determined, due to the different morphology of pores in different positions, the local density and young modulus of the coating will fluctuate to some extent, which will also cause the fluctuation of ultrasonic sound velocity [13]. Therefore, P-wave sound velocity is a multivariate function of porosity, pore morphology, pore distribution and so on. In order to avoid the introduction of multiple variables, this paper is different from the traditional regular pore model and random medium model [6], and adopts an in-situ SEM photo modeling method that preserves the actual morphology and distribution of pores in ZrO_2 coating. That is, a series of geometric models with the same pore morphology and pore distribution but different porosities are constructed based on SEM photos and image processing software. After the geometric model of materials is established, the model can be used to study the ultrasonic properties of different materials by assigning different elastic parameters to each region.

ZrO_2 belongs to cubic system and has three independent elastic constants: C_{11}, C_{12} and C_{44}. By the law of elasticity, stress and strain have the following relationship:

$$\begin{pmatrix} \sigma_{11} \\ \sigma_{22} \\ \sigma_{33} \\ \sigma_{12} \\ \sigma_{23} \\ \sigma_{31} \end{pmatrix} = \begin{pmatrix} C_{11} & C_{12} & C_{12} & 0 & 0 & 0 \\ C_{12} & C_{11} & C_{12} & 0 & 0 & 0 \\ C_{12} & C_{12} & C_{11} & 0 & 0 & 0 \\ 0 & 0 & 0 & C_{44} & 0 & 0 \\ 0 & 0 & 0 & 0 & C_{44} & 0 \\ 0 & 0 & 0 & 0 & 0 & C_{44} \end{pmatrix} \begin{pmatrix} \varepsilon_{11} \\ \varepsilon_{22} \\ \varepsilon_{33} \\ 2\varepsilon_{12} \\ 2\varepsilon_{23} \\ 2\varepsilon_{31} \end{pmatrix} \tag{4}$$

In the formula, σ is stress and ε is strain. From the wave equation, the elastic constant of the material and the density of the material are determined, and the sound velocity of the material can be calculated. In this paper, the microstructure characteristics of ZrO_2 coating are studied, and the Hudson model used in geophysical survey is used to modify the elastic parameters of ZrO_2 [14]. The parameters required in the numerical calculation are shown in Table 1.

Table 1. Material parameters in numerical calculation of ultrasonic testing

Medium	C_{11} (MPa)	C_{12} (MPa)	C_{44} (MPa)	density (kg·m^{-3})
ZrO$_2$ coating	142740	70039	36200	6500
20°C air	0.1467	0.1467	0	1.24

3 Simulation of Longitudinal Wave Velocity of ZrO2 Coating

3.1 Construction of Pore Models with Different Porosity in ZrO2 Coating

According to the in-situ modeling principle of SEM photos, a series of 24 physical models with the same pore morphology and pore distribution but different porosities are constructed. The porosity of the physical model varies from 5.7% to 28.5%, covering 7% to 25% of the porosity of common ceramic coatings. Figure 1 is a partial schematic diagram of physical models with porosity of 10.5%, 13.3% and 18.7%.

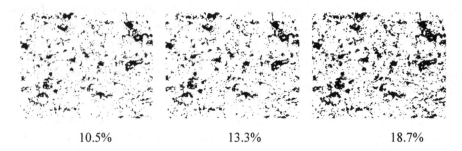

10.5%   13.3%   18.7%

Fig. 1. Partial schematic diagram of physical models with different porosity

The physical models constructed by this method have the following advantages: (1) compared with other methods, they are closer to the actual morphology of pores in ZrO$_2$ coating, which makes the calculation more reliable; (2) pore morphologies of different porosity models are basically equivalent. Following the principle of controlling a single variable, the longitudinal wave velocity only changes with porosity, and the research results are more targeted and rigorous.

3.2 Simulation of Ultrasonic Testing ZrO2 Coating

For the physical model of ZrO$_2$ coating, the ultrasonic numerical simulation is carried out by the FDTD method. The main parameters used in the simulation are that the probe chip size is 1 mm, the center frequency is 5 MHz, and that the water immersion method is coupled. The thickness of the coating is 288 μm, the longitudinal wave velocity of the water layer is 1497 m/s, and the longitudinal wave velocity of the dense ZrO$_2$ coating is 4686.15 m/s.

The pulse echo reflection method is used to simulate the narrow pulse sound source with a frequency of 5 MHz. The waveform of sound source is shown in Fig. 2(a), and the reflected echo waveform of ZrO_2 coating is shown in Fig. 2(b).

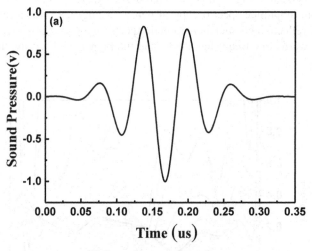

(a) Sound source waveform

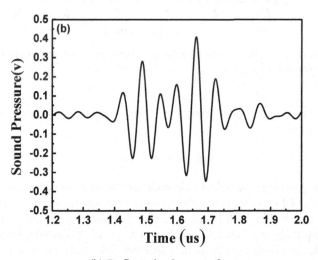

(b) Reflected echo waveform

Fig. 2. Sound source and reflected echo in numerical simulation

3.3 Results

The ultrasonic reflection coefficient amplitude spectroscopy of coating under different porosity is obtained by spectrum analysis of the time domain signal obtained by simulation, as shown in Fig. 3. It is observed that compared with the ultrasonic reflection coefficient amplitude spectroscopy of dense coating, the resonance frequency of the ultrasonic reflection coefficient amplitude spectroscopy of coating with pores deviates moves towards low frequency bands, and the higher the porosity, the greater the deviation.

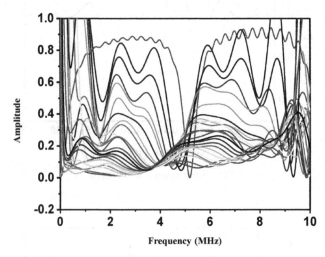

Fig. 3. Numerical simulation of ultrasonic reflection coefficient amplitude spectroscopy of ZrO_2 coating with different porosity

According to the resonance frequency formula of the sound pressure reflection coefficient spectrum:

$$f_n = n * c / 4d \tag{5}$$

The ultrasonic reflection coefficient amplitude spectroscopy of coating under different porosity is obtained by spectrum analysis of the time domain signal obtained by simulation, as shown in Fig. 3. It is observed that compared to the ultrasonic reflection coefficient amplitude spectroscopy of dense coating, the resonance frequency of the ultrasonic reflection coefficient amplitude spectroscopy of coating with pores deviates towards low frequency, and the higher the porosity, the greater the deviation.

3.4 Linear Relationship Between Porosity and Longitudinal Wave Velocity

It is observed that for porous materials, the higher the porosity, the lower the sound speed. According to the resonant frequency, change of longitudinal wave velocity under different porosity is calculated, and the calculated results are fitted. It is found that there is a good linear relationship between the change of longitudinal wave velocity and coating porosity, as shown in Fig. 4, which is similar to the conclusion of the relationship between porosity and sound velocity of ZrO_2 coating studied by B. Jeong et al. [16]. According to the simulation, the empirical formula of the change of longitudinal wave velocity with porosity of ZrO_2 coating prepared by APS method at 5 MHz can be expressed as:

$$c = c_0(1.05 - 2.09\,P) \qquad (6)$$

Where c_0 is the sound velocity of the dense ZrO_2 coating, and P is the porosity.

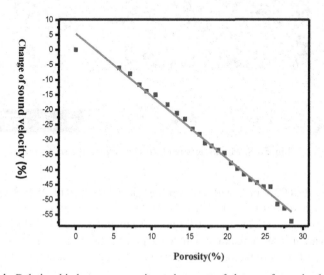

Fig. 4. Relationship between porosity and percent of change of sound velocity

3.5 Experimental Verification

The ultrasonic pulse reflection echo method is used to measure the longitudinal wave velocity of plasma sprayed ZrO_2 coating, and the experimental parameters are consistent with the numerical simulation parameters. The pulse reflection echo signal obtained in the experiment is analyzed by spectrum, and the ZrO_2 coating at the data acquisition point is observed by SEM. Figure 5 is an SEM photograph of ZrO_2 coating prepared by plasma spraying. It is observed that micro-cracks and pores are randomly distributed in the coating, and the microstructure is heterogeneous to some extent.

The result of spectrum analysis shows that there are two minima, whose resonance frequencies are $f_3 = 5.12$ MHz and $f_5 = 8.56$ MHz respectively. The porosity of ZrO_2 coating was measured by image analysis method [17], and the results showed that the average porosity of the coating was 22.9%. The average thickness of ZrO_2 coating is 380 μm automatically calculated by image processing of SEM photos and using mathematical software. The longitudinal wave velocity of the ZrO_2 coating can be calculated by formula (5). The velocities calculated from resonance frequencies f_3 and f_5 are

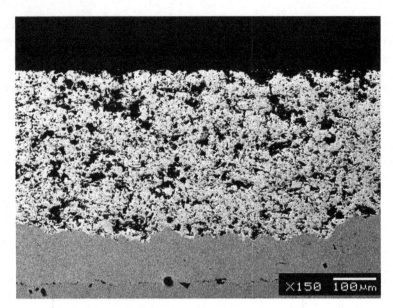

Fig. 5. SEM photograph of ZrO$_2$ coating cross section

2594.1 m/s and 2602.2 m/s, with an average sound velocity of 2598.2 m/s. According to the conclusion of numerical simulation, the longitudinal wave velocity of ZrO$_2$ coating with porosity of 22.9% is 2701.3 m/s, and the relative error is 3.82%. The simulation results are in good agreement with the experimental results, which further verifies the rationality of the simulation model construction and the validity of the simulation conclusions.

4 Conclusions

Given that the physical model construction is reasonable and that the credibility of the numerical simulation are fully verified, the following conclusions are obtained: the ultrasonic longitudinal wave velocity decreases with porosity, and there is a good linear relationship between the change of longitudinal wave velocity and coating porosity, and the empirical formula (6) of the change of sound velocity with porosity is obtained, which provides a basis for the determination of longitudinal wave velocity in actual testing work and better explains the inconsistency of acoustic parameters in different positions of coating in TBCs ultrasonic testing. This study reveals the interaction mechanism between ultrasonic wave and heterogeneous materials of TBCs and further discusses the numerical simulation theory of TBCs.

Acknowledgments. The authors are grateful to the financial support from the Key-Area R&D Program of Guangdong Province (No:2020B090925002), Science and Technology Innovation Commission of Shenzhen (No. ZDSYS20190902093209795), and the Science, Technology and Fundamental Research and Discipline Layout project of Shen Zhen (No. JCYJ20170818153048647).

References

1. Ma, Z., Zhang, W., Luo, Z., et al.: Ultrasonic characterization of thermal barrier coatings porosity through BP neural network optimizing Gaussian process regression algorithm. Ultrasonics **100**, 105981 (2020)
2. Kulkarni, A., Goland, A., Herman, H., et al.: Advanced neutron and X-ray techniques for insights into the microstructure of EB-PVD thermal barrier coatings. Mater. Sci. Eng. **426** (1–2), 43–52 (2006)
3. Portinha, A., Teixeira, V., Carneiro, J., et al.: Residual stresses and elastic modulus of thermal barrier coatings graded in porosity. Surf. Coat. Technol. **188**(1), 120–128 (2004)
4. Parthasarathi, S., Tittmann, B.R., Onesto, E.J.: Ultrasonic technique for measuring porosity of plasma-sprayed alumina coatings. J. Therm. Spray Technol. **6**(4), 486–488 (1997)
5. Rogé, B., Fahr, A., Giguere, J., McRae, K.I.: Nondestructive measurement of porosity in thermal barrier coatings. J. Therm. Spray Technol. **12**(4), 530–535 (2003)
6. Jeong, H., Hsu, D.K.: Quantitative estimation of material properties of porous ceramics by means of composite micromechanics and ultrasonic velocity. NDT and E Int. **29**(2), 95–101 (1996)
7. Zhao, Y., Lin, L., Ma, Z.Y.: Establishing TBC random pore model based on random media theory. China Surf. Eng. **23**(2), 78–81 (2010)
8. Ma, Z.Y., Luo, Z.B., Lin, L.: Characterization of ultrasonic longitudinal velocity evolution in TBCs with different porosity and void morphology based on random void model. J. Mater. Eng. **5**, 86–90 (2014)
9. Rogé, B., Fahr, A., Giguère, J.S.R. et al.: Nondestructive measurement of porosity in thermal barrier coatings. J Therm Spray Tech. **12**, 530–535 (2003)
10. Zhao, Y., Lin, L., Li, J., et al.: Study on the method of measuring the density of thermal barrier coating based on ultrasonic sound pressure reflection coefficient phase spectrum. Rare Metal Mater. Eng. **39**(1), 255–258 (2010)
11. Xiaoguang, Y., Geng, R., Changbing, X.: Study on the application of thermal insulation ceramic coating for hot end components of aeroengine. J. Aerospace Power **12**(2), 183–188 (1997)
12. Zhao, Y., et al.: Correlating ultrasonic velocity and porosity using FDTD method based on random pores model. In: Advanced Material Science and Technology, Pts 1 and 2. Tan, Y., Ju, D.Y., (eds) pp. 1221–1224 (2011)
13. Khor, K.A., Gu, Y.W., et al.: Surf. Coat Techn. **139**, 200 (2001)
14. Jianfeng, L., Xiaming, Z., Chuanxian, D.: Statistical analysis of porosity of plasma sprayed Cr_3C_2-NiCr coating. Chinese J. Aeronautical Mater. **01**, 33–39 (2000)

An Overview of Control Strategy and Trajectory Planning of Visual Servoing

Tan Wang[1,2], Weijun Wang[3], and Feng Wei[4(✉)]

[1] Xi'an University of Architecture and Technology, Xi'an 710055, China
tenney.wang@foxmail.com
[2] Guangzhou Institute of Advanced Technology,
Chinese Academy of Science, Guangzhou 511458, China
[3] Shenzhen CAS Derui Intelligent Technology Co., LTD, Shenzhen 518109, China
[4] Shenzhen Institutes of Advanced Technology,
Chinese Academy of Sciences, Shenzhen 518055, China
wei.feng@siat.ac.cn

Abstract. Visual servoing is a hot topic in the field of Robotics. Compared with other sensors, visual servo has the characteristics of high flexibility, high precision and strong robustness of calibration. In this paper, visual servoing are classified according to different types, and the research progress of visual servoing is reviewed from position-based visual servoing, image-based visual servoing, and hybrid visual servoing. Secondly, the feature extraction in the visual servoing process is introduced, and the current development situation is introduced from several feature extraction algorithms Then, the trajectory planning based on image interpolation, the trajectory planning based on optimization and the trajectory planning based on potential field are analyzed and discussed. Finally, based on the current research progress, the future research direction is prospected.

Keywords: Visual servoing · Control strategy · Feature extraction · Trajectory planning

1 Introduction

With the increasing demand for robots in production and life, the robot industry has developed rapidly, industrial robots have liberated productivity, and service robots have improved people's living standards. In the 1990s, computer hardware technology advanced by leaps and bounds, which greatly improved the processing speed of image information and greatly reduced the cost of visual sensors. Because of its large amount of information, high accuracy and good repeatability, it has been widely used in the field of robots. Therefore, robot visual servoing came into being and became a hot research topic in the field of robotics [1]. The robot's vision servo system is provided by the vision sensor to adjust the robot's posture to reach the target position. This is an interdisciplinary subject which integrates computer, image processing, signal processing, automatic control and other fields [2].

© Springer Nature Singapore Pte Ltd. 2020
M. Fei et al. (Eds.): LSMS 2020/ICSEE 2020 Workshops, CCIS 1303, pp. 358–370, 2020.
https://doi.org/10.1007/978-981-33-6378-6_27

2 Classification of Visual Servoing System

The visual servoing system can be divided into different types according to different classification methods. According to the installation position of the camera, it can be divided into eye-in-hand vision system, eye-on-hand vision system and hybrid vision system (see in Fig. 1) [3–5]. Eye-in-hand visual servoing system can get the precise position of the target and control the manipulator accurately. However, the field of view is so small that the target is easily out of the field of view. In addition, the eye cannot observe the robot arm itself, so the relationship between the target and the manipulator needs to be solved in kinematics. Eye-on-hand visual servoing system can observe the target and the robot arm at the same time, but it cannot obtain the detailed characteristics of the target, and the motion of the robot arm may block the target, causing the servoing task to fail. Hybrid vision system is proposed to solve the problems of the above two problems.

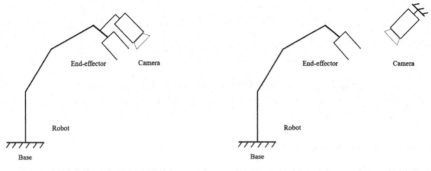

a) Eye-in-hand visual servoing system   b) Eye-on-hand visual servoing system

Fig. 1. Different camera installation locations

According to the number of cameras, it can be divided into monocular vision system, binocular vision system and multi-eye vision system [6]. The monocular vision system uses only one camera, which is simple in structure and can accomplish simple work in a single scene. Although depth information can be obtained through 3D models or motion compensation, the accuracy is low and the algorithm is complex. Compared with monocular vision system, binocular vision system does not need 3D model, but according to the difference of left and right two images, applying matching algorithm to obtain parallax map, and finally obtaining depth information. The multi-eye vision system can observe different parts of the target and obtain more information, but the design of the vision controller is relatively complicated. From the point of view of feedback information form, this paper introduces position-based visual servoing (PBVS), image-based visual servoing (IBVS) and hybrid visual servoing (HVS) [7–9].

2.1 Position-Based Visual Servoing

In PBVS, the initial given information and visual feedback signals are defined in the Cartesian coordinate system, so PBVS is also called 3D visual servoing. According to the image features and the spatial model of the target, the image collected by the camera is reconstructed in 3D to obtain the pose of the target in Cartesian space. Then the offset is calculated according to the error between the robot and the expected pose to control the robot end-effector to move to the expected pose [10]. The control strategy flow chart is shown in Fig. 2. The advantage of this method is that the controller directly controls the movement of the manipulator in the Cartesian space, so that the robot arm has better translation and rotation characteristics during the movement, and can avoid problems such as singularity and local minima [11]. However, due to the poor robustness of PBVS, robot kinematics model and camera calibration errors will directly affect the accuracy of pose estimation. Secondly, because there is no direct control over the image, the target may leave the camera's field of view.

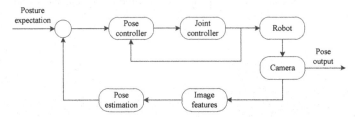

Fig. 2. Control flow of position-based visual servo system

2.2 Image-Based Visual Servoing

Different from the positional visual servo, the image-based visual servo does not require complex transformation of image features. Compare the desired image with the real-time image, and then the vision controller directly calculates the offset through the image feature error, and controls the robot end effector to move to the desired position and posture [12–15]. The control strategy flow chart is shown in Fig. 3. This method does not need to estimate the target pose, and the positioning accuracy is not sensitive to the calibration error of the camera [16]. In addition, since the image-based visual servoing is performed in the image space, it can be guaranteed that the image features are always in the camera's field of view. However, this method requires online calculation of the image Jacobian matrix and its inverse matrix, which requires a large amount of calculation, and there are singularity problems and local minimum problems in the solution process.

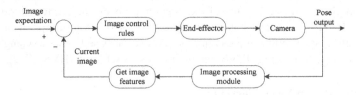

Fig. 3. Image-based control flow of visual servo system

2.3 Hybrid Visual Servoing

Combining the advantages of PBVS and IBVS, Malis et al. proposed a hybrid visual servoing method combining 2D information and 3D information [2]. First, the camera gets the image information of the target, and the homography matrix H is obtained through the calculation of the current image and the target image information. The homography matrix is decomposed into the translation component and rotation component of the system, corresponding to the translation and rotation of the camera. Its error signal e is defined as:

$$e = \left[u - u^*, v - v^*, \log\left(\frac{z}{z^*}\right) \theta u^T \right] \tag{1}$$

Where u and v are the current image coordinates of the feature, u^* and v^* are the ideal image coordinates of the feature, θ and u^T are the rotation angle and rotation axis obtained from the decomposed rotation matrix, z and z^* are Current depth and ideal depth information.

In this way, the translational motion is controlled in the image space, and the rotation control is controlled in the Cartesian space. The control strategy flow chart is shown in Fig. 4. Although this method inherits the advantages of PBVS and IBVS, it improves the stability of the system [17]. However, it is sensitive to image noise, and the homography matrix needs to be calculated and decomposed in real time, which requires a large amount of calculation.

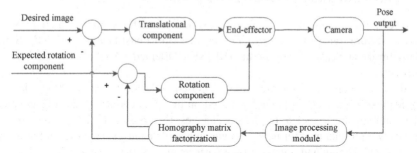

Fig. 4. Control flow of hybrid visual servoing system

3 Feature Extraction

3.1 Image Feature Classification

At present, image processing in visual servoing is mainly based on the selection of image features, and the performance of the visual servoing system depends on the image features used in the control loop. Simple geometric features are often used in the research of visual servoing, including points, lines, circles, rectangles, area areas, etc. and their combined features. The anti-interference ability of feature points is poor, the anti-interference ability of characteristic lines and circles is slightly stronger, and the anti-interference ability of rectangles and area areas is the strongest.

In 2005, Mahony et al. used the Plucker coordinates of parallel lines to form image features and realized image-based visual servoing of a 4-rotor UAV [18]. In 2008, Mahony et al. used centroid and optical flow characteristics to achieve visual servoing based on dynamic images [19].

In 2010, Lopez-Nicolas et al. used a three-viewpoint tensor to form an interaction matrix [20]. In 2011, Fomena et al. designed a visual servoing method based on spherical projection to separate position and attitude [21]. Among them, the Cartesian space distance of the spherical projection of 3 points is used to form 3 rotation invariants, which are used for position control; the images of two points are used to determine the rotation axis and the angle for attitude control. In 2012, Lin et al. defined 4 low-order moment invariants and realized model-based visual tracking [22]. In 2013, Dong et al. used the improved image moment to realize image-based visual servoing of a 6-DOF robot [23].

Among image features, feature points are relatively special points in the image and are used more frequently. Compared with other features, feature points have more obvious changes in texture, brightness, and grayscale. Image feature points include key-points and descriptors. Key-points refer to the location, size, direction and other information of the feature points in the image, and the descriptor is usually a vector, including key-point neighborhoods Pixel information. Establishing the point-to-point correspondence between images generally depends on an excellent image feature point descriptor.

3.2 Feature Extraction Algorithm Development

At present, the commonly used feature extraction methods mainly include Scale-invariant feature transformation (SIFT), Speeded Up Robust Features (SURF), Binary Robust Invariant Scalable Key points (BRISK), Oriented FAST and Rotated BRIEF (ORB), etc.

SIFT is a computer vision algorithm proposed by David Lowe in 1999 to describe local features in images [24]. SIFT is based on the local characteristics of the object, and it maintains invariance to brightness changes, rotations, and scale scaling, and has a relatively high tolerance for noise and viewing angle changes. However, it has a large amount of calculation and poor real-time performance. SURF is an accelerated version of the SIFT algorithm, which was proposed by Hebert Bay at the EVVC conference in 2006 [25]. Surf uses the Hessian matrix to calculate the scale and location information of the key points, and uses the integral graph to speed up the calculation. The speed of SURF algorithm is 3-7 times that of SIFT, and it has been applied in many scenarios. BRISK is a feature extraction algorithm with better performance in terms of rotation invariance, scale invariance and robustness proposed on ICCV in 2011 [26]. The BRISK algorithm performs best when registering images with large blur. ORB is a fast feature point extraction and description algorithm, proposed by Ethan Rublee et al. in 2011 [27]. The ORB feature combines the detection method of FAST feature points with the BRIEF feature descriptor, and adds the direction calculation method based on the BRIEF, which solves the problem of the invariance of the BRIEF in the rotation; In addition, BRISK's complex descriptor is simplified, and the calculation speed is greatly improved. Perform feature point extraction on the same data set, and the accuracy and speed comparison of each algorithm are shown in Fig. 5, 6.

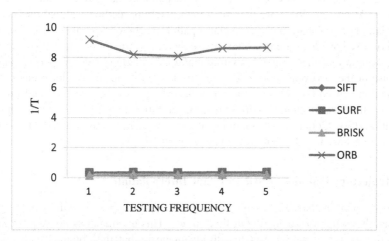

Fig. 5. SIFT, SURF, BRISK, ORB speed comparison

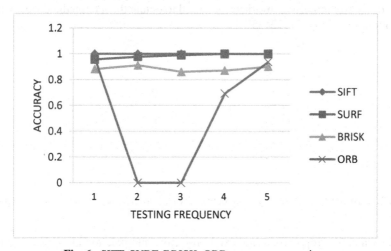

Fig. 6. SIFT, SURF, BRISK, ORB accuracy comparison

Tingtian Ma et al. reduced the original 128-dimensional feature descriptor to 24 dimensions by applying Gaussian weighting to the gradient modulus and gradient direction of each pixel, and introduced the relationship between data and nodes in each dimension to limit the search Range, which improves the calculation speed of SIFT [28]. Aiming at the problem of the weak rotation robustness of the SURF algorithm, Guo proposed an LBP-SURF image matching algorithm [29]. The algorithm integrates the information extracted by the improved rotation invariant LBP feature description method into the SURF feature description vector, and use multiplying distance instead of Euclidean distance for feature matching.

4 Plan Planning

With the development of visual servoing, path planning has attracted more and more attention. At present, there are two main path planning methods: global planning and local planning. Global planning is to plan a path from the start point to the end point for the robot based on the acquired global environmental information. The accuracy depends on the accuracy of the obtained environmental information, and it is poorly robust to the noise of the environmental model. Local planning relies on the local environment information sensed by the robot in real time, which is more real-time and practical. However, the lack of global environmental information may lead to local minima and oscillations [30]. Common path planning algorithms include image-based interpolation, optimization-based, and potential field-based methods [31].

4.1 Trajectory Planning Based on Image Interpolation

Trajectory planning based on image interpolation aims to obtain a path by interpolating between the initial image and the desired image. This method does not require camera calibration or target model. But the disadvantage is that the obtained path may not correspond to the only camera path, so the methods of epipolar geometry, projective homography and projective invariance are used to obtain the intermediate view without calibration (see in Fig. 7) [32–35].

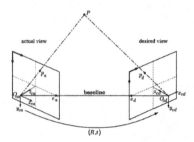

Fig. 7. Basic epipolar geometry setup defined by two different views of the same 3D point

Hosoda et al. proposed an improved projective interpolation algorithm to directly design a visual servo trajectory generator in image space. The algorithm can ensure the visibility of the observed target without the camera calibration parameters and target model [36]. J.A. Borgstadt proposed a method based on image feature extraction, which can insert the desired object path between two arbitrary object poses. This method fails to solve the following problems, such as the singularity of the internal working space and the discontinuity of the joint space [37].

Malis proposed a path planning method based on projection transformation defined in an invariant space [12]. The basic idea of projective invariance is to construct a task function that only depends on the position of the camera and the 3D structure of the observation target. This makes the feature vector path generated in the invariant space independent of the camera parameters, and the path of the corresponding camera in the working space is a straight line (see in Fig. 8). Gian Luca Mariottini et al. used an approximate input and output linearization feedback to zero the pole axis and align the robot with the target. In the case of calibration and partial calibration, the asymptotic convergence was proved [38].

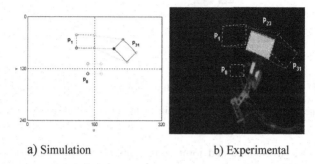

a) Simulation   b) Experimental

Fig. 8. Experimental results of method based on projection transformation in invariant space

4.2 Trajectory Planning Based on Optimization

Trajectory planning based on optimization aims to find a path from image boundary distance, robot path length, energy consumption and so on. However, this method is only suitable for simple scenarios, and the introduction of robotic arms or other physical constraints will increase the complexity of optimization problems. RRT is an algorithm proposed by Steven M. LaValle and James J. Kuffner Jr. to achieve a fast search for non-convex high-dimensional space by randomly constructing Space Filling Tree (see in Fig. 9) [39]. This method avoids the modeling of space, and can effectively solve the path planning problem of high-dimensional space and differential motion constraints.

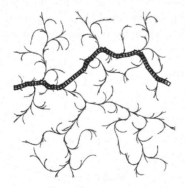

Fig. 9. A 2D projection of a 5D RRT for a kinematics and dynamics car

Graziano Chesi proposed a path planning method with the objective of optimizing cost functions such as spanning image area, trajectory length and curvature, considering

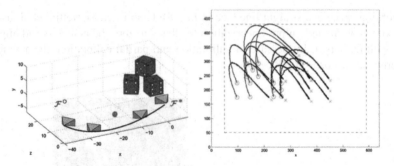

a) CAD model   b) The trajectory from the original center to the desired center

Fig. 10. Experimental results of path planning considering visibility and space constraints

visibility, obstacle avoidance and joint constraints [40]. Chesi also proposed a path planning framework based on homogeneous form and linear matrix inequality (LMI), which maximizes visibility and line similarity in a 3D space that satisfies all these constraints.

For non-holonomically constrained mobile robots, due to incomplete kinematics, it is also necessary to consider the visual field constraints and obstacle avoidance when planning the trajectory. Paolo Salaris proposed a path planning method for nonholonomic kinematics in a limited field of view (FOV) [41].

4.3   Trajectory Planning Based on Potential Field

In the field of robot trajectory planning, the potential field method has been widely used in real-time obstacle avoidance and smooth trajectory control (see in Fig. 11) [42]. The basic idea is to define the gravitational field in Cartesian space to attract the robot to its target attitude. In addition, the repulsive force field is defined in the image space to repel the object away from the edge of the image. Finally, trajectories in the image space are generated iteratively in the resultant field [43]. The potential field method is simple in structure, but its disadvantage lies in its local optimum and easy to lock up.

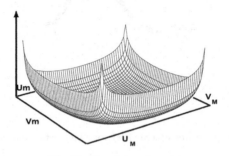

Fig. 11. Repulsive potential

Mezouar and Chaumette extended PFM to generate a visual servo path in the image space, especially, a 2D image path was constructed using the potential field method [44]. On this basis, the IBVS algorithm is used to enable the 6-DOF robot to track the generated path on the image plane, which is robust and insensitive to camera modeling errors. In addition, features can remain in the camera FOV, and other constraints such as joint limitations and obstacle avoidance can be considered at the same time.

Xuebo Zhang proposed a robust path planning switching method, which combines the idea of PFM with heuristics to avoid local minima [45]. On the premise of ensuring the characteristics of the camera's FOV, by introducing a gain function to adjust the weight of the translation and rotation parts of the repulsion, 2D and 3D paths are obtained (see in Fig. 12).

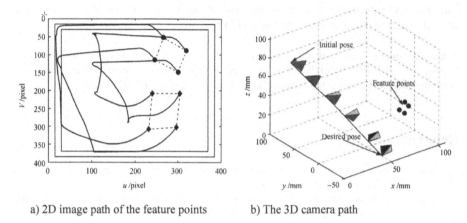

a) 2D image path of the feature points   b) The 3D camera path

Fig. 12. 2D and 3D paths

5 Research Prospects

With the advancement of technology, robots will be more widely used in various fields of social production and life, replacing humans for more complex and diverse labor tasks. Robot visual servo technology will become the main means of robot control. The future of visual servoing may be studied in the following directions:

First, ensure that the target is always within the camera's field of view. Both PBVS and IBVS have the problem that the target deviates from the camera's field of view. Once the target is lost, the servo directly fails. This kind of problem seriously hinders the development of visual servoing. In order to solve this problem, you can consider a mixed camera layout and use multiple cameras to collect image information. In addition, we can consider adding image space constraints to the trajectory planning of the robotic arm to ensure that the target is always in view.

Second, the real-time and robust research of visual servoing. The main reason for the poor real-time performance in visual servoing is the delay caused by image acquisition, image processing, feature extraction, etc. In the current control strategy, posture estimation, image Jacobian matrix, depth information estimation and other calculations are large, resulting in insufficient calculation speed. On the one hand, the prediction algorithm is used to estimate the motion of the target and reduce the delay; on the other hand, it is necessary to improve the relevant algorithms of visual servoing and simplify the algorithm process to increase the speed of visual servoing.

Third, research on calibration technology. The accuracy of most traditional visual servoing systems depends on the accuracy of calibration. However, under actual complicated conditions, it is difficult to complete accurate calibration of the system, and the high sensitivity of the calibration to the external environment also greatly increases the difficulty of calibration. In order to solve this problem, on the one hand, an uncalibrated visual servo system can be used to solve the image jacobian matrix to design a visual controller, which is insensitive to the camera internal and external parameters and robot parameters. On the other hand, the on-line calibration can offset the external interference and calibrate the parameters accurately and in real time.

Acknowledgment. This work was supported by the Key-Area R&D Program of Guangdong Province (No: 2020B090925002) and the Key Special Projects for Intergovernmental International Scientific and Technological Innovation Cooperation (No: YS2017YFGH001624).

References

1. Hill, J.: Real time control of a robot with a mobile camera. In: 9th International Symposium on Industrial Robots, pp. 233–246 (1979)
2. Malis, E., Chaumette, F.: 2-1/2 D visual servoing with respect to unknown objects through a new estimation scheme of camera displacement. Int. J. Comput. Vis. **37**(1), 79–97 (2000)
3. Feddema, J.T., Lee, C.S.G., Mitchell, O.R.: Automatic selection of image features for visual servoing of a robot manipulator. In: Proceedings 1989 International Conference on Robotics and Automation, pp. 832–837. IEEE (1989)
4. Tsai, R.Y., Lenz, R.K.: A new technique for fully autonomous and efficient 3D robotics hand/eye calibration. IEEE Trans. Robot. Autom. **5**(3), 345–358 (1989)
5. Leite, A.C., Lizarralde, F., Hsu, L.: Adaptive hybrid vision-force control with uncalibrated camera and uncertain robot dynamics. IX Simpósio Brasileiro de Automação Inteligente', Brasilia, DF, 1–6 (2009)
6. Yang, Y., Qin, R.: Research progress and challenges of robot visual servo control. J. Zhengzhou Univ. (Sci. Ed.) **50**(02), 41–48 (2018)
7. Zhong, X., Peng, X.: Robust Kalman filtering cooperated Elman neural network learning for vision-sensing-based robotic manipulation with global stability. Sensors **13**(10), 13464–13486 (2013)
8. Deng, L., Janabi-Sharifi, F., Wilson, W.J.: Hybrid motion control and planning strategies for visual servoing. IEEE Trans. Industr. Electron. **52**(4), 1024–1040 (2005)
9. Chaumette, F., Malis, E.: 2-1/2 D visual servoing. IEEE Int. Conf. Robot. Autom. **15**(2), 238–250 (1999)
10. Deng, L., Janabi-Sharifi, F., Wilson, W.J.: Stability and robustness of visual servoing methods. In: Proceedings 2002 IEEE International Conference on Robotics and Automation (Cat. No. 02CH37292), vol. 2, pp. 1604–1609. IEEE (2002)
11. Li, B., Fang, Y.: Visual servo stabilization control of mobile robot based on 2D trifocal tensor. Acta Automatica Sinica **40**(12), 2706–2715 (2014)

12. Malis, E., Rives, P.: Robustness of image-based visual servoing with respect to depth distribution errors. In: IEEE International Conference on Robotics and Automation (Cat. No. 03CH37422), vol. 1, pp. 1056–1061. IEEE (2003)
13. Conticelli, F., Allotta, B.: Nonlinear controllability and stability analysis of adaptive image-based systems. IEEE Trans. Robot. Autom. **17**(2), 208–214 (2001)
14. Cowan, N.J., Weingarten, J.D., Koditschek, D.E.: Visual servoing via navigation functions. IEEE Trans. Robot. Autom. **18**(4), 521–533 (2002)
15. Marchand, E., Comport, A., Chaumette, F.: Improvements in robust 2D visual servoing. In: IEEE International Conference on Robotics and Automation, Proceedings. ICRA 2004, vol. 1, pp. 745–750. IEEE (2004)
16. De Luca, A., Oriolo, G., Giordano, P.R.: On-line estimation of feature depth for image-based visual servoing schemes. In: Proceedings 2007 IEEE International Conference on Robotics and Automation, pp. 2823–2828. IEEE (2007)
17. Malis, E., Chaumette, F.: Theoretical improvements in the stability analysis of a new class of model-free visual servoing methods. IEEE Trans. Robot. Autom. **18**(2), 176–186 (2002)
18. Mahony, R., Hamel, T.: Image-based visual servo control of aerial robotic systems using linear image features. IEEE Trans. Rob. **21**(2), 227–239 (2005)
19. Mahony, R., Hamel, T., Pflimlin, J.M.: Nonlinear complementary filters on the special orthogonal group. IEEE Trans. Autom. Control **53**(5), 1203–1218 (2008)
20. Lopez-Nicolas, G., Gans, N.R., Bhattacharya, S., et al.: Homography-based control scheme for mobile robots with nonholonomic and field-of-view constraints. IEEE Trans. Cybern. **40**(4), 1115–1127 (2010)
21. Fomena, R.T., Tahri, O., Chaumette, F.: Distance-based and orientation-based visual servoing from three points. IEEE Trans. Rob. **27**(2), 256–267 (2011)
22. Lin, C.: Human behavior classification and anomaly detection based on video sequences. Harbin Engineering University (2012)
23. Dong, Z.: Visual servo control of 6-DOF manipulator based on image moment and vector product method. J. Univ. Shanghai Sci. Technol. **35**(3) (2013)
24. Ng, P.C., Henikoff, S.: SIFT: predicting amino acid changes that affect protein function. Nuclc Acids Res. **31**(13), 3812–3814 (2003)
25. Bay, H., Tuytelaars, T., Van Gool, L.: SURF: speeded up robust features. In: Leonardis, A., Bischof, H., Pinz, A. (eds.) ECCV 2006. LNCS, vol. 3951, pp. 404–417. Springer, Heidelberg (2006). https://doi.org/10.1007/11744023_32
26. Leutenegger, S., Chli, M., Siegwart, R.Y.: BRISK: binary robust invariant scalable keypoints. In: International Conference on Computer Vision. IEEE (2011)
27. Rublee, E., et al.: ORB: an efficient alternative to SIFT or SURF. In: IEEE International Conference on Computer Vision, ICCV 2011, Barcelona, Spain, November 6–13. IEEE (2011)
28. Ma, T.: Research on improved SIFT algorithm in vision-based target recognition. Mach. Manuf. Autom. **48**(002), 188–191 (2019)
29. Guo, J.: Research on image matching algorithm based on local features. Nanjing University of Posts and Telecommunications (2018)
30. Jia, Q., Wang, X.: An improved potential field method for path planning. In: 2010 Chinese Control and Decision Conference, pp. 2265–2270. IEEE (2010)
31. Li, G., Tamura, Y., Yamashita, A., et al.: Effective improved artificial potential field-based regression search method for autonomous mobile robot path planning. Int. J. Mechatron. Autom. **3**(3), 141–170 (2013)
32. Hosoda, K., Asada, M.: Versatile visual servoing without knowledge of true jacobian. In: Proceedings of IEEE/RSJ International Conference on Intelligent Robots and Systems (IROS 1994), vol. 1, pp. 186–193. IEEE (1994)
33. Zhang, Z.: Determining the epipolar geometry and its uncertainty: A review. Int. J. Comput. Vis. **27**(2), 161–195 (1998)

34. Loop, C., Zhang, Z.: Computing rectifying homographies for stereo vision. In: Proceedings. 1999 IEEE Computer Society Conference on Computer Vision and Pattern Recognition (Cat. No PR00149), vol. 1, pp. 125–131. IEEE (1999)

35. Hager, G.D.: Calibration-free visual control using projective invariance. In: Proceedings of IEEE International Conference on Computer Vision, pp. 1009–1015. IEEE (1995)

36. Hosoda, K., Sakamoto, K., Asada, M.: Trajectory generation for obstacle avoidance of uncalibrated stereo visual servoing without 3D reconstruction. J. Robot. Soc. Jpn. **15**(2), 290–295 (1997)

37. Borgstadt, J.A., Ferrier, N.J.: Visual servoing: path interpolation by homography decomposition. In: Proceedings 2001 ICRA. IEEE International Conference on Robotics and Automation (Cat. No. 01CH37164), vol. 1, pp. 723–730. IEEE (2001)

38. Mariottini, G.L., Oriolo, G., Prattichizzo, D.: Image-based visual servoing for nonholonomic mobile robots using epipolar geometry. IEEE Trans. Rob. **23**(1), 87–100 (2007)

39. LaValle, S.M.: Rapidly-exploring random trees: a new tool for path planning (1998)

40. Chesi, G., et al.: Homogeneous Polynomial Forms for Robustness Analysis of Uncertain Systems. Springer Science & Business Media, New York (2009)

41. Salaris, P., et al.: Shortest paths for a robot with nonholonomic and field-of-view constraints. IEEE Trans. Rob. **26**(2), 269–281 (2010)

42. Tang, L., et al.: A novel potential field method for obstacle avoidance and path planning of mobile robot. In: 2010 3rd International Conference on Computer Science and Information Technology, vol. 9, pp. 633–637. IEEE (2010)

43. Chesi, G., Vicino, A.: Visual servoing for large camera displacements. IEEE Trans. Rob. **20**(4), 724–735 (2004)

44. Mezouar, Y., Chaumette, F.: Path planning for robust image-based control. IEEE Trans. Robot. Autom. **18**(4), 534–549 (2002)

45. Zhang, X., Fang, Y.: A PFM-based global convergence visual servo path planner. Acta Automatica Sinica **34**(10), 1250–1256 (2008)

Intelligent Biology and Information System

A Higher Order Prediction Model of *Populus Simonii's* Net Photosynthetic Rate Based on Improved Gradient Boosting Method

Xiao-Yu Zhang[1](✉), Xinyue Ji[1], Yuepeng Song[2], Deqiang Zhang[2], and Qing Fang[3](✉)

[1] College of Science, Beijing Forestry University, Beijing 100083, People's Republic of China
xyzhang@bjfu.edu.cn
[2] College of Biological Sciences and Technology, Beijing Forestry University, Beijing 100083, People's Republic of China
[3] Faculty of Science, Yamagata University, Yamagata 990-8560, Japan
fang@sci.kj.yamagata-u.ac.jp

Abstract. We develop a net photosynthetic rate prediction model of *Populus simonii* by selecting 548 individuals as core populations that represent almost the entire geographic distribution of *P. simonii*. We measure photosynthetic characteristic data (net photosynthetic rate, stomatal conductance, intercellular CO2 concentration, water use efficiency) and leaf phenotypic data (leaf area, length, width, perimeter, length-width ratio, leaf shape factor) of these individuals. We first classify these individuals into three subpopulations by utilizing average linkage clustering and PAM clustering. Then we use different machine learning methods to predict net photosynthetic rate based on leaf phenotypic data. Especially in gradient boosting method, the criterion of shrinkage estimator and the iteration-stopping criterion are put forward to enhance the model. The cross-validated results show that our model has high prediction accuracy (90.87%, 88.34%, and 89.26%, respectively in three subpopulations) and improve overfitting in other machine learning method at the same time.

Keywords: *P. simonii* · Net photosynthetic rate · Gradient boosting · Iteration-stopping criterion

1 Introduction

For the past few years, *P. simonii* has become the main species of protection forest in China for its adaptability, drought resistance and cold resistance. It has important values in the fields of wind-break and sand fixation, civil architecture and papermaking [1]. As we all know, net photosynthetic rate represents the level of plant photosynthesis. Therefore, the research of *P. simonii's* net photosynthetic rate prediction is of practical significance in determining carbon fixation and promoting plant growth and development [2].

At present, the study of net photosynthetic rate prediction is mainly at the level of herbaceous plants [3], and the methods used are mostly linear regression. For example,

© Springer Nature Singapore Pte Ltd. 2020
M. Fei et al. (Eds.): LSMS 2020/ICSEE 2020 Workshops, CCIS 1303, pp. 373–385, 2020.
https://doi.org/10.1007/978-981-33-6378-6_28

in 1999, P.C. Nautiyal [4] researched the effect of leaf position on net photosynthetic rate based on a linear mathematical model. However, compared with herbaceous plants, few people have researched woody plants such as *P. simonii*. Considering a longer reproduction time and the more complicated relationship between various factors and net photosynthetic rate [5], it is no longer appropriate to use the linear model. Therefore, in this paper, net photosynthetic rate of *P. simonii* is predicted by using machine learning methods, including neural network, Gaussian process regression and gradient boosting. Then, we conduct numerical experiments to compare the prediction accuracy of these methods and establish an improved prediction model of *P. simonii's* net photosynthetic rate.

In this paper, we first divide *P. simonii* into three subpopulations based on average linkage clustering and PAM clustering [6, 7]. Then, in every subpopulation, we develop two prediction models based on photosynthetic characteristics and leaf phenotypic traits respectively. In the first model, stepwise method is utilized to select variables and k-fold cross-validation is selected to determine the equation parameters. Through numerical experiments, the prediction equations which are based on photosynthetic characteristics have accuracy of 92.01%, 95.32% and 90.87% in three subpopulations, respectively. In the second model, we utilize different machine learning method to predict net photosynthetic rate based on six leaf phenotypic traits (leaf area, length, width, perimeter, length-width ratio and leaf shape factor). In order to avoid overfitting, we improve the gradient boosting method by proposing the criterion of shrinkage estimator and the iteration-stopping criterion. After numerical verification of *P. simonii's* data, the validated-result show that the improved gradient boosting model not only has high prediction accuracy (90.87%, 88.34%, 89.26% respectively in three subpopulations), but also solves the problem of overfitting in other machine learning methods. Finally, we obtain the importance ranking and marginal effect of each phenotypic factor on *P. simonii's* net photosynthetic rate in each subpopulation.

2 Materials and Methods

2.1 Plant Material

In 2007, we collected 1233 local individuals in the whole natural distribution of *P. simonii* in China and selected 548 of them as representatives (from Shaanxi, Henan, Hebei, Qinghai, Ningxia) [8]. Root segments of these 548 individuals were used to establish a clonal arboretum in Guan county, Shandong province, China (36°23'N, 115°47'E). In this study, these individuals were used to explore the effect of photosynthetic characteristics and phenotypic traits on net photosynthetic rate.

2.2 Acquisition of Photosynthetic Characteristics and Phenotypic Traits

We used the portable laser leaf area device (CI-202) when measuring the six leaf phenotypic traits of leaf area, length, width, perimeter, length-width ratio and leaf shape factor. In order to obtain the maximum net photosynthetic rate, the photosynthetic photon flux density (PPFD) was set at 1600at 9:00 to 11:00 using the portable gas exchange system

(Li-6400xt, LiCor), with the concentration set at 400. The net photosynthetic rate: stomatal conductance (Gs), intercellular CO2 concentration (Ci) and water use efficiency (WUE) were recorded when the net photosynthetic rate became constant (). In order to obtain accurate data, we measured the photosynthetic parameters of the three leaves three times (from the fourth to the sixth of the top of the stem) and averaged the measured values of each photosynthetic parameter to provide the value of each individual.

3 Results

3.1 Classification of *P. simonii* Based on Average Linkage Clustering and PAM Clustering

Due to different geographical location and growth environment, *P. simonii* may have different characteristics so that they can be divided into different subpopulations. In our research, we utilize average linkage clustering and PAM clustering to classify these *P. simonii* individuals.

Average linkage clustering belongs to hierarchical clustering methods which require the user to specify a measure of dissimilarity between groups of observations [9], based on the pairwise dissimilarities among the observations in the two groups. Strategies for average linkage clustering divide into two basic paradigms: agglomerative (bottom-up) and divisive (top-down). In this paper, we select agglomerative clustering algorithms:

Step 1. Let x_{ij} ($i = 1, 2,..., n, j = 1, 2,..., p$) be the observation data of the jth index of the ith sample and let every observation represent a singleton cluster.

Step 2. Let G and H represent two such clusters. Define $d_{ii'} = \sqrt{\sum_{j=1}^{p} (x_{ij} - x_{i'j})^2}$ where one member of the pair i is in G and the other i' is in H.

Step 3. Define a measure of dissimilarity between two clusters. The dissimilarity $d(G, H)$ between G and H is computed from the set of pairwise observation dissimilarities $d_{ii'}$. $d(G, H) = \frac{1}{N_G N_H} \sum_{i \in G} \sum_{i' \in H} d_{ii'}$, where N_G and N_H are the respective numbers of observation in each group.

Step 4. Iterate steps 3 until all clusters are merged into a single cluster.

PAM Clustering belongs to K-medoids Clustering and is one of the earliest proposed K-medoids algorithms [9]. It improves the problem of lacking robustness against outliers in K-means clustering. The algorithm steps can be described as follows:

Step 1. For a given cluster assignment C find the observation in the cluster minimizing total distance to other points in that cluster: $i_k^* = \underset{\{i:C(i)=k\}}{\arg\min} \sum_{C(i')=k} D(x_i, x_{i'})$. Then $m_k = x_{i_k^*}$, $k = 1, 2,..., K$ are the current estimates of the cluster centers.

Step 2. Given a current set of cluster centers $\{m_1, \ldots, m_K\}$, minimize the total error by assigning each observation to the closest (current) cluster center: $C(i) = \underset{1 \le k \le K}{\arg\min} D(x_i, m_k)$.

Step 3. Iterate steps 1 and 2 until the assignments do not change.

Table 1. Three subpopulations of *P. simonii*.

Sub-populations	Area	Length	Width	Perimeter	Ratio	Factor
1	24.84	7.37	5.26	33.14	1.48	0.47
2	55.91	10.76	8.04	44.85	1.39	0.43
3	26.89	9.08	5.10	76.05	1.92	0.11

According to average linkage clustering and PAM clustering, we obtain three sub-populations (Fig. 1, Fig. 2 and Table 1). The first subpopulation members are GQ, LX, LY from Shaanxi, CC, CD from Hebei, XZ from Henan and JL from Ningxia. It covers the most provinces. The second subpopulation includes SX, YC from Henan and FX, XB from Shaanxi and is featured by large leaf area. That is, the median of leaf area reaches 59.91 cm^2. The third subpopulation includes HZ, W from Qinghai. Their leaf shape factors are significantly smaller than the other two subpopulations while their perimeters are significantly larger than the other two subpopulations.

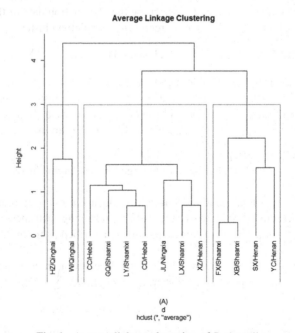

Fig. 1. Average linkage clustering of *P. simonii*.

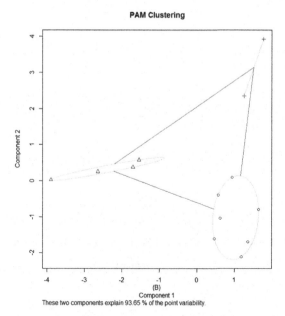

Fig. 2. PAM clustering of *P. simonii.*

3.2 The Non-linear Regression Model

In order to establish the optimal regression model, we use stepwise methods to select variables. It consists of variations on two basic ideas called Forward Selection and Backward Elimination. The equation can be described as:

$$y = a_1 \cdot x_1 + a_2 \cdot x_2 + a_3 \cdot x_3 + a_4 \cdot x_1 \cdot x_2 + a_5 \cdot x_1 \cdot x_3 + a_6 \cdot x_2 \cdot x_3 + a_7 \cdot x_1 \cdot x_2 \cdot x_3 + c$$

where y represents net photosynthetic rate (Pn), x_1 stomatal conductance (Gs), x_2 intercellular CO2 concentration (Ci) and x_3 water use efficiency (WUE). After using R software, we obtain the optimal variables of regression equation in three subpopulations:

$$y = 130.38x_1 + 0.04x_2 + 7.26x_3 - 0.41x_1x_2 - 24.94x_1x_3 - 0.02x_2x_3 + 0.09x_1x_2x_3 - 7.33 \quad (1)$$

Regression diagnostics of Eq. (1) are shown in Fig. 3, Fig. 4, Fig. 5, and Fig. 6. The residuals vs fitted plot shows that curvilinear relationship does not exist so that we do not need to add quadratic term to the equation. The normal QQ and scale-location plots show that the normality and homoscedasticity assumptions are satisfied. Thus, the above equations are reasonable. Furthermore, the residuals vs leverage plot provides the outliers and strong influence points in three subpopulations.

3.3 The Improved Gradient Boosting Prediction Model

Machine learning methods are considered to establish net photosynthetic rate prediction model based on leaf phenotypic traits since Pearson correlation coefficients between Pn and leaf phenotypic traits are not high (Fig. 7).

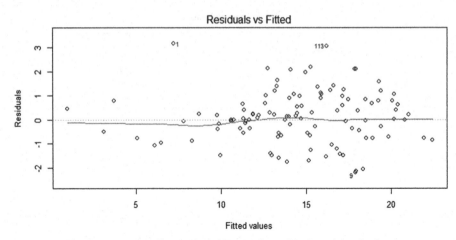

Fig. 3. Regression diagnostics of Eq. (1) for residuals and fitted values.

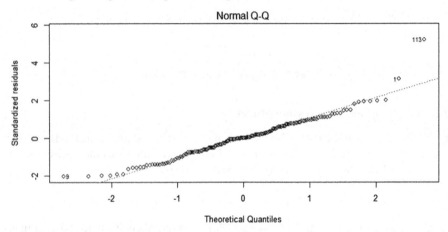

Fig. 4. Regression diagnostics of Eq. (1) for standardized residuals and theoretical quantiles.

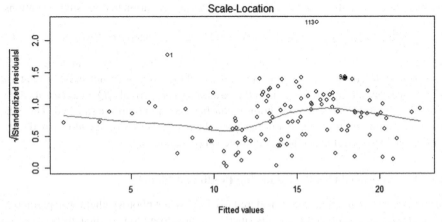

Fig. 5. Regression diagnostics of Eq. (1) for square root of standardized residuals and fitted values.

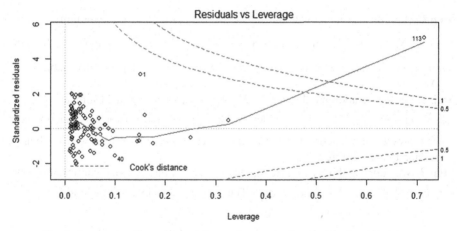

Fig. 6. Regression diagnostics of Eq. (1) for standardized residuals and leverage.

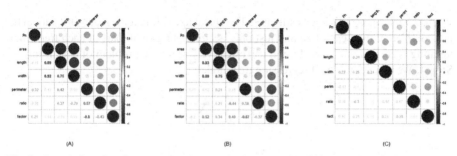

Fig. 7. Correlation plots between net photosynthetic rate and leaf phenotypic traits in three sub-populations. (A) In the first subpopulation, net photosynthetic rate is mostly related to leaf perimeter among leaf phenotypic traits and the correlation efficient between them is -0.32; (B) In the second subpopulation, net photosynthetic rate is mostly related to leaf shape factor and the correlation efficient between them is only -0.2; (C) In the third subpopulation, net photosynthetic rate is mostly related to leaf width and the correlation efficient between them is 0.27. Therefore, there is no obvious linear relationship between net photosynthetic rate and leaf phenotypic traits in three subpopulations.

After comparing gradient boosting method, neural network method and Gaussian process regression, we conclude that gradient boosting method has the highest prediction accuracy in net photosynthetic rate of *P. simonii*. Hence, in this paper, we choose gradient boosting and put forward two criteria to improve overfitting.

Step 1. Initialize the model with the constant value c, which needs to minimize $f_0(x) = \sum_{i=1}^{N} \frac{1}{2}[y_i - c]^2$, where $L(y, f(x)) = \frac{1}{2}[y_i - f(x_i)]^2$ is loss function, N is the number of observation data and y_i is observed as the value of dependent variable, $i = 1, 2...N$.

Step 2. For the iteration number $m = 1, 2...M$, compute the residual estimate $r_{mi} = -\left[\frac{\partial L(y_i, f(x_i))}{\partial f(x_i)}\right]_{f(x)=f_{m-1}(x)}$, which is the negative gradient for loss function, and $f_m(x)$ represents the function when we iterate over m times. Therefore, we can obtain a regression tree according to r_{mi}.

Step 3. We suppose that c_{mj} represents the constant value c which can minimize the equation $\sum_{x_i \in R_{mj}} \frac{1}{2}[y_i - (f_{m-1}(x_i) + c)]^2$. Thus, we can update the regression tree:

$$f_m(x) = f_{m-1}(x) + \lambda \cdot \sum_{x_i \in R_{mj}} c_{mj} \qquad (2)$$

where λ represents shrinkage estimator. The shrinkage estimator λ makes the model do not completely trust each tree and just only accumulate a small part of each tree when iterating is carried out. Therefore, it can effectively solve the problem of overfitting.

Step 4. We can get the final regression tree model:

$$f_M(x) = \sum_{m=1}^{M} \sum_{x_i \in R_{mj}} c_{mj} \qquad (3)$$

3.4 Net Photosynthetic Rate Prediction Analysis

Cross-validations are utilized in net photosynthetic rate prediction studies to estimate the accuracy with which predictions can be applied to other *P. simonii* individuals. We compare the performance of gradient boosting, neural networks and Gaussian process regression in three subpopulations respectively (Table 2, Table 3 and Table 4). Especially in gradient boosting model, we compare the results of different shrinkage estimators in regard to overfitting. According to theorem 1, we calculate the value of $\frac{t_{1-\frac{\alpha}{2}}(n-1) \cdot s}{\sqrt{n}}$ are 0.49, 0.54, 1.04 respectively. Since we have proved that the shrinkage estimator is negatively correlated with $\frac{t_{1-\frac{\alpha}{2}}(n-1) \cdot s}{\sqrt{n}}$, so the shrinkage estimator of gradient model in the first and second subpopulation ought to be smaller than the third one. Table 2 shows that even though the gradient boosting model has high prediction accuracy in training set when we take shrinkage estimator as 0.01, it performs badly in test set. Therefore, we take shrinkage estimator as 0.001. However in the third subpopulation, there is not much difference between prediction accuracy in training set and in test set when we take shrinkage estimator as 0.01 (Table 4). Thus, we take shrinkage estimator as 0.01 since it has higher prediction accuracy compared with the prediction accuracy when we take shrinkage estimator as 0.001 and 0.005.

Table 2. Cross-validation of different machine learning methods in the first subpopulation, where RE1denotes residual error in training set and RE2 denotes residual error in test set.

10-fold	Gradient boosting						Neural network		Gaussian process regression	
	0.01		0.005		0.001					
	RE1	RE2	RE1	RE2	RE1	RE2	RE1	RE2	RE1	RE2
1	1.44	2.46	1.97	2.59	1.78	2.52	0.78	3.89	3.39	4.89
2	1.06	1.16	1.43	1.33	1.66	1.34	1.43	4.21	3.51	4.14
3	1.47	2.25	1.96	2.31	2.19	2.35	2.30	3.74	3.40	3.95
4	1.46	2.00	1.96	2.37	1.74	2.38	1.95	4.44	3.37	4.23
5	1.40	1.84	1.93	1.91	2.21	1.97	1.89	3.79	3.43	3.99
6	1.43	2.05	1.94	2.11	2.22	2.45	1.04	4.24	3.34	4.82
7	1.44	1.93	1.94	2.12	2.10	1.59	1.15	4.58	3.37	4.30
8	1.44	2.31	1.96	2.49	2.19	2.01	1.42	4.22	3.48	4.46
9	1.34	1.83	1.87	1.65	1.81	2.42	1.64	3.32	3.38	4.02
10	1.43	1.84	1.90	1.44	2.12	2.12	0.75	3.43	3.35	4.63
Average	1.39	2.00	1.89	2.07	2.01	2.15	1.52	4.01	3.40	4.35

Table 3. Cross-validation of different machine learning methods in the second subpopulation, where RE1denotes residual error in training set and RE2 denotes residual error in test set.

10-fold	Gradient boosting						Neural network		Gaussian process regression	
	0.01		0.005		0.001					
	RE1	RE2	RE1	RE2	RE1	RE2	RE1	RE2	RE1	RE2
1	1.58	2.57	2.18	2.84	2.20	2.28	3.91	4.55	3.56	4.90
2	1.56	2.51	2.22	2.32	2.43	1.92	4.00	4.85	3.68	3.41
3	1.64	2.35	2.19	2.85	2.31	2.95	0.12	2.01	3.75	4.12
4	1.63	2.47	2.27	1.98	2.40	2.16	1.20	3.11	3.68	3.36
5	1.55	2.18	2.20	2.38	2.37	2.96	3.69	4.77	3.74	4.05
6	1.52	3.07	2.21	2.29	2.41	2.28	3.98	5.63	3.20	4.12
7	1.45	2.48	2.18	1.99	2.41	2.14	3.59	3.79	3.67	3.80
8	1.61	1.74	2.33	1.67	2.56	1.69	4.14	4.54	3.36	4.12
9	1.54	2.10	2.22	2.35	2.32	2.61	3.63	5.70	3.51	4.64
10	1.56	2.51	2.13	3.43	2.13	2.03	3.36	5.18	3.45	5.59
Average	1.56	2.42	2.21	2.46	2.36	2.33	3.42	4.55	3.56	4.26

Next, we compare the prediction results of the improved gradient boosting methods with neural networks and Gaussian process regression. As we can see, not only the

Table 4. Cross-validation of different machine learning methods in the third subpopulation, where RE1denotes residual error in training set and RE2 denotes residual error in test set.

10-fold	Gradient boosting						Neural network		Gaussian process regression	
	0.01		0.005		0.001					
	RE1	RE2	RE1	RE2	RE1	RE2	RE1	RE2	RE1	RE2
1	1.35	1.95	2.34	2.57	3.63	3.80	9.16E-14	3.20	1.12	3.91
2	0.60	1.26	1.78	2.70	3.67	4.70	5.28E-03	6.16	1.42	3.88
3	1.52	1.45	2.33	1.73	3.83	3.79	1.70E-12	7.23	1.88	1.11
4	2.07	2.23	2.75	2.79	3.69	3.74	5.93E-03	8.77	0.87	5.77
5	1.98	1.51	2.30	2.39	3.87	4.09	4.63E-15	6.51	1.84	2.93
6	1.13	1.26	2.16	1.91	3.82	3.14	1.47E-11	9.07	1.62	2.55
7	2.01	2.37	2.74	2.60	3.78	3.90	1.31E-05	3.59	1.40	2.83
8	0.72	1.29	2.03	2.22	3.81	3.14	4.38E-09	7.71	1.86	1.12
9	1.30	1.45	2.23	2.36	3.88	3.93	1.19E-06	4.94	1.70	0.61
10	1.43	1.51	2.33	2.45	3.67	3.78	2.15E-14	4.65	1.62	3.98
Average	1.49	1.67	2.32	2.39	3.77	3.82	2.51E-03	6.48	1.56	3.25

prediction accuracy is worse than our model, but also they exist obvious overfitting (Table 2, Table 3 and Table 4).

Finally, we choose model parameters which can minimize the residual error in the test set and set the iteration residual threshold for three subpopulations. In three subpopulations, the net photosynthetic rate predicted for each individual has an error of 1.34, 1.69 and 1.26, respectively. Figure 8 also provides the importance rankings of six leaf phenotypic traits on net photosynthetic rate. In conclusion, our model has effectively predicted net photosynthetic rate.

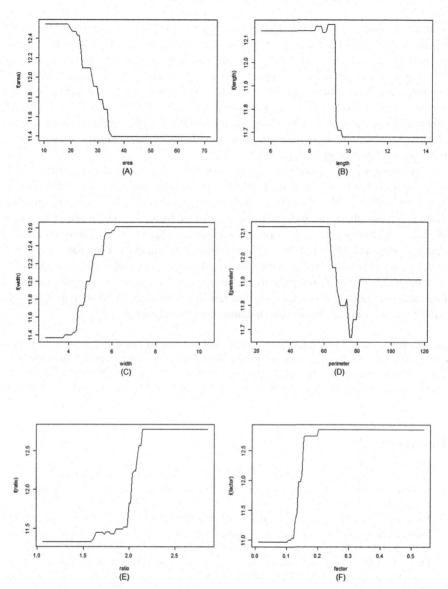

Fig. 8. Marginal effect of phenotypic factors on net photosynthetic rate in the third subpopulation. (A) leaf area (B) leaf length (C) leaf width (D) leaf perimeter (E) leaf length:width ratio (F) leaf shape factor.

4 Discussion

In this paper, we first utilized average linkage clustering and PAM clustering to divide the *P. simonii* into three subpopulations. Then, we discussed the effect of photosynthetic characteristics and phenotypic traits on net photosynthetic rate in each subpopulation. For three photosynthetic characteristics (Gs, Ci, WUE), non-linear regression models were established to predict net photosynthetic rate, which has the accuracy of 92.01%, 95.32% and 90.87% respectively. Afterwards, we used gradient boosting to analyze six phenotypic characteristics (leaf area, length, width, perimeter, length-width ratio and leaf shape factor). In order to avoid overfitting, we proved the criterion of shrinkage estimator and the iteration-stopping criterion. According to iteration-stopping criterion, we have the net photosynthetic rate prediction accuracy of 93.3%, 91.1%, 88.5% in each subpopulation respectively. Obviously, the model improved the prediction accuracy and effectively solved the overfitting problem in gradient boosting. Finally, we considered the marginal effect of phenotypic traits.

Through our paper, we can accurately predict *P. simonii's* net photosynthetic rate in different subpopulations just based on the leaf phenotypic data (leaf area, length, width, perimeter, length-width ratio and leaf shape factor), which effectively solves the problem that net photosynthetic rate measurement is restricted by weather conditions. Our results show that the first and second subpopulations have higher levels of photosynthesis and thus have higher levels of stability of photosynthesis productivity. Therefore, they generally have better performance as protection forest than the third subpopulation. However, in plateau climate where temperature differs greatly between day and night, rainfall is rare and concentrated and solar radiation remains strong, the third subpopulation has advantages over the other two subpopulations as protection forests.

Acknowledgments. This work is supported by the Fundamental Research Funds for the Central Universities (No. 2017ZY30), the Fundamental Research Funds for the Central Universities (No. 2015ZCQ-LY-01) and the National Natural Science Foundation of China (Grant No. 11501032). The last author is supported by Japan Society for the Promotion of Science (Grant No. 19K03613).

References

1. Zhang, C.X., Meng, S., Li, Y.M., Su, L., Zhao, Z.: Nitrogen uptake and allocation in *Populus simonii* in different seasons supplied with isotopically labeled ammonium or nitrate. Trees **30**(6), 2011–2018 (2016)
2. Shipley, B., Vile, D., Garnier, E., Wright, I.J., Poorter, H.: Functional linkages between leaf traits and net photosynthetic rate: reconciling empirical and mechanistic models. Funct. Ecol. **19**(4), 602–615 (2005)
3. Li, T., Ji, Y.H., Zhang, M., Sha, S., Li, M.Z.: Universality of an improved photosynthesis prediction model based on PSO-SVM at all growth stages of tomato. Int. J. Agric. Biol. Eng. **10**(2), 63–73 (2017)
4. Nautiyal, P.C., Ravindra, V., Joshi, Y.C.: Net photosynthetic rate in peanut: influence of leaf position, time of day and reproductive-sink. Photosynthetica **36**(1–2), 129–138 (1999)
5. Gao, H.L., Qiu, L.P., Zhang, Y.J., Wang, L.H., Zhang, X.C., Cheng, J.M.: Distribution of organic carbon and nitrogen in soil aggregates of aspen (*Populus simonii* carr.) woodlands in the semi-arid loess plateau of China. Soil Res. **51**(5), 406–414 (2013)
6. Ibrahim, L.F., Harbi, M.H.A.: Using modified partitioning around medoids clustering technique in mobile network planning. Int. J. Comput. Sci. **9**(6), 299–307 (2012)
7. Lei, D.J., Zhu, Q.S., Chen, J., Lin, H., Yang, P.: Automatic PAM clustering algorithm for outlier detection. J. Softw. **7**(5), 1045–1051 (2012)

8. Ci, D., Song, Y.P., Du, Q.Z., Tian, M., Han, S., Zhang, D.Q.: Variation in genomic methylation in natural populations of *Populus simonii* is associated with leaf shape and photosynthetic traits. J. Exp. Bot. **67**(3), 723–737 (2016)
9. Hastie, T., Tibshirani, R., Friedman, J.: The Elements of Statistical Learning. Springer, New York (2016)

Improving Metagenome Sequence Clustering Application Performance Using Louvain Algorithm

Yakang Lu[1,2], Li Deng[1,2(✉)], Lili Wang[1,2], Kexue Li[1,2], and Jinda Wu[1,2]

[1] School of Mechatronics Engineering and Automation, Shanghai University, Shanghai 200444,
China
dengli@shu.edu.cn
[2] Shanghai Key Laboratory of Power Station Automation Technology, Shanghai 200444, China

Abstract. Metagenomic assembly is a very challenging subject due to the huge
data volume of next-generation sequencing (NGS). The ability of clustering strategy to handle large amounts of data makes it an ideal solution to memory limitations. SpaRC (Spark Reads Clustering), a scalable sequences clustering tool
based on the Apache Spark, a distributed big data analysis platform, provides a
solution to cluster hundreds of GBs of sequences from different genomes. However, the Label Propagation Algorithm (LPA) used in SpaRC is usually unstable,
causing the clustering results to oscillate and contain too many tiny clusters. In this
paper, we proposed a method for clustering metagenomic sequences based on the
distributed Louvain algorithm to obtain more accurate clustering results. We performed experiments on two different datasets with millions of genome sequences
based on LPA and Louvain, respectively. The experimental results indicate that
this approach can effectively improve clustering performance. We hope that the
method applied in this paper can be widely used in other metagenomic clustering
studies.

Keywords: Metagenomic sequences · Clustering · Louvain algorithm · Apache
spark

1 Introduction

Metagenomics analysis is the key for researchers to fully understand the microbial community in nature, the relationship between the environment and humans, and the impact
on human health [1, 2]. At the end of 2019, a sudden outbreak of pneumonia named coronavirus disease 2019 (COVID-19) broke out in Wuhan, Hubei, China [3]. It quickly took
the lives of many people around the world and continued to spread. Rapid publication
of metagenomics analysis during COVID-19 outbreaks provided crucial public health
information. Next-generation sequencing (NGS) technology enabled metagenomics, but
at the same time it generates a large amount of sequence data, which puts forward
higher requirements for the analysis and mining of metagenomic sequences [4–6]. The
DNA/RNA segments from NGS are called reads, which can be seen as a string composed

© Springer Nature Singapore Pte Ltd. 2020
M. Fei et al. (Eds.): LSMS 2020/ICSEE 2020 Workshops, CCIS 1303, pp. 386–400, 2020.
https://doi.org/10.1007/978-981-33-6378-6_29

of four nucleotides (i.e., A, T, C, and G). Since many organisms in nature are unknown to humans, most of the sequences obtained by sequencing cannot determine their origin. Therefore, clustering sequence reads from the same species together is a crucial step in metagenomic analysis [7].

In the discipline of bioinformatics, we call a subsequence whose length is k of a sequence read as k-mer [8]. After the concept was proposed, it is often used in genome sequence analysis and research. It can be used to assemble DNA gene sequences, help biologists understand more organisms, and support medical personnel in the development of antiviral vaccines [7, 8]. Similarity clustering of the genome reads is a significant computational problem for studying bioinformatics and solving the huge data volume of NGS [6]. It groups similar sequences into families by the overlapping (number of k-mers) of reads. At present, most sequences read clustering tools are restricted by their efficiency and scalability, so they cannot process data with millions of reads or cannot obtain accurate clustering results [9, 10].

The basic idea of the label propagation algorithm (LPA) [11] is to add a unique label to each node to represent its community during initialization, and then randomly arrange all nodes. And the label with the largest number of neighbor nodes is used as the label of the node itself, and community detection is achieved through iterative operations [12]. Compared with other algorithms, the LPA algorithm has the advantages of near linear time complexity, simple calculation process, and fast speed. However, due to the randomness in the process of label propagation, its clustering stability is poor.

The clustering algorithm based on modularity regards the community network clustering as an optimization problem [13], and its target value is the modularity (Q). The modularity function is used to measure the quality of the results of the community detection algorithm. It can characterize the closeness of the communities found. Louvain is a multi-level optimization algorithm for community detection based on modularity. It can efficiently divide a large network into communities with high accuracy of division, and can clearly identify the hierarchical community structure [14, 15]. It is considered to be one of the best performing community detection algorithms. In the field of computer science and technology, Apache Spark introduces the concept of Resilient Distributed Dataset (RDD) [16], which is used to perform in-memory computing. Since intermediate results can be stored in memory, it is outstanding in iterative operations. The distributed Louvain strategy based on Apache Spark was applied to improve the clustering performance of SpaRC (Spark Reads Clustering) [17]. By this method, the clustering results of metagenomic reads sequences can be greatly improved.

2   The Proposed Method

2.1   Clustering Strategies

Since most organisms in the natural environment are not yet known to humans, the assembly of reads to restore the entire gene chain is of great significance to metagenomics analysis [4]. The strategy of most assemblers is based on assembling short reads into longer contigs and grouping contigs into genomes through a binning operation [1]. However, due to the large scale of data from NGS, the "assembly-then-cluster" approach faces huge computational challenges. In order to overcome this limitation, the "cluster-then-assembly" approach began to appear. It clusters reads first, and then assembles each cluster into draft genomes. Here, we aim to solve the clustering problem in Fig. 1 B1.

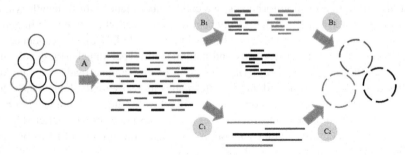

Fig. 1. Pipeline of clustering methods based on Louvain (A) sequencing and get short reads (B1)(B2) "cluster-then-assembly" (C1)(C2) "assembly-then-cluster".

2.2 Louvain Algorithm

Louvain algorithm is a graph clustering algorithm based on modularity. Modularity can be defined as the ratio of the sum of the number of edges in each community to the sum of the number of edges in the constructed graph minus a certain fixed value [18]. The fixed value is that when the graph is a random network, the ratio of the total number of edges in the community to the total number of edges in the community [14]. If the modularity of a network is higher, the node relationships intra each community in the network are closer, and the node relationships inter communities are sparse. Therefore, by taking the modularity as an optimization function and finding its global optimal solution [15], we can get the optimal community partition of a network.

The range of modularity is between [0,1). At the same time, the closer the Q is to 1, the higher the quality of the community obtained by clustering. The calculation formula is as formula (1):

$$Q = \frac{1}{2m} \sum_{i,j} [A_{ij} - \frac{t_i t_j}{2m}] \delta(v_i, v_j) \tag{1}$$

$$\delta(v_i, v_j) = \begin{cases} 1, & where \ v_i = v_j \\ 0, & else \end{cases} \tag{2}$$

where m is the sum of the number of edges in the undirected graph, A_{ij} is the edge weight between any node i and node j, $t_i = \sum_j A_{ij}$ represents the total number of all edges connected to node i, which is the sum of weights. Similarly, t_j is all the nodes connected to node j. v_i represents the community to which node i belongs.

Louvain iteratively clusters graphs formed by metagenomic sequences based on modularity increments. If a node joins a community, the modularity of the community will increase to the maximum extent [18], then the node should belong to the community.

Table 1. The iterative algorithm of Louvain

Louvain Algorithm

Input: Figure $G(V,E)$;

Output: The result of community partitioning on graph G;

  1: Initialize each node in the graph as a single independent community, calculate the modularity of the graph, and record it as Q_1;

  2: $Q_2 = Q_1$;

  3: **do**

  4:    **while** (i < n) # n represents the total number of nodes;

  5:      Calculate the modularity increment ΔQ of the graph after placing node i in the neighborhood of its neighbor node;

  6:      Find the neighbor node of max ΔQ and assign node i to the community where the node is located

  7:    **End while**;

  8: **while** (the community where all nodes are located remains unchanged);

  9: Calculate the modularity of the entire graph and record it as Q_1;

10: if $(Q_1 > Q_2)$

11:    Compress and merge all nodes in the same community into a single super node;

12:    Go to step 2;

13: Return the clustering results;

If it does not increase the modularity of other communities after joining them, it stays in its current community.

In the traditional Louvain algorithm, each iteration calculation uses a single-machine single-threaded method: each vertex checks its neighbor's community, and selects a new community based on the function that maximizes the modularity [14], and iterates continuously. However, this method is inefficient and is powerless for large data sets. Therefore, it is necessary to provide a strategy to enable distributed operation.

In order to implement the Louvain algorithm on a distributed cluster, the serial Louvain algorithm needs to be improved. In a distributed cluster, all vertices are no longer updated in order, but the node information is updated synchronously according to the information of the previous round of nodes [13], and the graph state is updated after each iteration.

In order to record the information of the nodes in the graph, the `VertexData` class is defined to store the state of the nodes. The `VertexData` class has five attributes: `community`, `internalWeight`, `nodeWeight`, `changed` and `communitySig-maTot`. The `community` stores the ID of the community where the node is located; the `communitySigmaTot` stores the sum of the degree values of the community where the node is located; the `internalWeight` stores the degree value of the edge contained in the node itself; the `nodeWeigh` stores the out-degree of the node; the `changed` attribute indicates whether the community to which the node belongs has changed, the default is False. The `VertexData` class should extends Java's `Serializable` interface and `KryoSerializable` to improve program performance.

In the iterative process, the community division of nodes mainly uses GraphX's `AggregateMessages` operation to aggregate messages [14]. The `map` phase sends messages to neighboring nodes (`sendMsg`), and the `reduce` phase merges messages received by nodes (`mergeMsg`). According to the information of all neighboring nodes received by the node, formula (1) can be used to calculate the modularity gain and decide to select the community to belong to. GraphX's `outerJoinVertices` operator is used to complete graph aggregation operations.

2.3 Pipeline

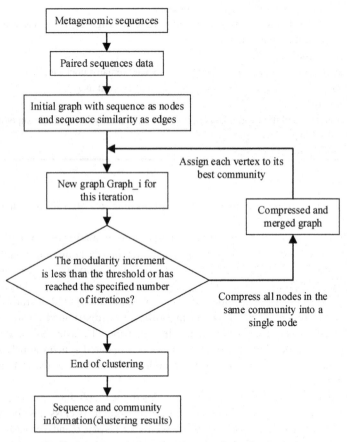

Fig. 2. Pipeline of clustering methods based on Louvain

During the execution of Louvain algorithm, the community division of the current step is used, and then iterative operations are performed continuously to achieve the expansion and merger of the community. It is based on the Apache Spark [16] platform and GraphX module to achieve distribution of Louvain. The pipeline of distributed [13] metagenomic sequences clustering method based on Louvain is shown in Fig. 2. It is divided into the following 3 parts:

1. Preprocessing largescale metagenomic sequences and measuring sequence similarity relationships to obtain paired sequences.
2. Filter out the paired sequences whose shared k-mer number is less than a certain threshold, and take the paired sequences as edges to construct an undirected graph representing the sequence relationship. The genome sequence in this graph is the vertex and sequence similarity is an edge.
3. This undirected graph is iteratively clustered. Calculate the modularity increment before and after each iteration. When the value is less than the predetermined threshold, it indicates that the community division has reached the optimal level, and the iteration can be stopped.

2.4 Evaluation Metrics

The experimental datasets used in this paper have the answers, that is, there is correct community partition results, which can be used to evaluate the metagenomic sequence analysis method. Here the clustering results are compared with known answers, and purity, completeness, Normalized Mutual Information (NMI), and F-Measure [19] are used to evaluate clustering performance and to compare the clustering effects of the two algorithms.

Purity (similar to precision) is usually defined as the percentage of the DNA sequence in each cluster in the clustering result that belongs to a known dominant genome. The purity of each cluster is calculated to obtain the purity set $P = \{p_1, p_2, \ldots, p_m\}$, where

$$p_j = \frac{\max a_{i1}}{\sum\limits_{i=1}^{n} a_{i1}}, \ 1 \leq j \leq m \tag{3}$$

Since sets are not easy to compare, the purity metrics can be simplified as the median purity(p_{med}), mean purity(p_{mean}) and the percent of the cluster with 100% purity ($p_{100\%}$).

Completeness (similar to recall) is generally defined as the percentage of DNA sequences of known classes captured by the clustering algorithm. Calculate the completeness of each known cluster in turn to obtain the completeness set $C = \{c_1, c_2, \ldots, c_n\}$, where

$$c_i = \frac{\max a_{1j}}{\sum\limits_{j=1}^{m} a_{1j}}, \ 1 \leq i \leq n \tag{4}$$

Similarly, the completeness metrics are represented by c_{med}, c_{mean} and $c_{100\%}$.

NMI is an important metric for community detection. Normalized Mutual information indicates the degree of correlation between two random variables:

$$I(U, V) = \sum_{u,v} p(u, v) log \frac{p(u, v)}{p(u)p(v)} \tag{5}$$

The standardized mutual information is normalized to 0~1 by dividing the mutual information by the maximum entropy.

$$NMI = 2\frac{I(U,V)}{H(U)+H(V)} \tag{6}$$

where

$$H(U) = \sum_{i=1}^{n} p(u_i)I(u_i) = -\sum_{i=1}^{n} p(u_i)log_b p(u_i) \tag{7}$$

F-Measure is a commonly used criterion for evaluating the quality of classification, clustering, and information search. Its calculation formula is:

$$F_\beta = \frac{(\beta^2 + 1)P \cdot C}{\beta^2 \cdot P + C} \tag{8}$$

where β is the parameter, P is the purity, and C is the completeness. A larger β indicates a higher recall weight (completeness). Here β takes 1 and 5.

3   Materials and Environment

3.1   Datasets

Two datasets (Mock, CAMI2) were selected for the experiments. Mock [17] is a sequence data set with a complete reference genome that has been identified in the community for benchmarking of metagenomic sequence analysis methods. The Mock dataset can be expressed as MBARC-26, which simulates a mixed community of bacteria and archaea. The number 26 represents that the data set contains 23 bacteria and 3 archaeal species with a complete genome. And CAMI2 [1] is a simulated dataset from the second CAMI Challenge (https://data.cami-challenge.org/participate). The data set contains 64 samples with different genome coverage, and it is simulated using genomes derived from mouse gut microbial communities. Here, sample 16 was selected as the second experimental dataset. Each read is a sequence of different lengths composed of nucleotides (four letters in the string perspective, i.e., A, T, G, and C). The dataset is described in Table 2.

Table 2. Datasets used in experiments

Dataset	Read length (bp)	Read number	Size (GB)
Mock	(90−150) * 2	10,756,542	3.6
CAMI2	150 * 2	6,800,165	2.0

3.2 Computing Environments

We performed the metagenomic sequences clustering experiments on Elastic MapReduce (EMR, emr-5.17.0) of Amazon Web Service (AWS) [1], which provides stable and reliable big data cloud computing services. Considering the size of the experimental dataset, we used several instances to test the algorithm applied in this paper. In the experiments, 21 r4.2x large instances (one as the master node and the others as worker nodes) were used to build a cluster. The details of computing environments and configuration are shown in Table 3.

Table 3. Computing environments of AWS

Parameter	Setting
Storage/node	300 GB SSD
Memory/node	61 GB
Cores/node	8
Master memory	40 GB
Master cores	5
HDFS Block Size	128 MB
# of executors/node	3
Memory/executor	16 GB
Cores/executor	2
Spark	2.3.2
Hadoop	2.8.4
Cluster mode	YARN

4 Results and Analysis

In order to verify the effectiveness of the community detection algorithm Louvain in largescale networks, we performed experiments on the Mock and CAMI2 datasets based on LPA used in SpaRC and Louvain applied in this paper respectively. The clustering results of them were compared. During the experiment, as the number of iterations increases, the modularity becomes closer to 1 (see Fig. 3). Analyzing only from the algorithm principle, this shows that the clustering effect is constantly approaching the optimal.

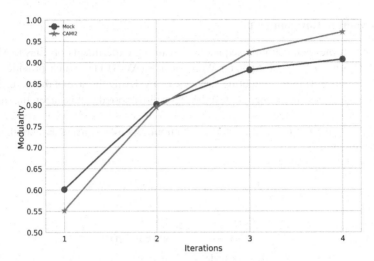

Fig. 3. Modularity increases with the number of iterations

Because the real community division results of this test dataset had been known, various external indicators (purity, completeness, NMI, F-Measure and cluster number) were also used to evaluate the performance of the community detection results.

The clusters in the LPA clustering results are generally small (under-clustering), so their purity is high, but the completeness of the genome classification corresponding to the answer is particularly low. Overall, this method performs less than ideal. However, the problem of low completeness can be largely solved by Louvain algorithm. As shown in Fig. 4, compared to the results of LPA, the distribution of purity and completeness of the clustering results produced by this algorithm is mostly concentrated in the area close to 100%, especially the completeness of clusters.

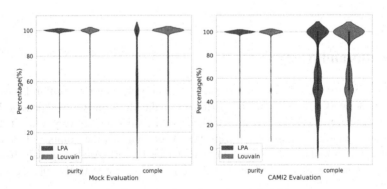

Fig. 4. Violin plots of Mock and CAMI2

In the evaluation of the Mock experimental results of the Louvain algorithm, when the median purity(100% vs 100%) has not changed and the proportion of 100% purity (87.75% vs 94.91%) has not decreased too much, its median completeness (100% vs 66.67%) and the proportion of 100% completeness (83.09% vs 27.79%) have been significantly improved. Similarly, Louvain algorithm outperforms on the CAMI2 dataset. The details of the results from the control experiments are shown in Table 4 and Fig. 5.

Table 4. Comparison of purity and completeness

Datasets	Algorithm	P_med	P_100%	C_med	C_100%
Mock	LPA (%)	100.00	94.91	66.67	27.79
	Louvain (%)	100.00	87.75	100.00	83.09
CAMI2	LPA (%)	100.00	89.51	66.67	36.50
	Louvain (%)	100.00	88.42	100.00	50.98

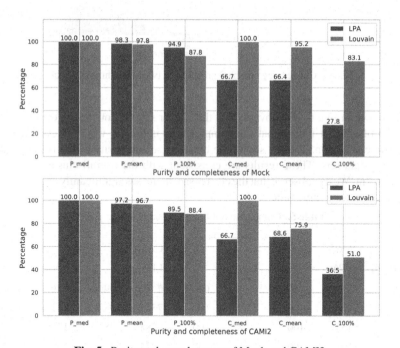

Fig. 5. Purity and completeness of Mock and CAMI2

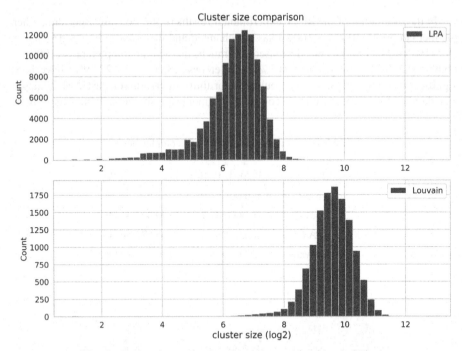

Fig. 6. Cluster size comparison between two algorithms of Mock

As shown in Fig. 6, it is obvious that the cluster size of Louvain algorithm is larger than that of LPA. It means that the completeness is higher, which is consistent with the above conclusion. Table 5 shows the cluster size and purity of the two algorithms in the two data sets. These results are consistent with the overall clustering accuracy measurement. We can observe that the cluster size increases significantly and it maintains a high purity at the same time.

Table 5. Comparison of top 20 cluster size and purity

Dataset	Mock				CAMI2			
Algorithm	LPA		Louvain		LPA		Louvain	
Top #	# read	purity	# read	purity	# read	purity	# read	purity
1	206	100	552	100	7356	98.41	14168	99.96
2	153	100	344	100	6585	98.31	12450	97.15
3	79	100	290	100	5974	100	10620	98.42
4	74	100	260	90.38	5542	98.34	9778	100
5	65	100	218	100	5276	100	9596	54.64
6	49	100	216	100	3239	95.8	9533	85.98

(continued)

Table 5. (*continued*)

Dataset	Mock				CAMI2			
Algorithm	LPA		Louvain		LPA		Louvain	
Top #	# read	purity	# read	purity	# read	purity	# read	purity
7	43	95.35	192	100	3224	96.74	9197	100
8	42	100	191	100	2643	84.83	8810	100
9	41	100	191	100	2239	93.17	8698	99.99
10	39	100	186	100	2178	93.89	8618	98
11	39	100	186	100	2059	99.37	8561	98.32
12	39	100	184	100	2017	93.95	8538	100
13	38	100	174	100	1946	92.34	8514	96.83
14	38	100	168	95.24	1880	96.65	8392	100
15	38	94.74	168	100	1877	100	8293	100
16	37	100	166	100	1845	94.8	8218	100
17	36	36.11	164	100	1843	98.16	8215	99.98
18	36	100	160	96.25	1822	93.96	8092	99.99
19	35	100	159	100	1785	58.77	8069	100
20	35	100	159	100	1631	86.02	7935	100

In addition, regardless of the case where the purity and completeness are 100%, in the impure and incomplete clusters (purity is not 100%, or completeness is not 100%), the purity and completeness of the cluster produced by Louvain are higher than those of LPA (see Fig. 7). It means that using this algorithm can produce better clustering results.

Considering that the single-type index evaluation may be one-sided, multiple methods were used to evaluate and compare the clustering results. NMI, F1 and F5 have all increased significantly. Besides, the number of clusters has also been greatly reduced. This solved the problem of under-clustering to some extent. The results of the two sets of experiments are described in Table 6.

Obviously, from the comparison of the experimental results, it can be easily observed that the Louvain algorithm can greatly improve the completeness, NMI and F-Measure on the basis of sacrificing a certain degree of purity. In practice, Louvain is popular with many data scientists due to its good efficiency and stability.

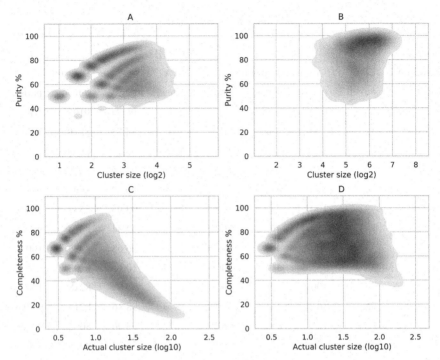

Fig. 7. Kernel density estimate plot of clustering results. (A) Cluster size and purity of LPA. (B) Cluster size and purity of Louvain. (C) Cluster size and completeness of LPA. (B) Cluster size and completeness of Louvain.

Table 6. Experimental results metrics

Datasets	Algorithm	NMI	F1	F5	# of clusters
Mock	LPA	0.7997	0.3138	0.1970	95,833
	Louvain	0.9287	0.5114	0.7033	12,675
CAMI2	LPA	0.6284	0.8044	0.6940	112,237
	Louvain	0.7183	0.8502	0.7650	58,430

5 Conclusion and Discussion

LPA has the advantages of simple logic and fast speed in solving community detection of large-scale networks [11], but its clustering results will produce many tiny clusters (under-clustering) and the results will be unstable. Here, we applied Louvain to the clustering problem of genome reads sequences. Although the purity was reduced slightly, more indicators (median completeness, proportion of 100% completeness, NMI and F-Measure) would be better much. All in all, by using distributed Louvain algorithm instead of LPA, the performance of SpaRC on clustering of metagenomic sequences can be greatly improved.

Some parameters in the clustering method based on distributed Louvain algorithm are need to be set manually. These parameters may have a large impact on clustering efficiency and clustering accuracy. If some parameter optimization models [20] are combined with the algorithm, it may be able to automatically select the optimal parameters, making it easier to get better clustering results.

Acknowledgments. This work is supported by National Natural Science Foundation (NNSF) of China under Grant 61802246.

References

1. Li, K., Lu, Y., Deng, L., Wang, L., Shi, L., Wang, Z.: Deconvolute individual genomes from metagenome sequences through short read clustering. PeerJ **8**, e8966 (2020)
2. Yan, W., Sun, C., Yuan, J., Yang, N.: Gut metagenomic analysis reveals prominent roles of Lactobacillus and cecal microbiota in chicken feed efficiency. Sci. Rep. **28**(7), 45308 (2017)
3. Dong, E., Du, H., Gardner, L.: An interactive web-based dashboard to track COVID-19 in real time. The Lancet Infectious Diseases, 19 February 2020
4. Hillmann, B., et al.: Evaluating the information content of shallow shotgun metagenomics. Msystems **3**(6), e00069–18, 30 October 2018
5. Sandhya, S., Srivastava, H., Kaila, T., Tyagi, A., Gaikwad, K.: Methods and tools for plant organelle genome sequencing, assembly, and downstream analysis. In: Legume Genomics, Humana, New York, NY, pp. 49–98 (2020). https://doi.org/10.1007/978-1-0716-0235-5_4
6. Compeau, P.E., Pevzner, P.A., Tesler, G.: Why are de Bruijn graphs useful for genome assembly? Nat. Biotechnol. **29**(11), 987 (2011)
7. Kelley, D.R., Salzberg, S.L.: Clustering metagenomic sequences with interpolated Markov models. BMC Bioinf. **11**(1), 544 (2010)
8. Onate, F.P., Batto, J.M., Juste, C., Fadlallah, J., Fougeroux, C., Gouas, D., Pons, N., Kennedy, S., Levenez, F., Dore, J., Ehrlich, S.D.: Quality control of microbiota metagenomics by k-mer analysis. BMC Genom. **16**(1), 1 (2015)
9. Zou, Q., Lin, G., Jiang, X., Liu, X., Zeng, X.: Sequence clustering in bioinformatics: an empirical study. Brief. Bioinform. **21**(1), 1 (2020)
10. Bao, E., Jiang, T., Kaloshian, I., Girke, T.: SEED: efficient clustering of next-generation sequences. Bioinformatics **27**(18), 2502–2509 (2011)
11. Jokar, E., Mosleh, M.: Community detection in social networks based on improved Label Propagation Algorithm and balanced link density. Phys. Lett. A **383**(8), 718–727 (2019)
12. Li, W., Huang, C., Wang, M., Chen, X.: Stepping community detection algorithm based on label propagation and similarity. Phys. A **15**(472), 145–155 (2017)
13. Chaudhary, L., Singh, B.: Community detection using an enhanced louvain method in complex networks. In: Fahrnberger, G., Gopinathan, S., Parida, L. (eds.) ICDCIT 2019. LNCS, vol. 11319, pp. 243–250. Springer, Cham (2019). https://doi.org/10.1007/978-3-030-05366-6_20
14. Blondel, V.D., Guillaume, J.L., Lambiotte, R., Lefebvre, E.: Fast unfolding of communities in large networks. J. Stat. Mech: Theory Exp. **2008**(10), P10008 (2008)
15. Ghosh, S., Halappanavar, M., Tumeo, A., Kalyanarainan, A.: Scaling and quality of modularity optimization methods for graph clustering. In: 2019 IEEE High Performance Extreme Computing Conference (HPEC), pp. 1–6. IEEE, 24 September 2019

16. Guo, R., Zhao, Y., Zou, Q., Fang, X., Peng, S.: Bioinformatics applications on apache spark. GigaScience, **7**(8), giy098, August 2018
17. Shi, L., Meng, X., Tseng, E., Mascagni, M., Wang, Z.: SpaRC: scalable sequence clustering using Apache Spark. Bioinformatics **35**(5), 760–768 (2019)
18. Chen, D., Yuan, Y., Zhang, R., Huang, X., Wang, D.: A smart weighted-louvain algorithm for community detection in large-scale networks. In: FSDM, pp. 273–281, 6 November 2019
19. Bascol, K., Emonet, R., Fromont, E., Habrard, A., Metzler, G., Sebban, M.: From cost-sensitive to tight f-measure bounds. In: The 22nd International Conference on Artificial Intelligence and Statistics, pp. 1245–1253, 11 April 2019
20. Wang, Y., Ni, X.S.: A XGBoost risk model via feature selection and Bayesian hyper-parameter optimization. arXiv preprint arXiv:1901.08433 (2019)

A Combined Method for Face and Helmet Detection in Intelligent Construction Site Application

Haikuan Wang[1], Zhaoyan Hu[1], Yuanjun Guo[2(✉)], Yuhao Ou[1], and Zhile Yang[2]

[1] Shanghai Key Laboratory of Power Station Automation Technology, Shanghai University, Shanghai, Baoshan 200444, China
[2] Shenzhen Institute of Advanced Technology Chinese Academy of Sciences, Shenzhen 518055, China
yj.guo@siat.ac.cn

Abstract. Environment of construction site is becoming more complicated and risky than ever before, due to the rapidly development of society. The traditional site management system is facing the challenges of manual supervision negligence as well as the inflexible attendance system, which cannot guarantee the life safety and legitimate interests of all employees. To address the above problem and reduce the risks, many construction sites utilize intelligent approaches such as effective safety helmet detection and face recognition. This paper propose a hybrid approach combining the popular YOLOv3 and Facenet model to detect the safety helmet wearing of construction workers and help them with attendance checking through camera simultaneously. At first, this method apply YOLOv3 to implement safety helmet detection. Then, Facenet is used to achieve face recognition with face detected by MTCNN model. Finally, combined with the above two modules, helmet detection and personnel information identification can be realized in site supervision system. The experimental results show the effectiveness of the proposed combined approach. As a result, database established by video and image process results can ensure the reasonable salary payment of construction workers, further improving the safety assurance measures and the management efficiency of the construction site.

Keywords: Construction safety · Safety helmet detection · Personnel information recognition

1 Introduction

In recent years, with the rapid development of construction industry, engineering construction projects can be seen everywhere across the city. Environment of construction site is becoming more complicated and risky than ever before, and accidents happen frequently under some severe construction environments. Wearing safety helmet when entering the construction site is a necessary protection measure for everyone, especially the workers who are more likely to be

© Springer Nature Singapore Pte Ltd. 2020
M. Fei et al. (Eds.): LSMS 2020/ICSEE 2020 Workshops, CCIS 1303, pp. 401–415, 2020.
https://doi.org/10.1007/978-981-33-6378-6_30

injured during working process. However, with the increased difficulty of personnel management by human, it is easy to cause safety accidents to the workers who are not wearing helmet or have non-standard operations. On the other hand, enough payment to many construction workers can not be guaranteed due to the confused and immature personnel management and non-transparent financial system.

Therefore, workers are forced to wear a safety helmet to enter the construction site. However, some workers enter the construction site without wearing safety helmets due to various reasons in many cases, which may cause a lot of potential safety problems. Consequently, it is hugely critical and meaningful to supervise the situation of construction personnel wearing safety helmets on the construction site. Meanwhile, face recognition is required at the same time, assisting workers to register their attendance as they enter the construction site so as to ensure the smooth progress of the construction project.

Some of the existing traditional video monitoring methods with regard to construction sites often adopt manual monitoring, which has many problems in this way. For example, human beings are unreliable because they are unable to stay focused for long time, easy to miss important picture information. In addition, the traditional methods can not analyze the actual situation on site, so that there is no safety function to provide the alarm for the personnel on duty when something dangerous occurs on the construction site.

Currently, with the rapid development of artificial intelligence, target detection and face recognition algorithm based on deep learning has attracted much attention for researchers at home and abroad. For example, Wei Liu et al. [1] presented a method named SSD for target detection in pictures with applying only one deep neural network in ECCV2016. With regard to the method of SSD, the of multistage feature maps is employed as the basis of classification and regression, achieving the multi-scale effect. Ross Girshick et al. [2] proposed a novel detection algorithm called R-CNN that is simple and scalable, achieving a mAP 53.3% in 2014. A point of view was put forward in the paper that when you lack a lot of labeled data, a more practical and effective approach is to perform neural network migration learning, employing network weight files trained with other data, and then fine tuning the parameters of the network structure for a small-scale specific datasets. In [3], the CNN network needs fixed size image input, so that the fixed length image representation was generated after that any size image passed through pooling layer in SPP-net, which improves the speed of R-CNN for detection by above 24 times.

This paper presents a combined method of helmet detection and face recognition for video surveillance on construction site. With the aim of evaluating the feasibility and the stability of the presented method, various visual conditions of construction sites were considered for the analysis of experimental results. The experimental results confirm that the novel method is efficient in the object detection on sites. The correct rate is all above 90% and the detection speed is 15fps, which achieves good performance for the real-time application on the construction sites. The structure of this paper is as follows. A review of literatures is provided in Sect. 2 regarding the development process of target detection on the construction sites. And then, our novel combined algorithm will be described elaborately in Sect. 3. In Sect. 4, the experimental results and analysis are put forward in detail. Finally, Sect. 5 shows the conclusion of the paper.

2   Literature Review

In recent years, massive research works have been done in the domain of target detection using video surveillance with regard to large-scale public places [4–12]. For example, a novel and functional method about safety helmet wearing detection method is put forward in [13], which is on the basis of traditional two-dimensional image feature extraction in 2017. The accuracy of detection for the proposed method is up to 80.7% and the frame rate is 7 fps. Dikshant Manocha et al. [14] presented a helmet detection method for two-wheeler riders with the assistance of machine learning and a user interface was provided to pay challans. This technology firstly captures the real-time image of road traffic and detect the two wheelers riders among all vehicles on the road, secondly it processed to recognize whether riders are wearing helmet or not. Rattapoom Waranusast et al. [15] presented a system which is able to recognize Motorcyclists and check whether they are wearing safety helmets or not automatically. The system utilizes the KNN classifier to extract features based on the regional features of the moving target and classify it as a motorcycle or other moving target. The experimental results of this method finally show excellent detection performance with the average correct rates are all above 70%.

The above methods for helmet detection are all based on traditional machine learning vision. The features used for safety helmet detection are artificially selected and designed, and the obtained feature are not robust enough. Consequently, target detection based on deep learning is developed rapidly and applied extensively on many fields. In 2014, the fast RCNN algorithm based on vgg16 [16] is nearly 9 times faster than RCNN in training speed, about 3 times faster than SPP-net, while in terms of test speed the fast RCNN is 213 times faster than RCNN, 10 times faster than SPP-net. The MAP calculated on voc2012 is about 66%. A Region Proposal Network is put forward in Faster RCNN [17] that shares full image convolutional features with the detection network, which has a frame rate of 5 fps on a GPU and achieves the detection accuracy of 73.25% mAP on PASCAL VOC2007.

In [18], a CNN model is utilized separately in YOLO algorithm to realize the purpose of end-to-end object detection. The input image is firstly resized to 448 × 448, after which the input is transmitted to CNN network for feature extraction, and the network prediction results are analysed and be dealt with to obtain final detect result. Compared with R-CNN algorithm, YOLO is a unified framework with faster speed, and its training process is also end-to-end. Joseph Redmon et al. [19] proposed YOlOv3 algorithm, which has no pooling layer and full connection layer in the whole V3 structure. In the forward propagation process, the tensor is transformed by changing the stride size of convolution kernel. It has clear structure and excellent real-time performance. Two methods are proposed in [20] to obtain better helmet detection performance, in which face detection was performed using features similar to haar and circle round transformations, respectively. Madhuchhanda Dasgupta et al. [21] presented a framework for helmet detection of riders travel on a motorcycle. In the proposed method, YOLOv3 model was employed for the detection of motorcycle riders at first stage, while a safety helmet detection system for motorcycle drivers based on Convolutional Neural Network (CNN) is presented. C A Rohith et al. [22] intended to establish an automated system to distinguish whether cyclists are wearing helmets and to impose fines on offenders as part of law enforcement. Yang Bo et al. [23] adopted YOLOv3 model to fine-tune the datasets for the electric power construction scene, in which the accuracy is over 90%. A hardhat detection framework based on machine learning and image processing technology is proposed in [23] by Zhong Kai et al.

In conclusion, previous methods are effective for target detection in close-range images, but there are many limitations when applied in far-field surveillance video. In contrast, YOLOv3 is robust enough to handle a variety of situations in terms of the detection performance.

3   Methodology

Currently, some achievements only focus on the detection of safety helmet or face recognition, making the function of intelligent construction system very limited in real practice. In this paper, the approach which combines the functions of safety helmet detection and personnel identification through surveillance video is proposed. Therefore, the information of any person appears in the camera is easily and quickly to be identified, including wearing safety helmets as well as face recognition for attendance check. The environment of the construction site is so complicated that many uncontrollable factors such as the long-term vision, different postures, and different degrees of occlusion are challenges for the detection and recognition task through video. Therefore, the novel method put forward in the paper not only focus on the accuracy of detection and recognition, but also the processing speed which is an important evaluation indicator in video target detection.

3.1   A Combined Detection Approach

This novel combined detection approach put forward in the paper contains two aspects: one is safety helmet detection module based on YOLOv3 and the other one is face recognition module based on Facenet. Which are efficiently applied in the monitoring part of intelligent construction site system. In this method, the above two modules are serially connected, which means that the original image is detected by helmet detection module, and then the output of detection module is used as the input of face recognition module. The overall framework of the presented method is shown in Fig. 1.

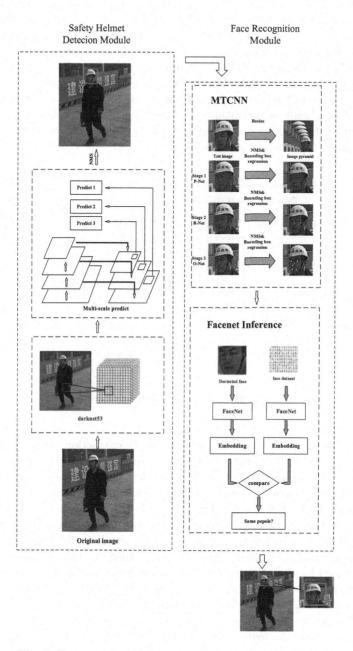

Fig. 1. Framework of the proposed hybrid detection method

YOLOv3 Network Structure. The algorithm of YOLOv3 adopted some good schemes for reference to integrate in YOLO, in practice, it actually performs better on real-time target detection and recognition. Under the premise of maintaining the speed of processing each frame, detection accuracy is effectively improved, moreover, the recognition performance of long-range objects is strengthened. Under GPU environment, YOLOv3 can achieve the frame rate of 20, so it can be applied to real-time safety helmet detection on construction site. YOLOv3 has three main improvements parts: tuning the parameters of network structure, employing multi-scale features for target detection and classification, using the algorithm for logistic instead of softmax. The YOLOv3 structure network involves three improvement as illustrated in Fig. 2.

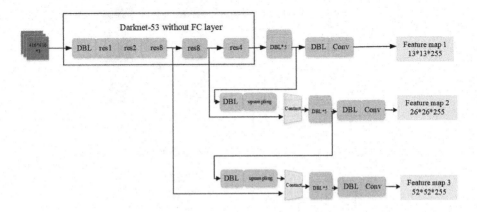

Fig. 2. YOLOv3 network structure

Firstly, in the aspect of basic image feature extraction network, YOLOv3 adopts the network structure of darknet-53 (including 53 convolution layers), which adopted the method of residual network for reference, and sets short connections between some layers. Secondly, passthrough structure was used in YOLOv3 to detect fine-grained features while three feature maps of different scales are further employed for object detection. In YOLOv3, the convolution layer generally has 79 layers, a scale detection result is obtained after several convolution layers. Compared with the input image of the backbone, the feature map used for detection here has 32 times down sampling, for example, the feature map is 13 * 13 if the input image is 416 * 416. Because of the high subsampling ratio, the receptive field of the feature image is large so that it is suitable for detecting large-scale objects in the image. In order to achieve fine-grained detection, the feature map of the 79th layer needs up-sampling first, and then concatenated with the 61th layer feature map. In this way, the 91th layer fine-grained feature map is obtained, and the 16 times down sampling feature map of the relative input image is obtained after several convolution layers, which has medium scale receptive field and is suitable for detecting medium scale objects. Finally, the feature map of the 91th layer starts to do up sampling again, then being concatenated with the 36th layer feature map so that the 8 times down sampling feature map of the relative input image is obtained, of which the receptive field is small so that it is suitable for detecting small-scale objects in the image. Thirdly, during the object classification process, the algorithm of softmax is replaced by logistic, which can support multi-label objects detection.

Facenet. Facenet can directly map face image to euclidean space, and the length of space distance represents the similarity of face image. For example, the Euclidean distance of different face images for the same person is smaller, while that of different face images in Euclidean space is larger. As long as mapping relationships are identified, some tasks such as face recognition, validation, and clustering can be easily accomplished. The model of Facenet is on the basis of deep convolution neural network, of which the detection accuracy is above 0.99 on LFW dataset and above 0.95 on YouTube Faces DB dataset.

The existing methods of face recognition algorithm on the basis of depth neural network adopts classification layer: the middle layer of this network architecture is the vector map of the face image, and then the classification layer is considered as the output layer of the network, of which the disadvantages are not direct and inefficient.

The Facenet architecture flow chart is presented in Fig. 3. Different from the current existed methods, facenet employs LMNN (maximum boundary nearest neighbor classification) as its loss function, which is based on triplets to train the neural network, from which 128 dimension vector space is output directly. The selected triplets loss include two matching face thumbnails and one non matching face thumbnail. The loss function aims to distinguish positive and negative classes by distance boundary.

3.2   Real Dataset of Construction Site

Since there are few open source dataset of safety helmet under the real construction site, we conduct a month's data collection from the cameras installed on a specific place. In order to maintain the richness and diversity of the dataset, the strategy of data collection is to employ ten cameras fixed on different locations across the construction site, recording videos with different range of vision covered, under various weather condition and separate time periods during a day. Meanwhile, in order to relieve the pressure of server cache needed, the video data was collected when many workers appear in video pictures instead of being collected all day. In this way, 36000 min of original video data was obtained and images were extract from the videos. In addition, crawler technology is adopted to obtain 8000 safety helmet image data from the network, and then the two kinds of data were fused together to make dataset for training YOLOv3 model in this paper.

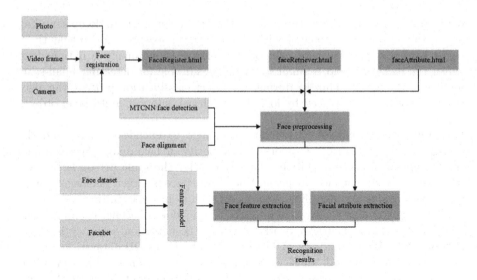

Fig. 3. The Facenet architecture flow chart

3.3   Performance Evaluation

Correctness. The correctness metrics were applied to measure the performance of safety helmet detection model. Two categories of result images can be get, including people who wearing safety helmet and others not. To describe the key indicator of precision, TP_{helmet}(true positive), FP_{helmet}(false positive) and FN_{helmet}(false negative) are used. In more concrete terms, TP means the number of people without wearing safety helmet in the monitored area on the construction site, and the detection results are accurate at the same time. FP indicates the number of those people who are actually wearing a safety helmet but are detected as people who are not. FN is the number of people without wearing helmet but being detected incorrectly. The metrics such as precision, recall and miss rate are defined as follows.

$$Precision_{helmet} = \frac{TP_{helmet}}{TP_{helmet} + FP_{helmet}} \tag{1}$$

$$Recall_{helmet} = \frac{TP_{helmet}}{TP_{helmet} + FN_{helmet}} \tag{2}$$

$$Missrate_{helmet} = \frac{FN_{helmet}}{TP_{helmet} + FN_{helmet}} \tag{3}$$

Speed. The speed of the proposed method refers to the time consumed by the two modules of YOLOv3 and Facenet for one picture frame. The computational complexity is much smaller than that of faster R-CNN algorithm and other previous methods. Considering that this method is to be applied in the monitoring system of the actual construction site, it is necessary to process the detection results in real time.

4   Experiments and Results

4.1   Construction Site Dataset Establishment

Experimental data set is a critical prerequisite for object detection in computer vision field. There is, however, no open-sourced dataset for helmet wearing detection and worker identification from construction plant. In this paper, we made a large-scale data set from ten cameras fixed on the specific construction site for detection and recognition. In the process of building the dataset, it mainly includes four aspects: data collection, data preprocessing, data filtering and data labeling.

Data Collection. The data source of helmet wearing data set is mainly composed of two parts: the practical monitoring data collected from site cameras and data crawled from the Internet. Among them, the practical monitoring data of construction site accounts for 70% of the overall dataset, while the data crawled from the internet is 30% of the total dataset.

In the paper, the practical dataset of a real construction site in a hospital is mainly utilized, in which massive video data in different time periods has been retrieved from ten installed cameras.

In the process of establishing the safety helmet wearing detection database, labeling tool is applied to mark the image. The annotations included the identification of workers without safety helmet. The annotations were saved as XML files in Pascal VOC data format.

Data Preprocessing. Data preprocessing includes two parts: data format conversion and redundant data filter.

Firstly, since the video monitoring data of the construction site is video format file, it is necessary to use OpenCV development library to convert the video format file into image files as the initial image data.

Secondly, since the initial image data were collected from the video surveillance frame by frame, many of the pictures have no pedestrians or helmets, most are backgrounds. Therefore, these redundant pictures need to be removed first. In this paper, the fast R-CNN algorithm is used for data filtering, and the pictures with pedestrians are selected as the target data set.

4.2   Experiment

Choice of Confidence Threshold. In YOLOv3, confidence threshold is an important parameter for the output of each bounding box, which is defined by two aspects: One is to indicate the probability of existing objects in the present bounding box. The other is to express the possible IOU value of the prediction box and the real box when the current box has objects. It is given as a mathematical form below,

$$C_i^j = P_r(\text{Object}) * IOU_{\text{pred}}^{\text{truth}} \tag{4}$$

where $P_r(\text{Object})$ means the possibility of objects in the current bounding box, and $IOU_{\text{pred}}^{\text{truth}}$ means the value of IOU.

In this paper, the target detection results have two categories, one is 'safety helmet' and the other is 'no safety helmet'. Therefore, the confidence here indicates the possibility that whether the tested object is wearing a safety helmet or not. For example, when the confidence value is set to 0.8, it means that the detected target is 80 % likely to wear a safety helmet. Positive samples are explicated that when the value of confidence exceeds the confidence threshold, as a result, the appropriate confidence threshold plays a critical role in the classification of positive and negative samples. Figure 4 shows a (P-R) graph, which contains different precision and recall values, given as below:

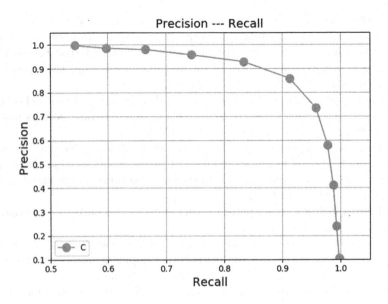

Fig. 4. The Precision-recall curve.

In order to obtain higher precision and recall values at the same time, confidence value is set as 0.85.

The Impact of Different Postures. Figure 5 shows the detection results of workers whether wearing safety helmet in different postures, from which Fig. 5(a) indicates the detection results of one people when he was running, and Fig. 5(b) shows the detection results when squatting, while the detection results for the same people when sitting is indicated by Fig. 5(c). Figure 5(d) shows that the detection results for people who is standing. In this experiment, different test results are marked with different colors, among which the detection for wearing safety helmet is marked with blue, and that of not wearing safety helmet is labeled with red. It can be seen that with regard to the four kinds of postures for the same worker, the test result expresses good performance of the detection for whether wearing safety helmet.

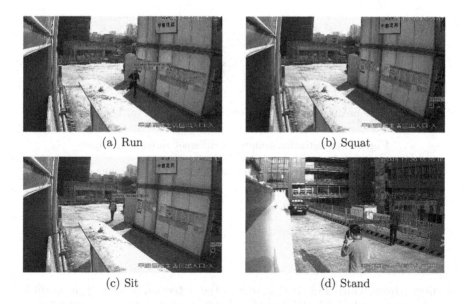

(a) Run

(b) Squat

(c) Sit

(d) Stand

Fig. 5. The detection results of different postures for people

The total number of pictures for different postures to be detected is 3000. Table 1 shows the number statistics of test results, including people with different postures in construction site.

Table 1. The number statistics of test results for different postures

Classes	TP_{helmet}	FP_{helmet}	FN_{helmet}	P_{helmet}	R_{helmet}	$Missrate$	Speed
Run	1561	81	93	95.0	94.3	5.7	0.059
Squat	1007	52	67	95.0	93.7	4.3	0.053
Sit	916	35	26	96.3	97.2	2.8	0.057
Stand	1208	68	54	94.6	95.5	4.5	0.051

The Impact for Different Range of Distance. In different range of distance, the detection results for workers whether wearing safety helmet are indicated in Fig. 6. Figure 6(a) shows that the result for safety helmet detection in close range of distance, from which the score is very accurate with 96.7% in long range. It also can be seen that the performance for safety helmet detection is excellent whether in moderate range or in long range of distance from Fig. 6(b) and 6(c).

The total number of pictures for different range of distance to be detected is 2000. Table 2 shows the number statistics of test results, including people at different range of distance.

The Impact for Different Degrees of Occlusion. Figure 7 shows the detection results of workers whether wearing safety helmet in different degrees of

(a) Close range   (b) Moderate range   (c) Long range

Fig. 6. The detection results for different range of distance

Table 2. The number statistics of test results for different range of distance

Classes	TP_{helmet}	FP_{helmet}	FN_{helmet}	P_{helmet}	R_{helmet}	$Missrate$	Speed
Close range	1003	57	42	94.6	96.0	4.0	0.067
Moderate range	1116	62	53	94.7	95.4	4.6	0.061
Long range	980	34	51	96.7	95.0	5	0.063

occlusion, from which Fig. 7(a) indicates the detection results of no occlusion between two people, with the score being 0.96 for safety helmet and 0.97 for no safety helmet, while Fig. 7(b) and 7(c) shows the good detection result for slight occlusion and serious occlusion between two people.

(a) No occlusion   (b) Slight occlusion   (c) Serious occlusion

Fig. 7. The detection results for different degrees of occlusion

The total number of pictures for different degrees of occlusion to be detected is 1000. Table 3 shows the number statistics of test results, including the detection results from different degrees of occlusion

Table 3. The number statistics of test results for different degrees of occlusion

Classes	TP_{helmet}	FP_{helmet}	FN_{helmet}	P_{helmet}	R_{helmet}	$Missrate$	Speed
No occlusion	1501	77	59	95.1	96.2	3.8	0.053
Slight occlusion	1037	47	32	95.7	97.0	3.0	0.061
Serious occlusion	977	53	41	94.8	95.9	4.1	0.057

5 Conclusions

The safety of the construction site is still a very severe challenge for both the construction unit and the workers, so it is particularly significant to improve the safety protection and supervision measures of the construction site. Some previous work about safety helmet detection is unsatisfactory in accuracy and speed so that those methods cannot be applied completely in the actual construction project. In addition, many of the previous site monitoring systems are only for the application of helmet detection module, which cannot identify the information of workers and then regulate the behavior of employees when they have negligent behavior.

In this paper, a novel method for site monitoring on the basis of combined object detection and face recognition is proposed, which is based on YOLOv3 and Facenet algorithm respectively. First of all, a large number of video data from the construction site was collected in the stage of preliminary data preparation, and then these video data are all divided into massive images so as to be made into the dataset of construction site. Secondly, the test dataset is divided into 12 categories according to the visual characteristics for the construction site, on which the performance of the helmet detection module can be evaluated. The test results demonstrate that the precision and recall rate are above 94% and 95% respectively, which shows that the method can be successfully applied to the monitoring system of the construction site. The facenet model is employed in the human information recognition module in this paper owing to its high performance. The face images and identity information of 30 workers on the construction site are collected in the paper, and then the face images are sent to the facenet model for face database. If employees are identified, their identity information will be returned to the monitoring platform, so as to assist them in video punch.

At present, the method that put forward in this paper can be successfully applied to the site monitoring platform. In the future, some other detection functions need to be added to the video monitoring system, such as the identification of intrusion behavior in dangerous areas, which can further improve the system monitoring function and ensure the safety of workers.

Acknowledgments. This work is supported by National Science Foundation of China (61473182, 61877065), China Post-doctoral Science Foundation (Grant 2018M631005), Key Project of Science and Technology Commission of Shanghai Municipality under Grant No. 16010500300, Natural Science Foundation of Shanghai (18ZR1415100), and Defense Industrial Technology Development Program(JCKY2017413C002).

References

1. Liu, W., et al.: SSD: single shot MultiBox detector. In: Leibe, B., Matas, J., Sebe, N., Welling, M. (eds.) ECCV 2016. LNCS, vol. 9905, pp. 21–37. Springer, Cham (2016). https://doi.org/10.1007/978-3-319-46448-0_2

2. Girshick, R., Donahue, J., Darrell, T., Malik, J.: Rich feature hierarchies for accurate object detection and semantic segmentation, pp. 580–587 (2014)
3. He, K., Zhang, X., Ren, S., Sun, J.: Spatial pyramid pooling in deep convolutional networks for visual recognition. CoRR abs/1406.4729 (2014)
4. Tan, H., Chen, L.: An approach for fast and parallel video processing on apache Hadoop clusters. In: 2014 IEEE International Conference on Multimedia and Expo (ICME), pp. 1–6 (2014)
5. Ghasemi, A., Kumar, C.N.R.: A video surveillance system methods at public zone. In: 2016 2nd International Conference on Applied and Theoretical Computing and Communication Technology (iCATccT), pp. 523–526 (2016)
6. Chen, S., Jia, K., Liu, P., Huang, X.: Taxi drivers' smoking behavior detection in traffic monitoring video. In: 2019 Asia-Pacific Signal and Information Processing Association Annual Summit and Conference (APSIPA ASC), pp. 968–973 (2019)
7. Luo, R., Li, L., Huang, W., Sun, Q.: Multi-strategy object tracking in complex situation for video surveillance. In: 2008 IEEE International Symposium on Circuits and Systems, pp. 2749–2752 (2008)
8. Wu, C.H., Ho, G.T.S., Yung, K.L., Tam, W.W.Y., Ip, W.H.: An RFID-based fallen object detection system: a case study of Hong Kong's light rail system. IEEE J. Radio Freq. Ident. 2(2), 55–67 (2018)
9. Li, X., Xue, Y., Malin, B.: Detecting anomalous user behaviors in workflow-driven web applications. In: 2012 IEEE 31st Symposium on Reliable Distributed Systems, pp. 1–10 (2012)
10. Su, C.C.: An open source platform for educators. In: Fifth IEEE International Conference on Advanced Learning Technologies (ICALT 2005), pp. 961–962 (2005)
11. Ghasemi, A., Kumar, C.N.R.: A novel algorithm to predict and detect suspicious behaviors of people at public areas for surveillave cameras. In: 2017 International Conference on Intelligent Sustainable Systems (ICISS), pp. 168–175 (2017)
12. Xu, Z., Zhang, J., Xu, Z.: Memory leak detection based on memory state transition graph. In: 2011 18th Asia-Pacific Software Engineering Conference, pp. 33–40 (2011)
13. Li, J., et al.: Safety helmet wearing detection based on image processing and machine learning, pp. 201–205 (2017)
14. Manocha, D., Purkayastha, A., Chachra, Y., Rastogi, N., Goel, V.: Helmet detection using ML & IoT (2019)
15. Waranusast, R., Bundon, N., Timtong, V., Tangnoi, C., Pattanathaburt, P.: Machine vision techniques for motorcycle safety helmet detection, pp. 35–40 (2013)
16. Girshick, R.: Fast R-CNN, pp. 1440–1448 (2015)
17. Ren, S., He, K., Girshick, R., Sun, J.: Faster R-CNN: towards real-time object detection with region proposal networks. IEEE Trans. Pattern Anal. Mach. Intell. 39(6), 1137–1149 (2015)
18. Redmon, J., Divvala, S., Girshick, R., Farhadi, A.: You only look once: unified, real-time object detection. In: 2016 IEEE Conference on Computer Vision and Pattern Recognition (CVPR), 779–788 (2016)
19. Redmon, J., Farhadi, A.: Yolov3: an incremental improvement. CoRR abs/1804.02767 (2018)
20. Doungmala, P., Klubsuwan, K.: Helmet wearing detection in Thailand using Haar like feature and circle Hough transform on image processing. In: 2016 IEEE International Conference on Computer and Information Technology (CIT), pp. 611–614 (2016)

21. Dasgupta, M., Bandyopadhyay, O., Chatterji, S.: Automated helmet detection for multiple motorcycle riders using CNN. In: 2019 IEEE Conference on Information and Communication Technology, pp. 1–4 (2019)
22. Rohith, C.A., Nair, S.A., Nair, P.S., Alphonsa, S., John, N.P.: An efficient helmet detection for MVD using deep learning. In: 2019 3rd International Conference on Trends in Electronics and Informatics (ICOEI), pp. 282–286 (2019)
23. Bo, Y., et al.: Helmet detection under the power construction scene based on image analysis. In: 2019 IEEE 7th International Conference on Computer Science and Network Technology (ICCSNT), pp. 67–71 (2019)

A Novel Approach of Human Tracking and Counting Using Overhead ToF Camera

Haikuan Wang[1,2(✉)], Haoyang Luo[1,2], Wenju Zhou[1,2], and Dong Xie[1,2]

[1] School of Mechanical Engineering and Automation, Shanghai University, Shanghai 200072, China
{hkwang,luohy,zhouwenju,xiedong}@shu.edu.cn
[2] Shanghai Key Laboratory of Power Station Automation Technology, Shanghai University, Shanghai 200072, China

Abstract. Human detection is a critical measure used for obtaining the information of flow density in a region, of which the data is conductive to the customer flow analysis for some businesses, avoiding human stampedes caused by excessive people simultaneously. Although previous research into human detection is generally based on a 2D camera, the data collected by it is vulnerable to pedestrian occlusion, complex background, shadow interference, and other uncontrollable factors. Meanwhile, human detection algorithms based on 3D are also studied currently by some researchers, however, these methods suffer from low efficiency and high error rate. Thus, a novel real-time human detection algorithm based on the ToF camera is proposed in this paper. Different from other detection algorithms running on the computer, our algorithm is implemented on an embedded 3D camera platform. This method is divided into two steps to go on: firstly, the novel detection algorithm determines the head position by finding the peak value for the region of interest (ROI). Secondly, combined with an improved Hungarian algorithm and a prediction of human movements using a Kalman filtering which facilitates the tracking rate significantly. The experimental results demonstrate that the accuracy of the novel algorithm in real-time people detecting and tracking is above 95%.

Keywords: People detection · Human tracking · Time-of-Flight · Hungarian algorithm

1 Introduction

Human counting is an important but challenging task in computer vision. It can be applied to a variety of video surveillance applications. Conventional methods are achieved by using 2D image features, such as popular human feature HOG which is proposed by DALAL et al. [1], and there are many applications combined HOG and SVM. Machine learning is widely used due to its outstanding performance, it is also applied in the field of people counting. For example, Huang et al. [2] use neural networks to count people.

Under the influence of external light, shadows, complex backgrounds, and other factors, the poor image quality which further increases the difficulty of human detection.

© Springer Nature Singapore Pte Ltd. 2020
M. Fei et al. (Eds.): LSMS 2020/ICSEE 2020 Workshops, CCIS 1303, pp. 416–429, 2020.
https://doi.org/10.1007/978-981-33-6378-6_31

As a result, another research direction is based on the 3D image captured by a depth sensor, such as the ToF camera. It can avoid unwanted problems due to illumination, color, and texture, etc. However, occlusion as the main problem in flow counting is still unsolved. In order to avoid mutual occlusion between pedestrians as much as possible, the camera is mounted perpendicular to the ground in the so-called top-view position, this enables an easier separation and tracking of people. In such images, pedestrians usually have only the head (especially the top of the head) being more complete.

In 2012, Dan et al. [3] developed a new robust people counting system based on the fusion of depth and visual data, which uses human models to extract human objects from preprocessed depth images. Then, the trajectory of the detected object is established by applying a bidirectional matching algorithm. Zhang et al. [4] use a vertical Kinect ®v1 [5] sensor to obtain a depth image. Since the head is always closer to the Kinect sensor than other parts of the body, people counting task equals to find suitable local minimum regions. The author proposed a novel unsupervised water filling method that can find head robustly, however, the computational cost of the water filling algorithm is high and not suitable in the embedded system. The proposal by Stahlschmidt et al. [6], where the authors present a system that also uses an overhead ToF camera. The proposal has two steps: first, they use a matched filter in order to distinguish people from different objects, which involves Mexican hat wavelet. Peaks indicate an upright standing person. Then, their location is used for the Kalman-filter based tracking procedure. Another interesting proposal is described in Del Pizzo and Carletti et al. [7, 8], they use background subtraction before blob detection. Then, the blob filter and split after calculating whether the pixels in blob belonging to the foreground surpass the threshold. The proposed method allows us to achieve comparable accuracy using very few computational resources, but it works well only in fewer interference scenarios.

In this work, we describe a real-time people counting system based on 3D depth images. Although many methods use the ToF camera that captures and makes use of both 2D and 3D images, our algorithm only uses depth information, because people counting based on sensor fusion may cause a heavy burden on the embedded processor. Besides, the ToF camera generated 3D image need preprocessing to improve detecting accuracy, including background subtraction and depth images recovery. After that, the sliding window is used to determine the peak value of the corresponding region for people segmentation. Finally, we introduce a tracking algorithm that combines improved Hungarian algorithm and prediction of object movement. Dataset from our laboratory is used to evaluate people counting accuracy

This paper is organized as follows. Section 2 introduces the camera system briefly. The image preprocessing is depicted in Sect. 3.1. Section 3.2 and 3.3 outline the specific process of segmentation and tracking algorithm. The experiment is described in Sect. 4, including experiment platform, result, and analysis. Finally, Sect. 5 draws a conclusion.

2 Overview of System

This paper mainly studies the movement (entering or leaving) of people in a specific area (such as doors, buses, etc.). The simulation diagram is shown in Fig. 1. The camera is suspended above and faces towards the ground.

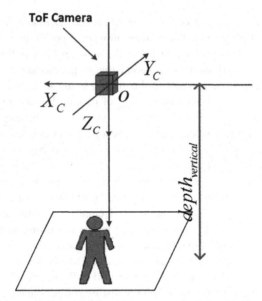

Fig. 1. Schematic diagram of tracking target and relative position of camera

The camera will capture every 320 * 240 raw image, then depth image will be sent to Arm-based embedded board, all image processing procedures are completed by the embedded system. Meanwhile, the camera can communicate with the computer by SOCKET, sending images and processing results to the host computer in real-time. As a result, we can see the original image, as well as the process and results of the tracking task on the computer. Real-time statistics of incoming and outgoing traffic will also be displayed on the computer program.

3   Proposed Method

Including three parts as mentioned ahead: image preprocessing, counting, and tracking.

3.1   Depth Image Preprocessing

Because the emission of infrared light will be absorbed by black objects according to the principle of ToF camera. Noise is especially profound in the dark region such as human hair and black tile on the floor. The depth image captured by using the ToF camera suffers from data loss when the surface of the object has poor reflectance for a similar reason. Therefore, the depth image of the object contains the lost data regions appearing as small holes as shown in Fig. 2 (a). Since our algorithm relies on find the head of people, these holds have a remarkable effect on recognition and tracking. And it is vital to improving the quality of acquired depth images in this case. As for our algorithm, background subtraction is the first step.

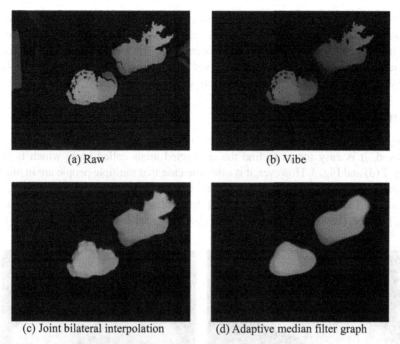

(a) Raw   (b) Vibe

(c) Joint bilateral interpolation   (d) Adaptive median filter graph

Fig. 2. Three steps of pretreatment. The image quality has been greatly improved by removing the background, repairing the holes, and smoothing the edges of the targets.

In order to avoid interference in subsequent processing, we decided to remove the irrelevant items in the image first, only people's body will be retained. ViBe [9] is a pixel-level video background modeling and foreground detection algorithm. Its principle is to store a sample set for all pixels. The sample values stored in the sample set are the pixel values of the frame before the pixel and the pixel values of the surrounding points. The historical value of the pixel value of the new frame and its sample pixel value are compared to determine whether it belongs to the background point. The vibe algorithm is simple and easy to implement. It is friendly to environmental changes and has high computational efficiency. It is also robust to stationary targets and shadow foregrounds.

After background subtraction, as we can see in Fig. 2 (b), there are many holes on the human body. Nowadays, image morphology processing tools can help us solve this problem, such as many upsampling algorithms. In this paper, we use the joint bilateral interpolation algorithm, because the algorithm adds a guide image to the image inter-polation process, it can better retain the edge information than the general interpolation algorithm. At the same time, the problems such as partial distortion, edge jaggedness, and edge blur that are easy to appear in the interpolated image are also solved.

Since different environmental noises will affect the image analysis, filtering and denoising the image is also an indispensable step in image preprocessing. Median filter-ing is a kind of non-linear digital filter technology, which is often used to remove noise signals in images, especially it has a strong suppression effect on isolated noise points such as pepper and salt noise. The basic principle is to select each pixel in the image in turn and replace the pixel with the median value of each pixel in its neighborhood.

After these three steps, the problems of missing depth values and regional noise in the image is mainly solved. The preprocessed image in Fig. 2 (d) has much better image quality, which does good to people counting in remaining steps.

3.2 Human Dectection

Through the preprocessing of 3D images, the image quality has been significantly improved. It is easy for us to find the connected areas called blob, which is shown in Fig. 2 (d) and Fig. 3. However, it is often the case that multiple people are in one blob because they are too close to others. And this is also a major problem when dealing with 3D images. The method proposed by [10] detects humans by region-growing. Lei and Wang [11] use the Adaboost algorithm to classifier head and shoulder profile.

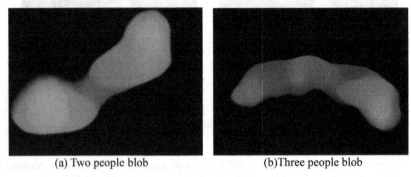

(a) Two people blob   (b)Three people blob

Fig. 3. When more than one person's body is too close, it can cause this kind of situation that bodies appear to be connected from the image.

Here we proposed a novel method, and make full use of the characteristics of 3D images. In 3D images, different gray values represent different heights. Head and shoulder is the most notable feature on top-of-view. Head is always higher than one's shoulder, in other words, the head is the closet organ to the camera in the human body (Fig. 4).

The outline of the edge of the head is round, while the outline of the shoulder surrounds the head and is oval. Generally, other parts of the body will be blocked by the shoulders. As a result, there will be a noticeable difference in height between the head and shoulders, as well as between the shoulders and the ground. And this feature basically exists from different perspectives, we use it for detection.

The minimum bounding rectangle of every blob was defined as ROI. The sliding window was introduced to distinguish different targets in ROI. When the target enters the camera's field of view, there will be obvious peaks and valleys in the depth histogram. When peak appears in the area of the sliding window, the highest point will be regarded as head at present. Other pixels surrounded peak in the sliding window will be taken into consideration in order to judge whether it meets the characteristic criteria mentioned above. Considering different person's height may vary widely, sliding window use adaptive threshold for detection.

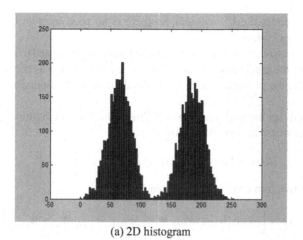

(a) 2D histogram

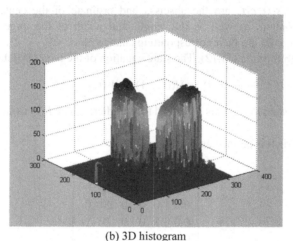

(b) 3D histogram

Fig. 4. The above two photos show two people entering the camera's field of view. There are obvious features as we can see from the histogram of the depth image.

3.3 Targets Matching

By searching in the ROI, we can find the head position which meets the conditions, but we also need an algorithm to associate the targets between different frames. So in this chapter, we first use the Kalman filter to estimate the possible areas of the target, and then use the Hungarian algorithm for correlation. Additionally, we took the computing power of the embedded platform into consideration.

3.3.1 Search Range Constraint Based on Kalman Filter

Kalman filtering can estimate the state of dynamic systems by calculating and analyzing a series of incomplete measurements. We only need to know the estimated value of the $(k-1)^{th}$ moment and the observation value under the kth moment state to calculate the estimated value under the kth moment, so there is no need to record the observed or estimated historical value.

Before target association matching, Kalman filtering can be used to narrow the search range. By using the eigenvalue of the ith target at the tth frame as the input parameter of Kalman filter, the predicted value is obtained and the corresponding search area is calculated.

It is assumed that the state variable of Kalman filter is an eight dimensional variable including the position and size of moving target:

$$R(t) = \left[x(t)\, y(t)\, s(t)\, c(t)\, x'(t)\, y'(t)\, s'(t)\, c'(t) \right]^T \tag{1}$$

Then the estimation error is as shown in (2):

$$W(t) = \left[W_{x(t)}\ W_{y(t)}\ W_{s(t)}\ W_{c(t)}\ W_{x'(t)}\ W_{y'(t)}\ W_{s'(t)}\ W_{c'(t)} \right]^T \tag{2}$$

where $x(t), y(t), s(t), c(t)$ are the abscissa and ordinate of the center of the mass of the target contour, as well as the length and width of the target contour. And $x'(t), y'(t), s'(t), c'(t)$ are their corresponding change rates.

According to formula (1)–(2). The state equation of the target is defined as follow:

$$R(t) = \delta R(t-1) + W(t-1) \tag{3}$$

where the latest state $R(t-1)$ is projected to a predicted state $R(t)$ using system matrix δ.

System matrix δ is defined as:

$$\delta = \begin{bmatrix} 1 & 0 & 0 & 0 & \Delta t & 0 & 0 & 0 \\ 0 & 1 & 0 & 0 & 0 & \Delta t & 0 & 0 \\ 0 & 0 & 1 & 0 & 0 & 0 & \Delta t & 0 \\ 0 & 0 & 0 & 1 & 0 & 0 & 0 & \Delta t \\ 0 & 0 & 0 & 0 & 1 & 0 & 0 & 0 \\ 0 & 0 & 0 & 0 & 0 & 1 & 0 & 0 \\ 0 & 0 & 0 & 0 & 0 & 0 & 1 & 0 \\ 0 & 0 & 0 & 0 & 0 & 0 & 0 & 1 \end{bmatrix} \tag{4}$$

Through defined system state equations and observation equations, Kalman filtering is used to estimate the target motion parameters. According to the center position of the search area, the boundary range of the search area can be obtained.

By $s(t), c(t)$, we can get the boundary of the search area:

$$W_x = s(t) \times 2.5,\ W_y = s(t) \times 2.5 \tag{5}$$

$$x_c = x_c^{prev} + \Delta x,\ y_c = y_c^{prev} + \Delta y \tag{6}$$

Then the search area is:

$$x_c - W_x/2 \le x \le x_c + W_x/2,\ y_c - W_y/2 \le y \le y_c + W_y/2 \tag{7}$$

3.3.2 Hungarian Tracking Algorithm

The Hungarian algorithm is a combinatorial optimization algorithm that solves task assignment problems in polynomial time. The assignment problem is actually the maximum matching problem for bipartite graphs.

However, the Hungarian algorithm treats the status of each matching object as equal, so the maximum match solved under this premise is often not our optimal solution, because the close target of adjacent frames should have a higher weight. Then we improved the Hungarian algorithm's matching strategy, so that it is more in line with the real situation.

The added loss function is defined as shown in below:

$$V(i,j) = \alpha D(i,j) + \beta S(i,j) + \lambda U(i,j) \tag{8}$$

$$D(i,j) = \frac{\sqrt{(x_t^i - x_{t+1}^j)^2 + (y_t^i - y_{t+1}^j)^2}}{Max_n \sqrt{(x_t^i - x_{t+1}^j)^2 + (y_t^i - y_{t+1}^j)^2}} \tag{9}$$

$$S(i,j) = \frac{|S_t^i - S_{t+1}^j|}{Max_n |S_t^i - S_{t+1}^j|} \tag{10}$$

$$C(i,j) = \frac{|C_t^i - C_{t+1}^j|}{Max_n |C_t^i - C_{t+1}^j|} \tag{11}$$

Among them $x_t^i, y_t^i.S_t^i, C_t^i$ are the abscissa, ordinate, area and perimeter of the center of mass surrounded by the contour of the ith target in the tth image sequence. ($1 \leq n \leq num(t+1)$), $num(t+1)$ is the target number in $(k+1)^{th}$ frame. $D(i,j)$ represents the size of the contour centroid of the ith target in the tth frame and the jth target in the $(k+1)^{th}$ frame. The smaller the value, the closer the centroids of the two targets. $S(i,j)$ and $C(i,j)$ indicates the deformation similarity of the area and perimeter of ith target in the tth frame and the jth target in the $(k+1)^{th}$ frame. $V(i,j)$ refers to proximity of surrounding contour of ith target in the tth frame and the jth target in the $(k+1)^{th}$ frame. The smaller the value, the more likely the target and the greater the weight. We define $\alpha = 0.7, \beta = 0.2, \lambda = 0.1$.

Suppose in the weighted bipartite graph $G = (X, Y)$, x and y correspond to the target set of frame t and frame t + 1. Give a mark to the vertices of $\forall i \subset X$ and $\forall j \subset Y$, denoted as $L(i)$ and $L(j)$ respectively. $L(i)$ is shown in following formula 7.

$$l(i) = \begin{cases} min_{\forall j \in Y}(\phi(i,j)), & i \in X \\ 0, & j \in Y \end{cases} \tag{12}$$

where $\phi(i,j)$ represents the value of the edge (i,j), and $\phi(i,j) = V(i,j) \geq L(j) - L(j)$. Meanwhile, the initial feasible vertex symbol G is marked to determine the equal subgraph G_L, and then the weighted Hungarian algorithm is used to find whether a perfect match exists.

If a perfect match that meets the requirements can be found in G_L, then the match is considered to be the minimum weight match as required, after the end of the search for the match, the target association is completed.

4 Experiments and Results

4.1 Hardware Platform

The ToF camera used in this paper is shown in Fig. 5 (a). After the depth data is collected by the ToF camera, it is processed by the embedded board (as shown in Fig. 5 (b), chip is Arm-based RK3399). The parameters of this camera are shown in Table 1. Many papers use Kinect cameras, but we use laboratory-made ToF cameras, which makes us more flexible in the process of use, because the camera has better performance and more adjustable parameters.

(a) Lab-made ToF camera   (b) Embedded board

Fig. 5. The lab-made ToF camera is used in experiment.

Table 1. Main parameters of the ToF camera.

Parameter	Value
Resolution	320×240
Frame rate	$35fps$
FOV	$72(h) \times 34.6(v)$
Range	$0 \sim 12$ m
Accuracy	10 mm

4.2 Experiment Strategy

The 3D image is divided into three regions which is shown in Fig. 6: region in, tracking region, and region out. The region is divided for the purpose of setting personnel entry and exit standards. When the target appears in the in area and out area, it indicates that the target may enter the tracking area. When the target crosses the boundary and is captured in the tracking area for three consecutive frames, the target is activated and tracked in the image, then retained in the tracking library including features and movement. When the target crosses the area line in the opposite direction, and the target is not captured in the tracking area for two consecutive frames, then the target is considered to have disappeared, the related trajectory will be cleared, and the historical information of the target is eliminated in the target tracking library.

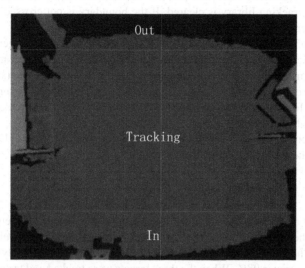

Fig. 6. Target entry and exit line. The picture is divided into three parts in order to count the flow of people accurately.

Assume that the distance of the center of the moving target between two frames is ζ. We define from the out area to the in area is positive direction (crossing the three area boundaries in order). When the algorithm starts running, the in and out counters are started. Here is a detailed tracking algorithm strategy:

(1) Three consecutive frames in an image sequence are collected to meet the standard that the centroid interval is less than ζ, which means that the target is activated in the image (prevents environmental interference), and the target's features and motion status are recorded in the target tracking library. If this condition is not met, activation fails.

(2) After activating the target and accepting the kth frame of the image. The kalman filter is used to search and predict the area where i^{th} active target may appear ($1 \leq i \leq sum$, sum is the number of targets that have been activated) according to the features and motion status in the target tracking library. After that, calculate the target to be matched with all the suspected active targets in the range predicted by the kth frame, and find the most appropriate corresponding target. Then judge if the distance d between j^{th} active target's centroid and the ith activate the target's centroid is greater than ζ.

(3) If $d \leq \zeta$, then the jth activated target is considered to be the subsequent movement of the ith activated target, then the information of the ith activated target in the tracking target library is updated with the current j^{th} activated target's motion and

feature status, and the activated target is marked. If $d > \zeta$, it indicates that the ith activated target has no subsequent state in the kth frame. Then determine whether the activated target has crossed the dividing line in the forward or reverse direction. If there is a positive crossing of the dividing line, then in +1. If there is a reverse crossing the dividing line, then out +1. The information of the activated target in the target tracking library is cleared. If the boundary is not crossed, the activated target is temporarily stationary, retaining its characteristics.

(4) After matching all the tracking targets, start searching for the target of the frame image to confirm whether there are any omissions. When there are no missing targets, it means that the connection between the kth frame and the $(k-1)^{th}$ frame have been established. If there is a target that has not been marked, temporarily retain the target features and wait for the $(k+1)^{th}$ and $(k+2)^{th}$ frame to determine if the target can be activated.

(5) After receiving the $(k+1)^{th}$ frame of image, continue with steps (1), (2), (3), and (4) until the end of the image frame.

4.3   Result

Figures 7, 8, 9 and 10 indicates that the complexity of the test experiment is gradually increasing. When tracking tests on two targets, the situation is relatively simple, the accuracy and the speed of the algorithm processing images are relatively high, and the overall test accuracy is 98%. When the three targets were tracked and tested, the overlap and occlusion of targets increased, but the accuracy remained at 98%. When four or five targets are tracked and tested, overlapping and occlusion of different targets is more serious, and to a certain extent will increase the probability of false detection and missed detection. However, we relies on multi-object segmentation algorithm and multi-object matching algorithm, and still achieved satisfactory results, with overall tracking accuracy of 97%. At the same time, it is worth noting that with the increase in tracking targets, the time required for algorithm processing has also increased relatively. The specific statistical results of the four groups of experiments are shown in Table 2.

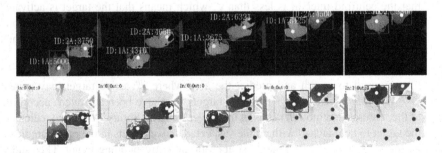

Fig. 7. Tracking two people.

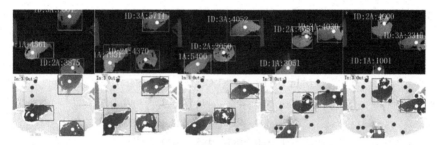

Fig. 8. Tracking three people.

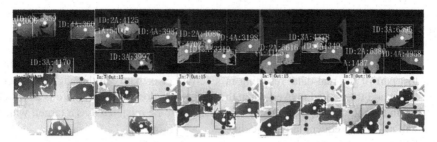

Fig. 9. Tracking four people.

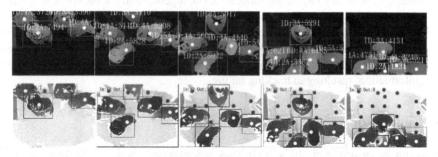

Fig. 10. Tracking five people.

4.4 Error Analyze

Based on the test and analysis of the laboratory platform and the bus platform, the error sources of the dynamic multi-target tracking algorithm proposed in this paper are mainly divided into the following points:

(1) Errors in the acquisition of depth data

The ToF depth camera uses the principle of infrared reflection for depth measurement, and near-infrared light is prone to multi-path reflection and scattering when it passes through the reflection to the collector. At the same time, the chip temperature will change with the use of time, which will affect the movement of the electrons. These factors will cause a certain degree of interference with the depth data.

Table 2. Test result statistics among four different Dataset.

Dataset	Real flow		Captured by camera		Frame rate	Accuracy
	In	Out	In	Out		
1	500	500	482	498	35.3	98%
2	500	500	487	493	33.8	98%
3	500	500	485	485	32.9	97%
4	500	500	477	483	32.1	96%

(2) Error during target segmentation

When the target is segmented, although the image has been pre-processed, a lot of noise is eliminated, and the holes in the image are filled. However, this inevitably lost part of the real data, resulting in a certain error between the edge of the cut target and the real edge. And when the targets are close together, there is a lack of depth data in the occlusion part, which is easy to be misjudged as a target, which will affect subsequent target tracking.

(3) Error during target tracking

When tracking the target, the target may be abruptly changed to a certain degree. For example, when the targets are suddenly close together and the backpack is opened, the target matching association algorithm is likely to cause misjudgment. Once the conditions of the tracking target are not met, it is easy to eliminate the information of the target in the tracking library. Then even if the late target is activated again, the complete motion trajectory cannot be formed due to the loss of the previous information, resulting in tracking failure.

5   Conclusion

Based on the principle of ToF's 3D images, we proposed an algorithm for image pre-processing, segmentation and tracking, and builds a multi-target detection and tracking system that can be used for people counting. Because our experiment is implemented on an embedded platform, it has a great constraint on the computing power of the algorithm, but in general, our algorithm meets the real-time requirements with very high accuracy.

Acknowledgments. This work is supported by National Science Foundation of China (61473182, 61877065), China Post-doctoral Science Foundation (Grant 2018M631005), Key Project of Science and Technology Commission of Shanghai Municipality under Grant No. 16010500300, Natural Science Foundation of Shanghai (18ZR1415100), and Defense Industrial Technology Development Program(JCKY2017413C002).

References

1. Dalal, N., Triggs, B.: Histograms of oriented gradients for human detection. In: 2005 IEEE Computer Society Conference on Computer Vision and Pattern Recognition (CVPR'2005), vol. 1, pp. 886–893 (2005). https://doi.org/10.1109/cvpr.2005.177
2. Huang, D., Chow, T.W.S.: A people-counting system using a hybrid RBF neural network. Neural Process. Lett. 18(2), 97–113 (2003). Doi: 10.1023/A: 1026226617974
3. Dan, B.K., et al.: Robust people counting system based on sensor fusion. IEEE Trans. Consum. Electron. 58(3), 1013–1021 (2012). https://doi.org/10.1109/TCE.2012.6311350
4. Zhang, X., Yan, J., Feng, S., Lei, Z., Yi, D., Li, S.Z.: Water filling: unsupervised people counting via vertical kinect sensor. In: 2012 IEEE Ninth International Conference on Advanced Video and Signal-Based Surveillance, pp. 215–220 (2012). https://doi.org/10.1109/avss.2012.82
5. Smisek, J., Jancosek, M., Pajdla, T.: 3D with kinect. In: Consumer Depth Cameras for Computer Vision, London, pp. 3–25. Springer London (2013)
6. Stahlschmidt, C., Gavriilidis, A., Velten, J., Kummert, A.: People detection and tracking from a top-view position using a time-of-flight camera. In: Dziech, A., Czyżewski, A. (eds.) MCSS 2013. CCIS, vol. 368, pp. 213–223. Springer, Heidelberg (2013). https://doi.org/10.1007/978-3-642-38559-9_19
7. Del Pizzo, L., Foggia, P., Greco, A., Percannella, G., Vento, M.: A versatile and effective method for counting people on either RGB or depth overhead cameras. In: 2015 IEEE International Conference Multimedia Expo Work. ICMEW 2015, pp. 1–6 (2015). https://doi.org/10.1109/icmew.2015.7169795
8. Carletti, V., Del Pizzo, L., Percannella, G., Vento, M.: An efficient and effective method for people detection from top-view depth cameras. In: 2017 14th IEEE International Conference Advanced Video Signal Based Surveillance, AVSS 2017, August 2017. https://doi.org/10.1109/avss.2017.8078531
9. Barnich, O., Van Droogenbroeck, M.: ViBE: a powerful random technique to estimate the background in video sequences. In: ICASSP, IEEE International Conference on Acoustics, Speech and Signal Processing - Proceedings, pp. 945–948 (2009). https://doi.org/10.1109/icassp.2009.4959741
10. Galčík, F., Gargalík, R.: Real-time depth map based people counting. In: B-T, J., Kasinski, A., Philips, W., Popescu, D., Scheunders, P. (eds.) ACIVS 2013. LNCS, vol. 8192, pp. 330–341. Springer, Cham (2013). https://doi.org/10.1007/978-3-319-02895-8_30
11. Zhu, L., Wong, K.H.: Human tracking and counting using the KINECT range sensor based on adaboost and kalman filter. In: Bebis, G., et al. (eds.) ISVC 2013. LNCS, vol. 8034, pp. 582–591. Springer, Heidelberg (2013). https://doi.org/10.1007/978-3-642-41939-3_57

A Novel Text Personalized Analysis Model Based on BP Neural Network

Zhongfeng Wang[1,2,3], Chu Wang[1,2,3]($\boxtimes$), Ligang Li[1,2,3], Zhudong Pan[4], and Yunfeng Zou[5]

[1] Key Laboratory of Networked Control Systems, Chinese Academy of Sciences, Shenyang 110016, China
wangchu@sia.cn
[2] Shenyang Institute of Automation, Chinese Academy of Sciences, Shenyang 110016, China
[3] Institutes for Robotics and Intelligent Manufacturing, Chinese Academy of Sciences, Shenyang 110169, China
[4] DongFang Eleceronics CO., LTD, Yantai 264001, China
[5] State Grid Jiangsu Marketing Service, Nanjing 210019, China

Abstract. In online interactive platform, text analysis has greatly changed people's communication, thinking, and promoted the explosive growth of user-generated information. A large number of texts generated by users have become one of the most representative data sources of big data in recent years. Mining and analyzing user-generated information has become essential part on research of social development. The sentiment analysis on social media text as an information processing technology for analyzing, processing, summarizing and reasoning subjective texts with emotions has received extensive attention in academia and industry in recent years, and has been used among many areas of social media and many applications. The traditional text sentiment analysis research work mainly focuses on analyzing emotions from texts, but ignores the individualized differences of users in emotional expression, thus affecting the quality of analysis results. To solve the problems, this paper is about solving the problem of personalized sentiment analysis of social media texts. Considering the wide application of BP neural network technology in social media was proposed to solve the possible challenges of social media text personalized sentiment analysis.

Keywords: Personalized · Sentiment analysis · Social media · BP neural network

1 Introduction

Text Sentiment Analysis (SA) is some process of summarizing, processing, analyzing and reasoning some sort of subjective texts with one or more kinds of emotional expression (eg, Weibo, online commentary and online news, etc.) [1]. The history of sentiment analysis research is not very long. It began to get widespread attention at around 2000. It developed rapidly and then gradually became a very popular area in the field of researching natural language and text mining. Sentiment analysis also has many aliases and similar tasks, such as Opinion Mining, Sentiment mining, Subjectivity analysis, etc., all of

© Springer Nature Singapore Pte Ltd. 2020
M. Fei et al. (Eds.): LSMS 2020/ICSEE 2020 Workshops, CCIS 1303, pp. 430–441, 2020.
https://doi.org/10.1007/978-981-33-6378-6_32

which can be summarized under sentiment analysis [2]. For example, for a film review, the user's evaluation of the movie is recognized, and analyzing of the product review of the digital camera, such as the emotional tendency of the "price", "size", "zoom", etc. At present, sentiment analysis has become interdisciplinary research area, for example, natural language processing, information retrieval, computational linguistics, data mining, machine learning, artificial intelligence and so on.

The existing text sentiment analysis algorithm is mainly about the user's viewpoints and opinions from the text. Due to the lack of user's characteristics to interpret the text, these algorithms are difficult to completely and accurately reflect the user's true emotional expression. By introducing the influence of users and even product features, a personalized text sentiment analysis method can be proposed to overcome the shortcomings of the current methods.

The rise of research on sentiment analysis is also gradually paid attention to with the development of the Internet, and it became popular in the academic circles at the beginning of this century. And gradually evolved from simple word sentiment analysis to complex text sentiment analysis, for example, specific to the text range, sentiment analysis can identify the user's praise and devaluation of the product from the product evaluation and obtain the user's public opinion hot spots from Weibo Emotional tendencies. In-depth study of these aspects, we can find that text sentiment analysis can mainly have the following two uses: (1) Text sentiment analysis technology can be used to obtain and monitor online public opinion. As social media plays an increasingly important role in people's lives, social media generally has the characteristics of openness, virtuality, concealment, divergence, permeability and randomness, which is The government and public relations companies have new requirements for grasping online public opinion. For example, in the Yinchuan 1.05 bus arson case in 2016 and the Yulin maternal crash in 2017, the government's response to and handling of public opinion was considered to be more Proper. This timely and appropriate treatment of hot public opinion is of great significance for maintaining social stability and safeguarding national security; (2) Text sentiment analysis technology can obtain market prediction information based on the analysis of existing user behavior. With the development of Internet technology, commercial companies have paid more and more attention to customers' feedback and evaluation of products and services. Metrini once wrote about the emotions of consumers in the book "Emotional Economics". The commercial value has been elaborated. He believes that consumer emotions such as joy, anger, sorrow, and happiness will greatly affect consumer behavior. This also confirms with the theory of marketing. The essence of commercial activities such as sales promotion is to create environment and objective conditions to mobilize consumers' emotions and thus affect their consumption behavior. As a whole, the sentiment analysis of social media texts, the current popular methods mostly stay at the qualitative level, and according to their different research granularity can be divided into different level according to the text length and attributes. Nowadays, sentiment analysis has become a common cross-cutting field shared by disciplines such as natural language processing, data mining, artificial intelligence, machine learning, and computational linguistics.

Emotional analysis of texts has been extensively researched and has achieved excellent performance in public evaluation tasks. However, there are few reports on the real usable text sentiment analysis tools, especially the personalized text sentiment analysis tools, which are neglected by academics and industry. The sentiment analysis based on the text on the real social media platform faces the real-time update of the data stream. How to dynamically capture the user's personalized preferences poses a challenge for implementing the sentiment analysis tool. Since personalized sentiment analysis introduces user information, it can cause a "cold start" problem to recommendation systems, which includes two situations: inactive users and unobserved users. Inactive users usually have very little historic record, so it is difficult to accurately get the impact of the users or products through personalized emotional modeling; while new users completely have not documented in the system, which can result in the trained personalized sentiment analysis models totally unavailable. For product review data with evaluation objects, the product also has a "cold start" problem similar to the user.

2 Related Work

Emotions are closely related to individuals, so similarity emotion can be expressed through many words, and the same expression can also carry different emotional polarity or emotions [3]. Most of current sentiment analysis methods ignore this fact and focus on building non-personalized models at a huge population level [2, 4]. Emotional analysis can be used in many areas, including robotic autonomic cognition [5]. However, the particularity and diversity of people's communication and expression often make the global sentiment analysis model incompetent or produce the wrong perspective mining results. For example, a globally shared statistical classification model is difficult to identify a user's personalized emotional expression in a restaurant review. The word "dear" usually contains negative emotions, but sometimes it can also reflect some users' satisfaction with the quality of the restaurant. Therefore, a personalized sentiment analysis model needs to implement a fine-grained emotional interpretation of dynamic, personally different perspectives, and ultimately to mine relevant perspective applications (eg word-of-mouth analysis, emotional search, personalized recommendations, etc.) provide more accurate classification and prediction results. With the continuous deepening of text sentiment analysis research, more and more work begins to consider the influence of user characteristics on sentiment analysis, and focuses on capturing the differences in users' expression. For example, Gao et al. [6] designed effective user-related features to capture the influence of tolerant users on sentiment classification; Li et al. [7] introduced topic information of text and user-word factors to topic models for sentiment analysis; Tan Etc. [8] and Hu et al. [9] used user-to-text and user-to-user relationships to do sentiment analysis on some text of tweets.

A personalized sentiment classification model is proposed by Song et al. [10] based on matrix decomposition for Sina Weibo data, predicting the polarity of emotions by internal product calculation of user and microblog text feature vectors, and by introducing social relations and text grammar units. To enhance the capture of personalized preferences of users and the fine-grained modeling representation of Weibo text, respectively, and finally achieve a significant improvement in the effect of text sentiment analysis. Li et al. considered the attributes of users in social media to help individualized emotional modeling, such as age, gender and location information, and then proposed a sentiment classification algorithm based on graph model [11].

These researches mainly use the characteristics of artificial design to perform sentiment analysis modeling. However, the quality of feature selection may affect the final sentiment analysis results. For example, Song et al. [10] pointed out the long-distance relationship between Weibo users would bring noise.

In addition, some researchers suggest that users' data modeling could solve the sparsity of user data. However, previous studies have not been able to achieve personalized emotional modeling and data sparseness by establishing a unified emotion model or a personal emotion model. For solving the problem, some recent studies have regarded personalized sentiment analysis as multi-task learning, and built an adaptive personalized sentiment analysis model by sharing global models, thereby reducing the number of parameters and alleviating the sparsity of data. Wu et al. [12] combined personalized text sentiment analysis with multi-task learning under a unified architecture, and achieved low-complexity and high-accuracy performance in big data.

Gong et al. [13] adopted the concept of social theory, that is, people's views are different and influenced by changing social norms, and then a personalized emotional classification model MT-LinAdapt based on shared model adaptation is proposed, which captures people. Emotional differences can be learned online efficiently. The method uses a two-layer adaptive strategy, followed by Gong et al. [14] to further consider the three-layer adaptive strategy, which considers personalized emotion modeling at the user group level. This approach introduces social contrast theory [15] (that is, people with similar perspectives and abilities form a group) and cognitive coherence theory [16] (that is, people within the group interact and ultimately agree). The models based on deep neural network are different from traditional machine learning method. It does not need to use well-designed features and dictionaries, but automatically learns the feature representation with discriminantness and can capture deep semantic information. The neural network method has a lot of research work on sentence/document level sentiment classification [17, 18]. Recently, Tang et al. [19] considered the user's preference influence on words and passed the user preference matrix and word vector. The combination is used to capture the local influence of the user, and the personalized emotion is modeled by a combination of word vectors containing user preference information; Tang et al. [20] then consider the product feature impact to further extend the sentiment classification model, first Consensus assumptions related to users and products are proposed and verified, and then the user/product preference matrix and user/product vector are introduced to capture the emotional impact of users and products from the local and global perspectives, and finally in the product review. Chen et al. [21] used the two-layer Long Short-Term Memory of words and sentences to generate vector expressions of sentences and documents, respectively, and introduced attention mechanisms related to users and products (Attention) captures important words from sentences and captures important sentences from the document. The final model achieves good emotional classification performance.

3 BP Neural Network

We use BP neural network-based methods during this paper to process text sentiment analysis. Specifically, we first briefly introduce the basic BP neural network method.

BP neural network method was first proposed by Rumelhart D E and Mcclelland J L.[22] It is a multi-layer feedforward neural network trained according to the error back propagation algorithm and is the most widely used neural network, and has been widely used and developed in many area which is showed in Fig. 1 [23].

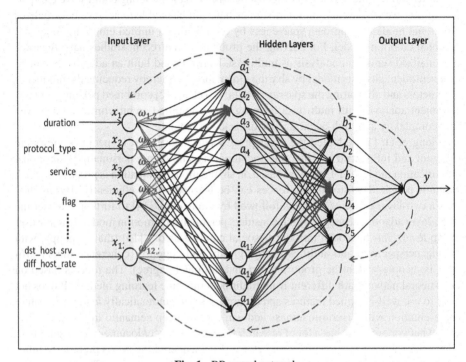

Fig. 1. BP neural network

4   Algorithm

For proving the validity and advantage, in this section we give the algorithm.

4.1   TF-IDF

Figure 2 shows the text word level semantic representation of a 3-dimensional vector space. Each piece of text is represented by the weight of the three word attributes. After expanding to n-dimensional space, the document is represented as $D_i = (W_1,i, W_2,i,..., W_n,i)$. In this paper, the weight of the word is calculated using the TF-IDF method, which is calculated as follows: TFIDF = TF*log(Ndoc/DF) TF is the frequency of the entry, IDF is the number of times a term appears in all text in the text set, and Ndoc is the total number of text sets.

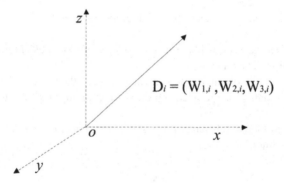

Fig. 2. TF-IDF method

5   Word2vec Model (Skip Gram)

We use word2vec word training method to get the word vector. This paper first select a word in the middle of the sentence as the input word, and then define a parameter of skip_window, which represents the number of words selected from the side (left or right) of the current input word. The other parameter is num_skips, which represents how many different words we choose from the entire window as our output word. Based on these training data, the neural network will output a probability distribution that represents the probability that each word in our dictionary is an output word.

If two different words have similar text "contexts" (the window words are very similar), then the embedded vectors of the two words will be very similar through the Word2Vec model training. Thus, we can better get the sentiment of the word level of the text. The words are independent relationships, and the semantic features of the word level cannot be prepared. The model is showed in Fig. 3.

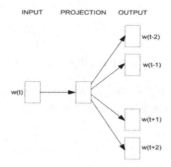

Fig. 3. Word2vec Model

6   Personalized Text Sentiment Analysis Model Based on User Semantic Integration

The shortage of current text Emotional classification and personalization methods:

1) The current text sentiment analysis method ignores the individual differences between users.

2) Ignores potential personalization factors for users, such as language habits, user personality, opinion bias, and so on.

3) How to accurately obtain the dependence between the text and the user, and grasp the potential personality of the user, which is of great significance to the emotional classification.

To solve the problems, this paper proposed a neural network model named UBPNN that based on the emotion of users.

$$x = \sum_{i=1}^{l} e_{(x_i)}/l \tag{1}$$

where l is length of sentence x, $e(x_i)$ is word embedding (word vector) of x_i.

We apply the activation function $g(.)$ to perform a nonlinear transformation projecting,

$$b = g(W_b x + t_b) \tag{2}$$

$$y = g(W_y b + t_y) \tag{3}$$

$$s = soft\,max(W_s y + t_s) \tag{4}$$

where W_b, t_b, W_y, t_y, W_s and t_s are the trainable parameters, $g(.)$ is tanh activation function.

$$b = g(W_b[x,\ u] + t_b) \tag{5}$$

$$y = g(W_y b + t_y) \tag{6}$$

$$s = soft\,max(W_s[y,\ u] + t_s) \tag{7}$$

where u is user embedding, W_b, t_b, W_y, t_y, W_s and t_s are the trainable parameters, $g(.)$ is $tanh$ activation function. The whole method is showed in Fig. 4.

7 Training Method

Input layer: The word embedding of all words of the input sentence is taken as the semantic representation of the sentence. At the same time, the user of the corresponding sentence is randomly initialized into a vector of a certain dimension, and the joint sentence represents the effect of the user information on the semantic level of the word in the common input model. Hidden layer: For the semantic representation of the input, the linear operation of matrix multiplication and the nonlinear activation function are used to obtain the hidden layer semantic representation.

Output layer: Input the semantic representation of the hidden layer, and use the dimensionality reduction operation to get the semantic representation of sentence. The representation joint user representation is entered into the classification layer, incorporating the role of user information for the sentence level.

Classification layer: The obtained vector is mapped into the two-dimensional emotion space, and the softmax method is used for emotion classification.

The whole process is showed in Table 1.

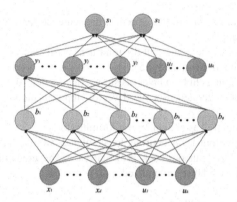

Fig. 4. BP Neural network

8   Experiment

8.1   Dataset

We choose two public datasets to verify our method, at last we selected two public product review datasets: Yelp 2013 and Yelp 2014, which have been showed in Table 2, and divided them into training set, validation set, and testing sets in terms of 80/10/10. The text is scored from one star to five stars, and the number of stars from low to high also indicates that the evaluation is increasingly from negative to positive. (multi-classification problem).

9   Experimental Description and Metric

Due to the complexity of the cold-start problem in personalized sentiment analysis, this article only considers sentiment analysis for users who have never seen before. Based on the UBPNN model mentioned above, we obtained the vector representation of all users. In the test, for the users who never appeared, the arithmetic mean of the user vector was used as the representation of such users, and the combined text was used for emotion classification.

In this paper, we use precision, recall, accuracy rate, F-score and Root Mean Squared Error (RMSE) methods to evaluate our model.

Table 1. The process of our method

Input: training set D = $\{((x_{1k} ..., x_{dk}), (y_{1k} ..., y_{5k}), u_k)\}$, $k = 1, ..., M$, M is the number of training data.
Output: A well-trained personalized UBP neural network with parameters determined.
Function UBPNN(D)

Randomly initialize parameters in the network with uniform distribution
Repeat
For all $((x_{1k},...,x_{dk}), (y_{1k},...,y_{5k}), uk) \in D$
According to the word vector in the sentence, the sentence representation is obtained, as shown in equation (1).
The sentence represents the input, and the joint user word vector is spliced together to the input layer, as shown in equation (5).
Input layer output input as hidden layer, as shown in equation (6)
Hidden layer output, federated user word vector, input to output layer, as shown in equation (7)
Calculate the loss function as shown by the formula (loss)
The gradient of the output layer neurons is calculated, and the adadalta optimization method is used to carry out the back propagation of the gradient, and the network parameter values of each layer are updated.
End for
Until reaches the break condition
End function

Table 2. Dataset of yelp 2013 and 2014

Dataset	category	amount of users	Amount of comments	Average words of comments
Yelp14	5	4,818	231,163	196.9 (words)
Yelp13	5	1,631	78,966	189.3 (words)

Precision:

$$Precision = TP/(TP + FP)$$

Recall:

10   Recall = TP/(TP + FN) = TP/P

The F–score:

11   F–score = 0.5*(Precision + Recall)

The Accuracy:

$$Accuracy = (TP + TN)/(P + N)$$

Generally speaking, the Accuracy rate is higher, the better the classifier;

12 Rmse:

$$RMSE(X, h) = \sqrt{\frac{1}{m} \sum_{i=1}^{m} (h(x^{(i)}) - y^{(i)})^2}$$

13 Experimental Result

The experimental results of the model proposed in this paper are showed in Table 3 and Table 4.

Table 3. Experimental results in Yelp 2013

Yelp2013	Precision	Recall	F-score	Accuracy	RMSE
UBPNN	0.5429	0.5479	0.5430	0.5954	0.7766
UBPNN-coldstart	0.5407	0.5501	0.5389	0.5747	0.7893
BPNN	0.5272	0.5030	0.5095	0.5704	0.7981
K-NN	0.3352	0.31625	0.3216	0.3940	1.2574
SVM	0.2273	0.2063	0.1324	0.4200	1.0934
Naive Bayesian	0.3169	0.3510	0.3221	0.3766	1.634
Decision Tree Model	0.2660	0.2685	0.2671	0.3489	1.668

Table 4. Experimental results in Yelp 2014

Yelp2014	Presion	Recall	F-score	Accuracy	RMSE
UBPNN	0.5583	0.5707	0.5632	0.5956	0.7596
UBPNN-coldstart	0.5433	0.5571	0.5486	0.5752	0.7673
BPNN	0.5161	0.5345	0.5195	0.5535	0.8011
K-NN	0.3457	0.3305	0.3351	0.39824	1.2758
SVM	0.2773	0.2896	0.2806	0.4380	1.0211
Naive Bayesian	0.3212	0.3704	0.3269	0.3699	1.8741
Decision Tree Model	0.2764	0.2786	0.2774	0.3477	1.7224

The experimental results show that the UPBNN method proposed in this paper is superior and effective on these two data sets than other baseline methods.

14   Conclusion

In the paper, we proposed a method based on the neural network BPNN, and then improved it as UBPNN model. The experimental results show that the proposed method is superior to other comparison methods in many aspects and has certain advantages. It is also suitable for the measurement and acquisition of some commodity website evaluation indicators, which can better understand the psychology of users and customers. It should improve their own services and product quality, which is conducive to the expansion and strengthening of the brand.

In the following work, this paper needs to further consider the various situations of the cold start problem, and at the same time improve the model to make it suitable for situations other than Binary Classification and after that we look forward to finding more comparison methods so that our method will gain more recognition in different fields.

Acknowledgments. Project supported by Key Technology of Intelligent Robot Application System in Electricity Business Hall (5210EF18001X), Science and Technology Project of State Grid Corporation Headquarters.

References

1. Zhao, Y., Qin, B., Liu, T.: Sentiment analysis. J. Softw. **21**(8), 1834–1848 (2010)
2. Liu, B.: Sentiment analysis and opinion mining. Morgan and Claypool, pp. 7–8 (2012)
3. Wiebe, J., Wilson, T., Cardie, C.: Annotating expressions of opinions and emotions in language. Lang. Resour. Evaluat. **39**(2), 165–210 (2005)
4. Pang B, Lee L. Opinion mining and sentiment analysis. Found. Trends® Inf. Retrieval, **2**(1–2), 1–135 (2007)
5. Fei, L.U., Yuan, J., Guohui, T.: Autonomous cognition and personalized selection of robot services based on emotion-space-time information. Robot (2018)
6. Gao, W., Yoshinaga, N., Kaji, N., Kitsuregawa, M.: Modeling user leniency and product popularity for sentiment classification. In: IJCNLP, pp. 1107–1111 (2013)
7. Li, F., Wang, S., Liu, S., Zhang, M.: SUIT: a supervised user-item based topic model for sentiment analysis. In: AAAI, pp. 1636–1642 (2014)
8. Tan, C., Lee, L., Tang, J., Jiang, L., Zhou, M., Li, P.: User-level sentiment analysis incorporating social networks. In: KDD, pp. 1397–1405 (2011).
9. Hu, X., Tang, L., Tang, J., Liu, H.: Exploiting social relations for sentiment analysis in microblogging. In: WSDM, pp. 537–546 (2013).
10. Song, K., Feng, S., Gao, W., Wang, D., Yu, G., Wong, K.: Personalized Sentiment Classification Based on Latent Individuality of Microblog Users. In: IJCAI, pp. 2277–2283 (2015).
11. Li, J., Yang, H., Zong, C.: Sentiment classification of social media text considering user attributes. In: NLPCC/ICCPOL, pp. 583–594 (2016).
12. Wu, F., Huang, Y.: Personalized microblog sentiment classification via multi-task learning. In: AAAI, pp. 3059–3065 (2016).
13. Gong, L., Al Boni, M., Wang, H.: Modeling social norms evolution for personalized sentiment classification. In: ACL, pp. 855–865 (2016).
14. Gong, L., Haines, B., Wang, H.: Clustered model adaption for personalized sentiment analysis. In: Proceedings of the 26th International Conference on World Wide Web, pp. 937–946 (2017).

15. Bruhn, J.: The concept of social cohesion. The Group Effect, pp. 31–48 (2009).
16. Newcomb, T.M.: The acquaintance process: looking mainly backward. J. Personal. Soc. Psychol. **36**(10), 1075 (1978).
17. Socher, R., Pennington, J., Huang, E.H., Ng, A.Y., Manning, C.D.: Semi-supervised recursive autoencoders for predicting sentiment distributions. In: EMNLP, pp. 151–161 (2011).
18. Socher, R., Huval, B., Manning, C.D., Ng, A.Y.: Semantic compositionality through recursive matrix-vector spaces. In: EMNLP-CoNLL, pp. 1201–1211 (2012).
19. Tang, D., Qin, B., Liu, T., Yang, Y.: User modeling with neural network for review rating prediction. In: IJCAI, pp. 1340–1346 (2015)
20. Tang, D., Qin, B., Liu, T.: Learning semantic representations of users and products for document level sentiment classification. ACL (1), 1014–1023 (2015)
21. Chen, H., Sun, M., Tu, C., Lin, Y., Liu, Z.: Neural sentiment classification with user and product attention. In: EMNLP (2016)
22. Mcclelland, J.L., Rumelhart, D.E.: Parallel Distributed Processing, Explorations in the microstructure of Cognition, Volume 2: Psychological and Biological Models. MIT Press (1986)
23. Zhang, R., Li, W., Mo, T.: Summary of deep learning research. Inf. Control, **47**(04), 5–17+30

Dynamic Pedestrian Height Detection Based on TOF Camera

Fulong Yao[1], Tianfang Zhou[2], Meng Xia[1], Haikuan Wang[1], Wenju Zhou[1(✉)], and Johnkennedy Chinedu Ndubuisi[1]

[1] School of Mechanical Engineering and Automation, Shanghai University, Shanghai 200444, China
{yaofl,SHU_xiameng,hkwang,zhouwenju}@shu.edu.cn, Kennedite4real@yahoo.com
[2] The Faculty of Engineering, Architecture and Information Technology, The University Of Queensland, Brisbane 4072, Australia
tianfang.zhou@uqconnect.edu.au

Abstract. Pedestrian height is a significant factor in many scenarios, such as behavior analysis or virtual reality. The paper develops a novel height detection method based on the time-of-flight (TOF) camera for dynamic pedestrian. Firstly, the prior error and lens distortion caused by the hardware setups are corrected to provide a more accurate result. Secondly, a new detection method, including a background difference - level set (BD-LS) denoising algorithm and a target extraction based on the maximally stable extremal regions (MSER), is proposed for getting the height of the dynamic pedestrian. Finally, the VICON system has been used as the ground truth to confirm the robustness and practicality of the proposed method. The experimental results show that our method can detect accurately the height of the dynamic pedestrian. In addition, the proposed detection method also can be applied to other related application fields.

Keywords: Height detection · Image denoising · Level set · TOF camera

1 Introduction

Height is a vital parameter that can be used for criminal cases such as suspect tracking or behavior analysis and prediction. It is also useful for the deployment of the 3D virtual reality technology. In the last decade, many detection methods related to the pedestrian height are proposed, such as Cao et al. and Du et al. adopted learning multilayer channel features for active recognition of pedestrian characteristics [1,2], Chen et al. proposed a new action-based pedestrian identification algorithm that can roughly detect the pedestrian height [3], and Shin et al. proposed a motion recognition-based 3D pedestrian navigation system that can be used to get the height information [4]. However, the above papers do not make the height detection as the main content, which leads to the relatively low accuracy of height detection. Up to now, accurate detection of the pedestrian heights is still a big challenge.

© Springer Nature Singapore Pte Ltd. 2020
M. Fei et al. (Eds.): LSMS 2020/ICSEE 2020 Workshops, CCIS 1303, pp. 442–455, 2020.
https://doi.org/10.1007/978-981-33-6378-6_33

In addition, there is a critical issue that is often overlooked. When pedestrians are walking, their heights cannot be kept on the same level, which further increases the difficulty of the height detection for dynamic pedestrians. Although some motion tracking systems (MTS), such as VICON (one of the most accurate MTS in the world) [5], can accurately detect the dynamic changes of height, the devices require high costs for installation and maintenance. In the paper, depth images in the continuous seconds are addressed to get the heights of dynamic pedestrian. The fluctuation of the pedestrian height during the movement is also investigated through analyzing the sequences of images.

Because the ToF camera has a compact structure and is not sensitive to light, it is frequently used in vision fields, such as object detection [6,7], material classification [8,9], positioning and tracking [10–13]. Thus, ToF camera is used in this paper to obtain the pedestrian's stature information. When the TOF camera is used, there are two knotty problems need to be discussed in advance. The first problem is how to correct the prior error caused by the equipment and environment. To face the problem, a pre-correcting test is conducted in the paper to get the prior error. The other problem is how to reduce the impact of lens distortion. A calibration model proposed in our previous work [14] is used to calibrate the depth images grabbed by the TOF camera. After that, a novel detection method based on the depth images is proposed to get the height of dynamic pedestrians. Firstly, a normalization algorithm is designed to convert the depth images into grey images. Secondly, a novel BD-LS algorithm is proposed to eliminate the complex background information. Then, the MSER algorithm and circularity & area are combined to extra the head region. Finally, the data in head region is adopted to calculate the pedestrian height. Furthermore, the VICON is used to verify the feasibility of our method.

In summary, the main contributions of this paper are listed below: 1) A novel accurate height detection method based on TOF camera is created for dynamic pedestrians. It takes into account the changes of pedestrian height during the movement, which is hardly considered in the existing papers. 2) A new target extraction, consists of a BD-LS denoising algorithm and a MSER-based target segmentation, is proposed to get the region of interest (ROI).

The rest of this article is organized as follows. Section 2 introduces the relevant research background. Section 3 shows the core framework and algorithms. Section 4 presents some experiments to show the feasibility and practicability of the proposed Method. A brief conclusion and further work are finally given in Sect. 5.

2   Research Background

2.1   Camera Correction

In this paper, the TOF camera is used to get depth images. The TOF camera determines the distance information by measuring the time of light between the sensor and the surface of the object. In TOF cameras, photons are emitted by high frequency modulated LEDs or Laser Diodes, also known as the modulated infrared light [15]. The modulated infrared light will be reflected by the forward objects, and produce a phase shift that can be used to calculate the distance [16]. The modulation frequency of the used TOF camera is 12 MHz. From the physical characteristics of the camera, the maximum reach distance of the TOF camera is 12.5 m, and the corresponding maximum depth value in depth image is 30000.

The physical distance between the TOF camera and the object can be recovered from the depth data using (1),

$$D = \frac{12.5 * d_i}{30000} \tag{1}$$

where d_i is the depth value of one point from the depth image, and D is the physical distance between the object corresponding to d_i and the TOF camera.

However, due to the difference in camera performance, test equipment and operators, there will inevitably be prior errors [17]. Fortunately, we can obtain the prior error in the pre-correcting test since the prior error is constant and won't change until the camera moves. The correction equation is as follows,

$$E_p = D' - D \tag{2}$$

where E_p represents the constant prior error, D' is the actual physical distance between the camera and the target object, which is obtained by manual measurement, D is the physical distance calculated by (1).

Once the prior error is got, the transfer Eq. (1) can be improved to get (3),

$$D_{dis} = E_p + \frac{12.5 * d_i}{30000} \tag{3}$$

where D_{dis} is the physical distance after camera correction.

In this paper, the resolution of the depth image acquired by the TOF camera is 320 * 240. The subsequent experiments are based on depth images with this resolution.

2.2 Height Fluctuation of Dynamic Pedestrians

The height cannot be kept on the static level while the pedestrian is moving. And the height changes in different people are different because of the various walking habits. Unfortunately, this issue is hardly considered in the existing literature. Moreover, the static height was always used as the criterion in existing paper to verify the accuracy of height detection methods, which may lead to inaccurate results. The VICON system whose space measurement accuracy achieved 0.01 mm is used in the paper to study the fluctuation of moving individuals in different heights. To further improve the credibility, the height data captured by the VICON system is adopted as the ground truth to verify the feasibility of the proposed method.

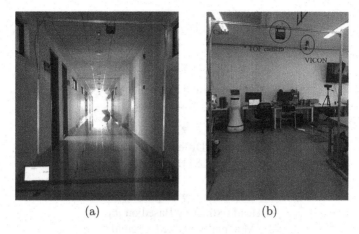

(a)   (b)

Fig. 1. Experimental Setups. (a) Campus corridor, (b) Classroom.

In the paper, we test our method in two different sites. The first site is the campus corridor, as shown in the Fig. 1(a). The second site is the classroom, and the VICON system fixed in the classroom is used to compare with the proposed method. Figure 1(b) shows the experimental setup in the classroom, only a part of the VICON system is presented in the figure. As can be seen from Fig. 1, the depth images captured by the TOF camera are sent to the personal computer to obtain the height information, and the internal algorithm will be explained in Sect. 3.

3 Height Detection Method for Dynamic Pedestrians

In this paper, head data is adopted to determine the height of the pedestrian. The depth images in the continuous seconds are addressed by the proposed method to obtain the height. The main framework of this method is shown in Fig. 2, and the detailed steps are shown below.

3.1 Calibration and Normalization

TOF camera has the same imaging principle as a normal two-dimensional camera. It collects light through a lens and generates an image on the sensor core. Therefore, the depth image acquired by the TOF camera also has a problem of lens distortion, and the degree of distortion is related to the position. The distortion at the center of the imager is zero, and the more severe the distortion becomes as it moves towards the edge. The calibration model proposed in our previous work [14] is adopted here and the results are shown in Fig. 3, in which (a) (b) are the raw depth images and (c) (d) are the corresponding depth image after calibration. For clarity, the images are represented by HSV (Hue, Saturation, Value) format.

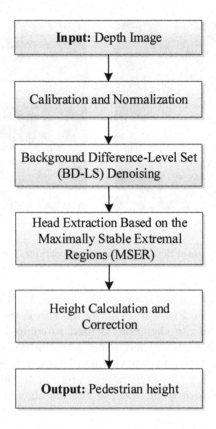

Fig. 2. Overall framework of the proposed method.

According to transfer Eq. (3), the depth value among depth image will be large because of the surprised conversion ratio between the physical distance and depth value. And most mature algorithms are designed for regular images with pixel values between 0 and 255. When we adopt mature algorithms to process depth images, it may not achieve the ideal results. Therefore, a normalization algorithm, as shown in (4), is designed in the paper to convert the depth images into grey images. Take Fig. 3(c) and Fig. 3(d) as the example, the corresponding grey images are shown in Fig. 3(e) and Fig. 3(f).

$$p_i = \frac{255 * (d_i - d_{min})}{d_{max} - d_{min}} \tag{4}$$

where d_i represents the depth value in a depth image corresponding to the pixel value p_i in a grey image, d_{max} represent the maximum depth value in a depth image, $d_{min} = 0$.

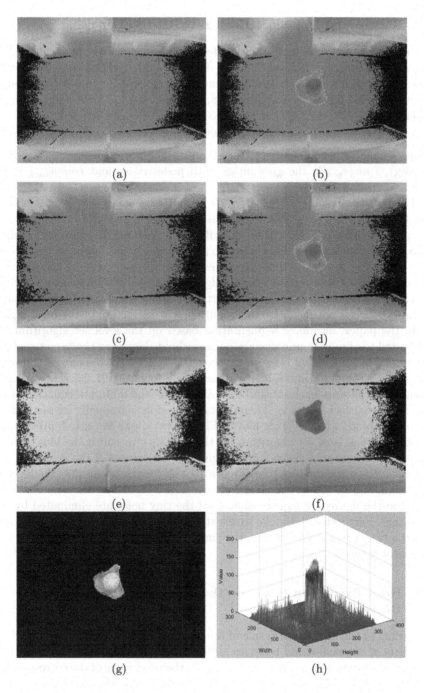

Fig. 3. (a–d) show the depth images represented by HSV format: (a) represents a background depth image captured in advance, (b) represents a depth image with pedestrian, (c) represents the background depth image after calibration, and (d) represents the depth image with pedestrian after calibration; (e) and (f) are the grey images corresponding to (c) and (d), respectively; (g) is the grey image after background difference; (h) is the 3D format of (g).

3.2 Background Difference-Level Set (BD-LS) Denoising

In this paper, we aims to use the pedestrian head data to obtain the height information. However, it is very hard to directly obtain the head region of pedestrians due to the complicated background. To remove the complex background, the background difference algorithm [18,19], as shown in (5), is adopted in the paper,

$$Image_{dif} = Image_{back} - Image_{ped} \tag{5}$$

where $Image_{dif}$ is the result of background difference algorithm, as shown in Fig. 3(g), $Image_{ped}$ is the grey image with pedestrians, and $Image_{back}$ is the grey background image.

To show the result more clearly, 3D format of the Fig. 3(g) is presented in Fig. 3(h), in which Z-axis (Value) represents the pixel value. It is clear that the background difference algorithm can effectively remove complex background. However, the background difference also generates a lot of unwanted tiny noises, as shown in Fig. 3(h). And these unwanted noise will greatly reduce the accuracy. Although some traditional denoising algorithm such as Gaussian and Median filtering can reduce the noise to a certain extent, they may blur the target contour, which is not facilitate to the subsequent extraction of the head region.

In the paper, an image segmentation based on the level set algorithm [20] is adopted to eliminate unnecessary noise without damaging the target contour. The main idea of level set is to embed the curved surface of the moving deformation as a zero-level set into a higher-dimensional function [21,22]. The evolution equation of the zero-level set can be obtained from the evolution equation of the closed hypersurface function. Then the target region is gradually extracted from the background as the level set moves. Figure 4(a) shows the search process and result using the level set algorithm for the Fig. 3(g), in which the blue curve is the initial contour, the yellow curves represent the process of the evolution, and the red curve is the final contour. Figure 4(b) shows the result after the level set algorithm, and its 3D format is also shown in the Fig. 4(c).

From the figures we can see almost all the tiny noise are eliminated by this algorithm. Furthermore, thanks to the strong contrast between the background and the target in the grey image, the time cost in the evolution of the level set is greatly reduced.

3.3 Head Extraction Based on the Maximally Stable Extremal Regions (MSER)

According to the imaging principle of the TOF camera, the depth value in different parts of the pedestrian body are different. The farther away from the TOF camera, the larger the depth value, the larger the pixel value of the corresponding grey image. Based on this point, the MSER algorithm is adopted to get the head region. In MSER algorithm, the stable extremal regions reflects that the grey value of the pixel in the set is always greater or smaller than the grey value of the pixel in the neighborhood region [23,24]. When performing a local threshold set operation on an image, the area with the smallest change in the number of pixels is called the maximum stable regions [25]. In this paper, we continuously binarize the grey image got by BD-LS algorithm (such as Fig. 4(b)), and the binary threshold is incremented from 0 to 255. The image obtained after each threshold increase is evaluated by the (6),

$$\triangle v = \frac{dR}{dT} \tag{6}$$

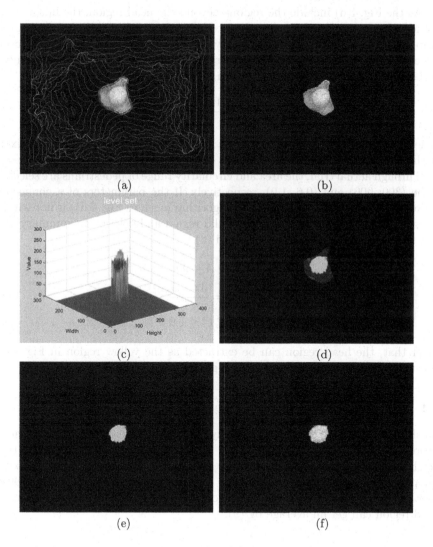

(a)

(b)

(c)

(d)

(e)

(f)

Fig. 4. (a-d) show the extraction process and result of the level set algorithm, (c) is the 3D format of (b), (d) shows the image got by the MSER-based segmentation along with (b), (e) is the result after constraint conditions, and (f) is the final image corresponding to (e). (Color figure online)

where R represents the number of pixels in a closed region; T represents the selected grey value, it is incremented from 0 to 255; Δv represents the region change rate. If the number of pixels in a closed regions change little or not within a wide range, this closed region is the maximum stable extreme region.

Figure 4(d) shows the result got by the MSER algorithm along with Fig. 4(b). For ease of display, the different maximum stable extreme region in Fig. 4(d) are marked with different colours. It shows the MSER algorithm can get hierarchically connected regions.

As the Fig. 4(d) include the regions beyond the head region, the image after the MSER need to be further processed. No matter where the pedestrian is, the shape of the head region is the relatively stable ellipse. Therefore, we used both the circularity and the area as the constraint conditions to remove the unwanted parts. The circularity equation is shown below,

$$C = \frac{4\pi * A}{l^2} \tag{7}$$

where A is the number of pixels in the closed region, l is the number of pixels at the boundary of the region, and C is the region circularity.

Through actual trial, the area and circularity range of pedestrians are selected to be (300, 900) and (0.6, 1.0), respectively. If the parameters of a connected region in the image got by the MSER algorithm (such as Fig. 4(d)) is not within the given range, it is marked as an invalid region and deleted. In summary, the constraint conditions are summarized in (8).

$$\begin{cases} 300 \leq A \leq 900, \\ C = \frac{4\pi * A}{l^2}, \\ 0.6 \leq C \leq 1.0. \end{cases} \tag{8}$$

After that, the head region can be extracted as the yellow region in Fig. 4(e), and the final image corresponding to Fig. 4(e) is shown in Fig. 4(f).

3.4 Height Calculation and Correction

As shown in Fig. 3(h), there is inevitably some tiny noise in the extracted head region. Thus, the method of taking the top of the head data as the pedestrian height is unconvincing. In the paper, the average of the head region is used to get the pedestrian height. The grey average p_{ave} can be calculated by the improved average algorithm (9). While the image contains more than one pedestrian, each head region can get an average by (9),

$$p_{ave} = \frac{5}{4n}(S - S_{max} - S_{min}) \tag{9}$$

where n is the number of pixels in the head region, S represents the sum of all pixel values in the region, S_{max} or S_{min} represent respectively the sum of the maximum or minimum $\lceil \frac{n}{10} \rceil$ pixel values in the head region, $\lceil \frac{n}{10} \rceil$ represents the largest integer less than $\frac{n}{10}$. The improved average algorithm can effectively reduce the effect of possibly sporadic noise in the head region.

Then, the corresponding depth average d_{ave} can be got by the deformation of (5) and (4) along with p_{ave}.

Considering the prior error, the physical distance between the pedestrian head and the TOF camera, defined as D_{dis}, can be obtained by (3) along with d_{ave}.

After that, the pedestrian height H can be finally calculated by (10),

$$H = H_{tof} - D_{dis} \tag{10}$$

where H_{tof} is the distance between the ground and the TOF camera.

4   Experiments and Analysis

To verify the robustness of the proposed algorithm, we conduct some experiments on the images containing more than one pedestrian. The partial results are shown in Fig. 5, where two pedestrians appear at the same time. The figures show that when the amount of people increases, the proposed method still gets a desired result.

To increase credibility, the VICON system is used as the ground truth to confirm the accuracy of our method. The VICON has 12 high-precision infrared cameras that are evenly mounted on the walls around the room, part of the infrared cameras is shown in Fig. 1(b). It can accurately measure the heights of pedestrians no matter where they are in the room. Four volunteers are invited in the paper to participate in the experiments. The volunteers are allowed to walk alone or with others at the usual speed.

The experimental results obtained by using VICON alone in several consecutive seconds are shown in Fig. 6, and the legend shows the static height of the four volunteers. It can be seen from the picture that the height cannot be kept on the static level while the pedestrian is moving. Thus, it is necessary to study the pedestrian height under the dynamic condition.

In the paper, many experiments are carried out based on the four volunteers, and a part of the experiments is shown here to shows the feasibility of our method. The depth images captured in five consecutive seconds are taken as the example in the section. Considering the practical application, we take five images evenly per second and use the proposed method to calculate the pedestrian height in each image. The results are then compared to the height data collected by VICON at the same moment. Figure 7 shows the comparison results in the five consecutive seconds. As shown in the legend, the solid curve represents results of our method, the dotted curve represents results of VICON system. It can be seen from Fig. 7 that the heights measured by the method proposed in this paper is feasible and effective.

To analyze the error of the proposed algorithm, we sort out the data shown in Fig. 7. Figure 8 shows the average error of every second in the five consecutive

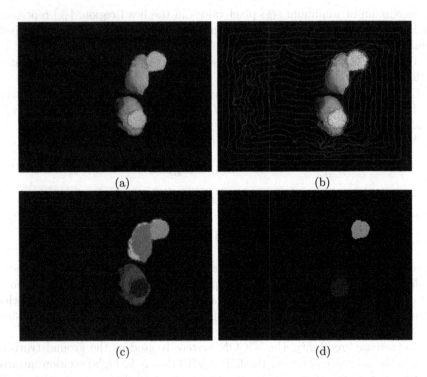

(a)   (b)

(c)   (d)

Fig. 5. The processes and result obtained by the proposed algorithm along with an image including two pedestrians.

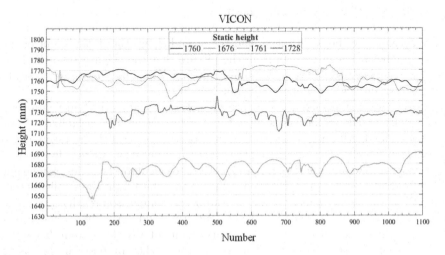

Fig. 6. Experimental results obtained by VICON alone from four volunteers with different heights.

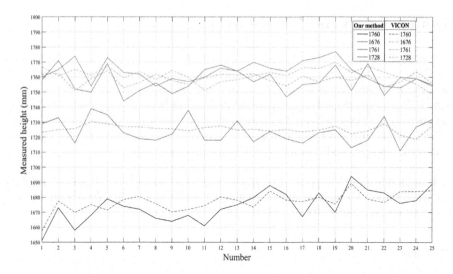

Fig. 7. Experimental results of four volunteers in five consecutive seconds.

seconds with the four volunteers. According to the figure, the results obtained by the proposed algorithm can accurately detect the height of the dynamic pedestrian, which proves the feasibility of our method.

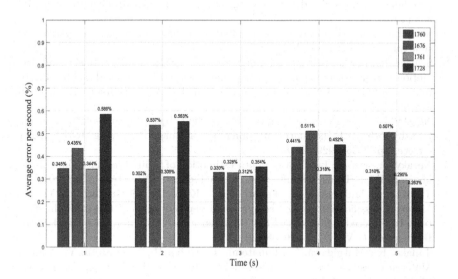

Fig. 8. Average error per second of the four volunteers in five consecutive seconds.

5   Conclusion and Future Work

A height detection method based on the continuous depth images is developed for dynamic pedestrian. To reduce the impact of the complicate background, a new BD-LS algorithm is proposed in the paper. A target extraction based on the MSER is also developed to extract the pedestrian head region. The comparison between our method and VICON shows the great feasibility of the proposed algorithm. Our future work will continue to explore the characteristics of dynamic pedestrian height, and gradually realize height prediction.

Acknowledgement. Supported by Natural Science Foundation of China (61877065) and Key Project of Science and Technology Commission of Shanghai Municipality under Grant No. 16010500300.

References

1. Cao, J., Pang, Y., Li, X.: Learning multilayer channel features for pedestrian detection. IEEE Trans. Image Process. **26**(7), 3210–3220 (2017)
2. Du, X., El-Khamy, M., Lee, J., Davis, L.: Fused DNN: a deep neural network fusion approach to fast and robust pedestrian detection. In: IEEE Winter Conference on Applications of Computer Vision (WACV), Santa Rosa, California, USA, pp. 953–961 (2017)
3. Chen, S.B., Xin, Y., Luo, B.: Action-based pedestrian identification via hierarchical matching pursuit and order preserving sparse coding. Cognit. Comput. **8**(5), 797–805 (2016)
4. Shin, B., et al.: Motion recognition based 3D pedestrian navigation system using smartphone. IEEE Sens. J. **16**(18), 6977–6989 (2016)
5. Schlagenhauf, F., Sreeram, S., Singhose, W.: Comparison of kinect and vicon motion capture of upper-body joint angle tracking. In 2018 IEEE 14th International Conference on Control and Automation (ICCA), Anchorage, AK, USA, pp. 674–679 (2018)
6. Noraky, J., Sze, V.: Depth estimation of non-rigid objects for time-of-flight imaging. In 2018 25th IEEE International Conference on Image Processing (ICIP), Athens, Greece, pp. 2925–2929 (2018)
7. Noraky, J., Sze, V.: Low power depth estimation of rigid objects for time-of-flight imaging. IEEE Trans. Circuits Syst. Video Technol. **30**(6), 1524–1534 (2019)
8. Tanaka, K., Mukaigawa, Y., Funatomi, T., Kubo, H., Matsushita, Y., Yagi, Y.: Material classification using frequency-and depth-dependent time-of-flight distortion. In Proceedings of the IEEE Conference on Computer Vision and Pattern Recognition (CVPR), pp. 79–88 (2017)
9. Tanaka, K., Mukaigawa, Y., Funatomi, T., Kubo, H., Matsushita, Y., Yagi, Y.: Material classification from time-of-flight distortions. IEEE Trans. Pattern Anal. Mach. Intell. **41**(12), 2906–2918 (2018)
10. Buigas, L.K.P., Jimenez, H.: Tracking object using a TOF camera using the intensity and depth channel. In: 2019 International Conference on Electronics Communications and Computers (CONIELECOMP), Cholula, Mexico, pp. 129–133 (2019)

11. Plank, H., Egger, T., Steffan, C., Steger, C., Holweg, G., Druml, N.: High-performance indoor positioning and pose estimation with time-of-flight 3D imaging. In: 2017 International Conference on Indoor Positioning and Indoor Navigation (IPIN), Sapporo, Japan, pp. 1–8 (2017)
12. Behrje, U., Himstedt, M., Maehle, E.: An autonomous forklift with 3D time-of-flight camera-based localization and navigation. 2018 15th International Conference on Control. Automation, Robotics and Vision (ICARCV), Singapore, Singapore, pp. 1739–1746 (2018)
13. Alkhawaja, F., Jaradat, M., Romdhane, L.: Techniques of indoor positioning systems (IPS): a survey. In: 2019 Advances in Science and Engineering Technology International Conferences (ASET), Dubai, United Arab Emirates, United Arab Emirates, pp. 1–8 (2019)
14. Le, W., Yu, L., Hai, W., Min, F.: Measurement error correction model of TOF depth camera. Chinese J. Syst. Simul. **29**(10), 2323–2329 (2016)
15. Shim, H., Lee, S.: Recovering translucent objects using a single time-of-flight depth camera. IEEE Trans. Circuits Syst. Video Technol. **26**(5), 841–854 (2015)
16. Corti, A., Giancola, S., Mainetti, G., Sala, R.: A metrological characterization of the Kinect V2 time-of-flight camera. Robot. Auton. Syst. **75**, 584–594 (2016)
17. Jung, J., Lee, J.Y., Jeong, Y., Kweon, I.S.: Time-of-flight sensor calibration for a color and depth camera pair. IEEE Trans. Pattern Anal. Mach. Intell. **37**(7), 1501–1513 (2014)
18. He, L., Ge, L.: CamShift target tracking based on the combination of inter-frame difference and background difference. In: 2018 37th Chinese Control Conference (CCC), Wuhan, China, pp. 9461–9465 (2018)
19. Filonenko, A., Jo, K.H.: Unattended object identification for intelligent surveillance systems using sequence of dual background difference. IEEE Trans. Industr. Inf. **12**(6), 2247–2255 (2016)
20. Zhang, K., Zhang, L., Song, H., Zhang, D.: Reinitialization-free level set evolution via reaction diffusion. IEEE Trans. Image Process. **22**(1), 258–271 (2012)
21. Osher, S., Sethian, J.A.: Fronts propagating with curvature-dependent speed: algorithms based on Hamilton-Jacobi formulations. J. Comput. Phys. **79**(1), 12–49 (1988)
22. Khadidos, A., Sanchez, V., Li, C.T.: Weighted level set evolution based on local edge features for medical image segmentation. IEEE Trans. Image Process. **26**(4), 1979–1991 (2017)
23. Matas, J., Chum, O., Urban, M., Pajdla, T.: Robust wide-baseline stereo from maximally stable extremal regions. Image Vis. Comput. **22**(10), 761–767 (2004)
24. Rahmany, I., Arfaoui, B., Khlifa, N., Megdiche, H.: Cerebral aneurysm computer-aided detection system by combing MSER, SURF and SIFT descriptors. In: 2018 5th International Conference on Control Decision and Information Technologies (CoDIT), Thessaloniki, Greece, pp. 1122–1127 (2018)
25. Akula, A., Ghosh, R., Kumar, S., Sardana, H.K.: WignerMSER: pseudo-wigner distribution enriched MSER feature detector for object recognition in thermal infrared images. IEEE Sens. J. **19**(11), 4221–4228 (2019)

Quantitative Regression Modeling of Cocoa Bean Content Based on Gated Dilated Convolution Network

Yayu Chen[1], Wenju Zhou[1], Minrui Fei[1(✉)], Haikuan Wang[1], Xiaofei Han[1], and Huiyu Zhou[2]

[1] School of Mechanical Engineering and Automation, Shanghai University, Shanghai, China
{chenyayu,zhouwenju,hkwang}@shu.edu.cn, mrfei@staff.shu.edu.cn, hanxiaofei@i.shu.edu.cn
[2] School of Informatics, University of Leicester, LE1 7RH Leicester, UK
hz143@leicester.ac.uk

Abstract. By analyzing the near-infrared spectrum, we can determine the quantitative relationship model between the spectral data of different cocoa beans and the target components. This paper proposes a predictive regression model based on 1D-CNN. Based on the traditional convolutional neural network, gating mechanisms and dilated convolutions are combined. The particle swarm optimization method is used to optimize the hyper-parameters of one-dimensional convolution. The end-to-end near-infrared predictive regression model does not require wavelength selection. It is convenient to use and has a strong promotional value. Taking the public cocoa beans near-infrared data set as an example, the method can predict the water and fat content in cocoa beans, and the effectiveness of the method is verified. Comparing the improved one-dimensional convolution with traditional one-dimensional convolution results and partial least squares regression, it shows better prediction accuracy and robustness.

Keywords: Convolutional neural network · Infrared spectroscopic data · Gating mechanisms · Dilated convolution

1 Introduction

Near-infrared spectroscopy analysis technology has a wide range of applications due to its fast analysis speed, no need for chemical reagents, reduced environmental pollution, low cost, and easy online analysis. Near-infrared spectroscopy detection technology will not destroy the substance that predicts the information of a sample by establishing a correlation model between the near-infrared spectroscopy data of the sample and the true value of the attribute to be measured.

In the past few decades, due to the complexity of NIR spectrum and the diversity of analysis objects, NIR quantitative analysis generally uses multivariate information processing technology, such as PLSR [1–3], MLR [4], PCA. PLSR is the most commonly used linear stoichiometric model. However, with the deepening of the application, these

© Springer Nature Singapore Pte Ltd. 2020
M. Fei et al. (Eds.): LSMS 2020/ICSEE 2020 Workshops, CCIS 1303, pp. 456–468, 2020.
https://doi.org/10.1007/978-981-33-6378-6_34

methods also expose many problems, such as MLR will encounter collinearity problems and the limitation of the number of input variables, PCR can not distinguish noise or valid information, and the running speed is slow, and PLS cannot be effective Handling nonlinear problems.

The study of machine learning method is a significant task to resolve the problem of accurate quantitative measurement in chemometrics. Especially in the industrial field, The machine learning algorithm of fast near-infrared detection has established several mathematical models to estimate the fidelity of measured data, such as ANN [5], SVR [6] and so on. However, they all produce more parameters and complicated operation, which is not conducive to the promotion of NIR analysis technology.

Neural network has powerful modeling capabilities and has achieved great success in image processing and natural language processing for multi-large-scale data. At the same time, due to its deep network structure and nonlinear activation ability, the application of neural network in modeling and analysis of near-infrared spectroscopy has been reported. Chao Ni et al. proposed to quantify the nitrogen content of pinus massoniae seedling leaves by using variable weight convolutional neural network [7]. Muhammad Bilal et al. proposed to reduce the eigenvector size by training L2 regularized sparse auto en-coder [8]. Salim Malek et al. used a 1D-CNN for feature extraction [9].

In view of the characteristics of deep learning's powerful expression ability, we improve the dilation gate convolution model and introduce it into the prediction regression of infrared spectrum. In order to optimize the convolution kernel and some other parameters, we used the PSO, which aims to study a non-destructive and effective method to identify the cocoa fat content and the moisture content and explore a new model applied to the learning of spectral characteristics of foods.

2 Neural Network

2.1 CNN

The convolution calculation layer is to extract feature on the original input. There can be multiple convolutional layers. We specify the size and weight of the convolution kernel. In the input matrix, select the data convolution calculation with the same size as the convolution kernel. Continuously extract and compress features, and then obtain relatively advanced features for convolution. The parameters of the convolution kernel of each convolution layer are the same, which is weight sharing, which significantly reduces the number of parameters.

In the excitation layer, the activation function can smooth the results of the operation. Meanwhile, the activation function can remove the redundant data, and retain the main features of the data.

After the excitation layer, the pooling layer specifies the size of a window and replaces it with a value. It is commonly used to maximize the pool and average the pool.

The output layer outputs multiple neurons. For regression problems, the fully connected layer should be located before the regression layer at the end of the network. We use linear regression.

CNN training can be seen as two stages of forward propagation and back propagation. Back propagation uses the BP algorithm. The difference between CNN and traditional multilayer neural networks is the addition of convolution and pooling operations, so CNN is in convolution The training process of the layer and the pooling layer is different from the neural network. Before the CNN network starts training, first, the network must be initialized, the connection weights are assigned random numbers in $(-1, 1)$, and the calculation accuracy and maximum learning are given. Number of times. Second, randomly select input samples and corresponding expected outputs. Finally, start the training of the network.

2.2 Spectral Regression Model Based on CNN

The wavelength range of NIR is 780 nm to 2526 nm, and a complete nir spectrum sequence is relatively long. Therefore, dilated convolution and gating mechanisms are introduced.In order to identify larger features, all pooling layer uses maximum pooling.

Block Structure. The gating mechanism will control the flow of information to make it multi-path circulation. Aäron van den Oord [10] proved that it could model more complex interactions with LSTM-style gating mechanism.

$$y = \tanh(x * W_1 + b_1) \odot \sigma(x * W_2 + b_1). \tag{1}$$

W and b represent kernel and deviation, which represents the sigmoid function, $\odot$ is element-wise multiplication, and * is the convolution multiplier. The gradient of gating in LSTM style is

$$\nabla y = \tanh'(x * W_1 + b_1)\nabla(x * W_1 + b_1) \odot \sigma(x * W_2 + b_2)$$
$$+ \sigma'(x * W_2 + b_2)\nabla(x * W_2 + b_2) \odot \tanh(x * W_1 + b_1). \tag{2}$$

Among them, $\tanh'(x*W_1+b_1)$, $\sigma'(x*W_2+b_2) \in (0, 1)$. Under normal circumstances, the problem of disappearing gradients appears as the network depth increases.

We constructed a set of block structures (as shown in Fig. 1), which is equivalent to two gating mechanisms in a set of block structures.

Dilated Convolution. Because the complete sequence of a near-infrared spectrum is relatively long, dilated convolution is used to enable the CNN model to capture a longer distance without increasing model parameters. Dilated convolution was first developed from the wavelet decomposition algorithm [11] called "convolution with dilated filter", and later applied to image and semantic segmentation. Dilated convolution modifies the convolution operator so that it can use filter parameters in a variety of ways. In a two-dimensional convolution,

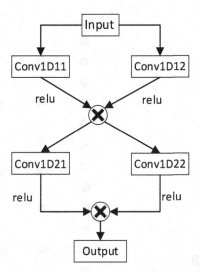

Fig. 1. Block structure

$F : Z^2 \to R$ is a discrete function, $\Omega_r = [-r, r]^2 \cap Z^2$, $k : \Omega_r \to R$ is a discrete filter of size $(2r + 1)^2$ [12].The two-dimensional convolution operator * is defined as

$$\left(F^*k\right)(p) = \sum_{s+t=p} F(s)k(t). \tag{3}$$

Then r represents a factor of dilation, $*_r$ is defined as

$$\left(F^*_rk\right)(p) = \sum_{s+rt=p} F(s)k(t) \tag{4}$$

So we can extend to one-dimensional expansion convolution,

$$\left(F^*_rk\right)(p) = \sum_{s+rt=p} F(s)k(t), \quad t \in [-m, m] \cap Z \tag{5}$$

It can expand the vision of convolution kernel without increasing computation. Figure 2 shows two convolutional neural networks with three layers, and the size of the convolution kernel is 3. Traditional convolution obtains a feature through a three-layer convolutional layer. The first layer has 7 inputs. After three-layer expansion convolution, a feature is obtained. The first layer has 15 inputs. In the dilated one-dimensional convolution, if the dilated rate of the kernel increases exponentially, the size of the receiving domain has an exponential relationship with the number of the layers.

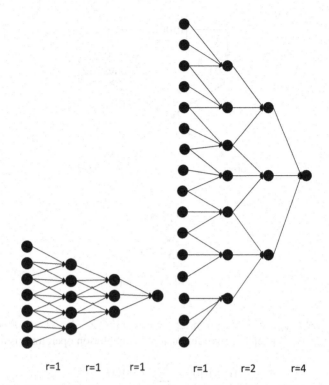

r=1   r=1   r=1   r=1   r=2   r=4

Fig. 2. Left: a 1-D CNN with three conventional convolutional layers. Right: a 1-D CNN with three dilated convolutional layers

3   PSO Algorithm

Through the combination of local perception, weight sharing and pooling, CNN reduces the number of parameters in network training. However, CNN's hyper-parameters, such as the size of convolution kernel and the number of convolution kernel,all need to be set manually.

PSO was proposed at the end of the 1990s [13, 14]. It has become a research hotspot due to its fast convergence and few parameters.The idea comes from simulating the behavior of birds flying for food in a cluster. Scattered birds then fly for food. There is an indirect mechanism to let them know the distance between food and the current position. Each particle has different velocity attributes and position attributes. The value of these attributes can be determined by the fitness function of the problem to be solved, and then iterated. Each particle can represent a bird. We know that the global optimal solution has been found, otherwise it will always calculate the new flight speed and position for each example through the historical optimal solution. The search speed is fast, the algorithm is easy to implement, and the high precision is the advantage of the algorithm

Suppose we have a search space with dimension d. There are m particles in each generation group. Then the i-th particle can be expressed as $x_i = (x_{i1}, x_{i2}, \ldots, x_{id})$,

$i = 1, 2, \ldots, m$. The position of the i-th particle in the d-dimensional search space is x_i, The velocity of the i-th particle can be expressed as: $v_i = (v_{i1}, v_{i2}, \ldots, v_{id})$. The optimal position found in the current iteration is $p_i = (p_{i1}, p_{i2}, \ldots, p_{id})$, and the optimal position found in the entire population is $p_i = (p_{g1}, p_{g2}, \ldots, p_{gd})$. The particle velocity and position update formula in the standard PSO is as follows:

$$v_{id}^{k+1} = wv_{id}^k + c_1 r_1 (p_{id}^k - x_{id}^k) + c_2 r_2 (p_{gd}^k - x_{id}^k) \tag{6}$$

$$x_{id}^{k+1} = x_{id}^k + v_{id}^{k+1} \tag{7}$$

In the above formula, v_{id}^k and x_{id}^k respectively represent the d-th element of the velocity and position of the i-th particle in the k-th iteration. p_{id}^k and p_{gd}^k respectively represent the d-th element of the local optimal value of particle i and the global optimal value in the entire population. k represents the current iteration times, the learning factor is represented by c_1, c_2, r_1, r_2 are random numbers, the value range is [0, 1], the inertia weight is w.

Assuming that the optimization goal is to find the minimum value, the update methods of p_{id}^k and p_{gd}^k are:

$$p_{id}^{k+1} = \begin{cases} x_i^{k+1} & f(x_i^{k+1}) < f(p_{gd}^k) \\ p_{gd}^k & f(x_i^{k+1}) \geq f(p_{gd}^k) \end{cases} \tag{8}$$

$$p_{gd}^k = \arg \min_{1 \leq i \leq N} (f(p_{gd}^k)) \tag{9}$$

Where $f(\cdot)$ is the objective function and N is the number of particles in each iteration (Fig. 3)

The steps of particle swarm optimization for CNN hyper-parameters are:

(1) Determine the number and value range of CNN hyper-parameters.
(2) Set the parameters of PSO, such as inertia factor, learning factor, and maximum iteration number.
(3) Initialize the particle swarm, and each particle swarm represents a group.
(4) CNN was trained according to hyper-parameters, the fitness function value of each particle was calculated, and the advantages and disadvantages of particles were sorted according to the fitness function value.
(5) Find the best position of the current particle and particle swarm according to the value of the fitness function.
(6) Update the position and speed of each particle to generate a new group of particles.
(7) Add 1 to the current number of iterations.
(8) If the maximum current iteration number is exceeded, the optimization is terminated; otherwise, step 4 is returned.
(9) According to the optimal position of PSO algorithm, the optimal value of CNN hyper-parameters is obtained.

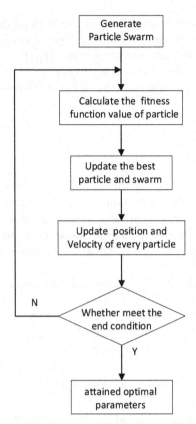

Fig. 3. Particle swarm optimization algorithm (PSO) optimizes CNN hyper-parameters flowchart

4 Experimental Methods

4.1 Dataset Description

The feasibility and effectuality of the method proposed is based on the public cocoa bean near-infrared spectral dataset [15]. The cocoa bean near-infrared data set has 72 samples. Near-infrared spectroscopy is an electromagnetic wave with a wavelength in the range of 780 nm to 2526 nm, Fat and water are the ingredients in cocoa beans samples. They correspond to the chemical information in cocoa beans, Near-infrared light will absorb the frequency doubled and combined frequency of the vibration of the hydrogen-containing group, which contains the information of fat and water. Near-infrared spectroscopy is a carrier for obtaining information. and the content of fat and water will be reflected by the absorption intensity of the near-infrared absorption band.

The wavelength range of near infrared spectrum in this data set is 1000 nm–2500 nm. Training sets can be trained to establish a quantitative analysis model, and the test set is used to evaluate the reliability of the trained model. The spectra of each sample between 1000 nm and 2000 nm is more dispersed. The spectrum of each sample between 2000 nm and 2200 nm is relatively messy (Fig. 4 and 5).

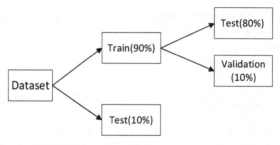

Fig. 4. Modeling the dataset for training and testing the model

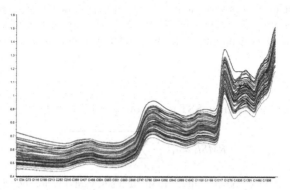

Fig. 5. Cocoa beans near infrared spectrum

4.2 Spectral Pretreatment

Noise will influence the model during spectral measurement, and data preprocessing is required. The data preprocessing method is usually used to preprocess the spectral data before the model is established. Common pre-processing methods include derivative, multivariate scattering correction and vector normalization. Derivative preprocessing can reduce the impact of low-frequency noise and improve resolution. Although neural networks have powerful feature extraction, preprocessing the spectrum and selecting the wavelength of the spectrum can make the prediction results more accurate.

Experiments have shown that the cocoa beans have a high degree of spectral overlap after SNV, MSC, and derivatives. They can not clearly reflect the differences between the samples, which is not conducive to the regression prediction of the neural network. The spectral preprocessing in this paper first adopts Gaussian smoothing to eliminate the dimensional influence between the indicators and the influence of the size of the variable itself and the value of the variable, then standardizes the spectral data to speed up the training of neural networks.

4.3 Network Structure

The network structure we propose is as follows, we construct a group of block structures, which is equivalent to two gating mechanisms in a group of block structures, and the subsequent groups repeat the same pattern. Zero padding is applied to the one-dimensional convolution to make the input and output size the same. Use maximum pooling to take the largest feature point in the neighborhood. Add dropout to the pooling layer to make some neurons in the pooling area randomly set to 0 with a probability of 0.1, which can reduce the probability of over-fitting.

4.4 Model Evaluation Criteria

We used the root mean square error (RMSE) and coefficient of determination (R2) to evaluating the quality of our model. The RMSE is very sensitive to the large or small error in a group of measurements. Coefficient of determination (R2) represents that the proportion of the change in variable y that the regression model can explain to the total change in the controlled independent variable x. The higher the goodness of fit, the higher the degree of explanation of the independent variable to the dependent variable.

$$R2 = 1 - \frac{SSE}{SST} \tag{10}$$

SSE is sum of squares for error, SST is sum of squares fortotal.

$$RMSE = \sqrt{\frac{1}{N} \sum_{i=1}^{N} (y_i - \hat{y}_i)^2} \tag{11}$$

N is the number of cocoa bean samples in the test set, y_i and $\hat{y}_i$ are the true and predicted values of x_{it} samples. Gain measured using accuracy can provide information about how much our method has improved [16], the formula is:

4.5 Parameter Setting

The improved 1DCNN architecture includes some parameters: the convolution layer number L and the sample number m selected for one training and Particle swarm optimization parameters. As shown in Table 1 and Table 2 are the set value of parameters. The parameter setting is greatly reduced, and it is easier to operate and use for users who lack professional knowledge. It can be seen from the set number of layers that for the regression model of the near infrared spectrum, the deeper the structure, the better the result. Since the network is not deep, there is no problem of the gradient disappearing.

Table 1. Parameter values of the improved 1D-CNN

Dataset	L	m
Moisture	1	4
Fat	2	2

Table 2. Particle swarm optimization parameters

Parameter	Value
Inertia factor	0.5
Learning factor	0.5
Number of iteration	10
Number of particles	5

Figure 6 is the comparison between the model calculated value and the real value of the cocoa bean moisture content test set (Fig. 7 and 9)

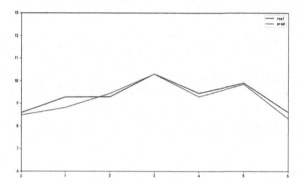

Fig. 6. The comparison between the model calculated value and the real value of the cocoa bean moisture content test set.

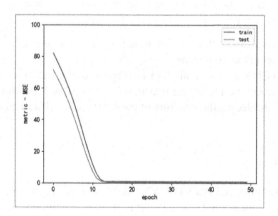

Fig. 7. Training error and verification error of water content

Figure 8 is the comparison between the model calculated value and the real value of the cocoa bean fat content test set.

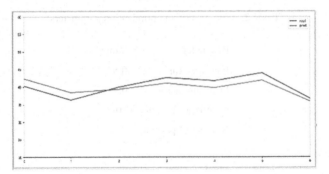

Fig. 8. The comparison between the model calculated value and the real value of the cocoa bean fat content test set.

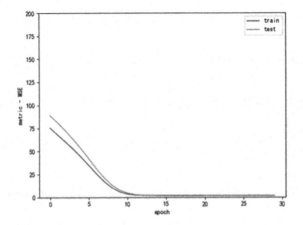

Fig. 9. Training error and verification error of fat content.

The figure above shows that the improved one-dimensional CNN can predict the water and fat content of cocoa beans.

The improved CNN uses several blocks to superimpose, and its effect is not as good as blocks with few layers. Indicating that in the near infrared spectroscopy prediction regression model, the deeper the structure of the neural network does not mean the better the effect.

5   Discussion

When we evaluate Improved PSO-1DCNN, we choose two regression methods, PLSR and traditional one-dimensional CNN. PLSR is the most commonly used linear stoichiometric model, and one-dimensional CNN is a common algorithm in machine learning. Table 3 and Table 4 are the water and fat content of cocoa beans, and their R2 and RMSE are calculated by three methods. It can be seen that in predicting the moisture and fat content of cocoa beans, the improved PSO-1DCNN has the highest R2 and the lowest RMSE.

Table 3. Results for moisture content dataset

Method	R2	RMSE
Improved PSO-1DCNN	0.876	0.223
PSO-1DCNN	0.744	0.794
PLSR	0.861	0.472

Table 4. Results for fat content dataset

Method	R2	RMSE
Improved PSO-1DCNN	0.832	0.801
PSO-1DCNN	0.554	1.197
PLSR	0.825	0.846

6 Conclusion

We proposed a new method for chemometric data analysis based on CNN, improved the traditional 1D-CNN architecture, and added dilated convolution and gate mechanisms. Experimental results in cocoa beans near infrared spectroscopy show that the improved 1D-CNN can predict the water and fat content of cocoa beans. Compared with the traditional 1D-CNN, the gate mechanism and expansion volume The integrated convolutional neural network has higher accuracy and better fault tolerance. Our contribution is that it simplifies the process of quantitative analysis of infrared spectroscopy and is also more suitable for users who lack professional knowledge and is easy to promote.

Acknowledgments. This research is financially supported by Natural Science Foundation of China (61877065), the National Key Research and Development Program of China (No. 2019YFB1405500) and Key Project of Science and Technology Commission of Shanghai Municipality under Grant (No. 16010500300).

References

1. Wold, S., Sjöström, M., Eriksson, L.: PLS-regression a basic tool of chemometrics. Chemometr. Intell. Lab. Syst. **58**(2), 109–130 (2001)
2. Nicolai, B.M., Theron, K.I., Lammertyn, J.: Kernel PLS regression on wavelet transformed NIR spectra for prediction of sugar content of apple. Chemometr. Intell. Lab. Syst. **85**(2), 243–252 (2007)
3. Arrobas, B., et al.:Raman spectroscopy for analyzing anthocyanins of lyophilized blueberries. In: 2015 IEEE Sensors. IEEE (2015)
4. Galvao, R.K.H., et al.: Multivariate analysis of the dielectric response of materials modeled using networks of resistors and capacitors. IEEE Trans. Dielectr. Electr. Insul. **20**(3), 995–1008 (2013)

5. Verikas, A., Bacauskiene, M.: Using artificial neural networks for process and system modelling. Chemometr. Intell. Lab. Syst. **67**(2), 187–191 (2003)
6. Devos, O.R., Cyril, D.A., Duponchel, L., Huvenne, J.-P.: Support vector machines (SVM) in near infrared (NIR) spectroscopy: focus on parameters optimization and model interpretation. Chemometr. Intell. Lab. Syst. **96**(1), 27–33 (2009)
7. Ni, C., Wang, D., Tao, Y.: Variable weighted convolutional neural network for the nitrogen content quantization of m asson pine seedling leaves with near-infrared spectroscopy. Spectrochim. Acta Part A: Mol. Biomol. Spectro. **209**, 32–39 (2019)
8. Bilal, M., Ullah, M., Ullah, H.: Chemometric data analysis with autoencoder neural network. Electron. Imaging, 679–681 (2019)
9. Malek, S., Melgani, F., Bazi, Y.: One dimensional convolutional neural networks for spectroscopic signal regression. J. Chemometr. **32**(5), e2977 (2018)
10. Van den Oord, A., et al.: Conditional image generation with pixelcnn decoders. Adv. Neural. Inf. Process. Syst. **29**, 4790–4798 (2016)
11. Shensa, M.J.: The discrete wavelet transform: wedding the a trous and Mallat algorithms. IEEE Trans. Signal Process. **40**(10), 2464–2482 (1992)
12. Yu, F., Koltun, V.: Multi-scale context aggregation by dilated convolutions. arXiv preprint arXiv:1511.07122 (2015)
13. Kennedy, J., Russell E.: Particle swarm optimization. In: Proceedings of ICNN 1995-International Conference on Neural Networks, vol. 4. IEEE (1995)
14. Shi, U., Eberhart, R.:A modified particle swarm optimizer. In: 1998 IEEE International Conference on Evolutionary Computation Proceedings. IEEE World Congress on Computational Intelligence (1998)
15. Agussabti, R., Purwana S., Munawar, A.A.: Data analysis on near infrared spectroscopy as a part of technology adoption for cocoa farmer in Aceh Province, Indonesia. Data Brief 29 (2020)
16. AlHichri, H., Bazi, Y., Alajlan, N., Melgani, F., Malek, S.Y., Ronald, R.: A novel fusion approach based on induced ordered weighted averaging operators for chemometric data analysis. J. Chemometr. **27**(12), 447–456 (2013)

A Novel Hybrid Forecasting Model Based on Time Series

Yu Pu, Jun-jiang Zhu$^{(\boxtimes)}$, and Tian-hong Yan

College of Mechanical and Electronic Engineering, China Jiliang University,
Hangzhou 310018, China
zjj602@yeah.net

Abstract. The forecast of cigarette sales is crucial for tobacco companies to
formulate long term development policies and optimize the inventory control system. Considering the long-term trend and seasonal fluctuation characteristics of
cigarette sales sequence, a hybrid method including wavelet decomposition, auto-
regression and fusion of several intelligent algorithms is proposed. The original time series of cigarette sales are first decomposed into two components of
low-frequency component and high-frequency component which simulate overall trends and seasonal fluctuation, by using wavelet decomposition. Then they
are predicted by auto-regression and fusion of several intelligent algorithms, and
the final result is the combination of the predictions of the two components. The
experimental results show that the hybrid forecasting model performed well, and
its minimum MAPE (mean absolute percentage error) is 3.58%. Compared with
BP (back-propagation neural network), SVM (support vector machine), and ELM
(extreme learning machine) algorithms, the hybrid model proposed in this paper
decreases MAPE by 2.01%, 1.58%, and 0.93%, while the stability of the model
increased by 16.92%, 22.85%, and 56.09%, respectively. The proposed hybrid
model provides a valid way for improving the accuracy and stability of time series
forecasting.

Keywords: Forecasting · Artificial intelligence · Cigarette sales forecast ·
Seasonal fluctuation · Decomposition

1 Introduction

The forecast of cigarette sales is an important start for cigarette marketing, which affects
the formulation of tobacco companies' purchasing plan and inventory decision [1–3].
Reducing human intervention and establishing a scientific sales forecasting model based
on market trends is necessary, it is not only beneficial to organize goods reasonably for
tobacco companies and provide marketable cigarette products for retailers [4], but also
helpful for tobacco companies to reduce the difficulty of monopoly management and
offend 'sham-private-none' market space [5]. Therefore, it has great significance for the
development of the tobacco industry.

There are two representative methods for forecasting of cigarette sales, namely a
physical-based model and a time-series-based model. The method based on physical

© Springer Nature Singapore Pte Ltd. 2020
M. Fei et al. (Eds.): LSMS 2020/ICSEE 2020 Workshops, CCIS 1303, pp. 469–482, 2020.
https://doi.org/10.1007/978-981-33-6378-6_35

model is used to predict the actual demand for cigarettes [6], it establishes regression model between macro information or micro individual information and cigarette sales. The physical model shows the advantages of physical explanation, but there are many factors that affect cigarette sales, and it is difficult to find them all. By contrast, the method based on time series [7] can be more convenient and intuitive to predict the cigarette sales sequence [8]. The typical methods include regression model (AR) [9], back-propagation neural network algorithm (BP) [10], support vector machine (SVM) [11] and extreme learning machine (ELM) [12, 13]. Especially BP, SVM, and ELM can fully consider the nonlinear and non-stationary time series [14] of cigarette sales, so these typical methods based on time series are more likely to obtain higher prediction accuracy than the physical model [15].

Many factors such as population, season, market, and economic level will have an impact on cigarette sales [16], which results in two characteristics of cigarette sales: the overall trend and seasonal fluctuation trend [17]. For this complex forecast system, previous studies have shown that the physical models, and single linear or nonlinear forecasting models [18] cannot capture characteristics of the time series deeply, so the accuracy and stability are not ideal. Wavelet decomposition [19, 20] as a preprocessing method was used in the field of wind power forecast to extract ill-behaved time-series features, and structure different frequency components into input-output series [21, 22]. A combination of wavelet transform and time-series modeling [23, 24] has been used to improve forecasting accuracy and stability [25].

In this paper, wavelet decomposition is used as a preprocessing method to fit the various characters of different sequence components. By wavelet decomposition, the cigarette sales sequence is decomposed into two components. Both of them have clear physical significance, but the high-frequency component is still the result of multiple factors. In theory, choosing the best result from several prediction methods can ensure that the prediction accuracy is higher than or equal to one method. However, it is hard to determine which method can obtain better prediction results of cigarette sales when limiting a specific region and time. Therefore, for the character of the high-frequency component, we propose a combined forecast model based on the literature [26] research. The aim is to acquire optimal forecasting results by various single methods and combine the results with a hybrid forecast model, thus improving the stability of the prediction results and ensuring accuracy.

To be specific, we first describe the collection and normalization of actual cigarette sales sequence in Sect. 2. Then the AR forecast model of low-frequency components and various single forecast models of high-frequency components are introduced. On this basis, a combined forecast model is presented. The forecast results of different methods are shown in Sect. 3. In Sect. 4, the performance evaluations of the different frequency components are discussed using real-time series. The last section demonstrates a conclusion on the advantages of the hybrid forecast model in this study.

2 Methodology

The difference between the overall trend, seasonal fluctuation, and stochastic volatilities in the cigarette sales sequence is that they have different fluctuation frequencies [15].

The discrete wavelet transform uses the two-channel decomposition filter bank to achieve the decomposition iteratively, and it is easy to realize the adaptive frequency band division of the signal. Therefore, in this study, the wavelet decomposition and reconstruction are used as a preprocessing method [24]. Due to the instability of the intermediate frequency component and residual component, the wavelet decomposition method is used to divide the signal sequence into two parts: low and high frequency components. The sales sequence is decomposed into 4 layers by Mallat algorithm, where the mother wavelet is DB6 (Daubechies wavelet with the order of 6). Then, the contour of the fourth layers will be labeled as low-frequency components, and then the overall trend of cigarette sales is simulated by using low-frequency components. The sum of the partial reconstruction of the first to fourth layer as the high-frequency component is used to simulate the fluctuations of cigarette sales. Low-frequency components change slowly, which is a steady-state process, and thus it can be predicted the overall trend of cigarette sales by AR model. Due to drastic non-stability characteristics of the high-frequency component, a fusion method based on SVM, BP algorithm, and ELM is used to predict sales fluctuation. The implementation process will be illustrated in Fig. 1.

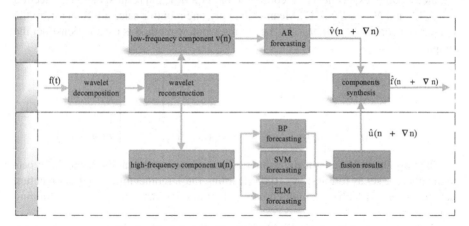

Fig. 1. Flowchart for the Hybrid forecasting model

2.1 Forecasting Method of Low-Frequency Component

The low-frequency components of the sales sequence are predicted by AR model. There is an assumption that s data: $\{v(m-s), v(m-s+1),\ldots, v(m-1)\}$ have been known before the establishment of the $v(n)$, then the AR model uses the linear combination of these p data to predict the value of $v(m)$:

$$\hat{v}(m) = \sum_{k=1}^{s} a_k v(m-k) \tag{1}$$

The steps to predict low-frequency components using the AR model are as follows. First of all, the selection of training data by minimizing the error between $\hat{v}(m)$ and

$v(m)$ can get the coefficient $\{a_k\}$. Secondly, the p-value is calculated by Akaike Information Criterion. Finally, the coefficients are taken into formula (1), and according to the new data sequence of the $\{v(m - s + 1), v(m - s + 2), v, \hat{v}(m)\}$, forecast the amount $\hat{v}(m + 1)$ of cigarette sales next month.

2.2 Forecasting Method of High-Frequency Component

Prediction Method. (1) Neural network: Due to the strong ability of self-adaptation and self-organization, BP is an effective prediction tool [27]. At the same time, it has a strong generalization ability and nonlinear mapping ability. In theory, it can approximate any nonlinear continuous function [28].

In the paper, a three-layer feedforward neural network is used to implement BP. The input sample $\{u_L(m - s), u_L(m - s + 1), v, u_L(m - 1)\}$ is the known sales for the s month before the month that will be forecasted, and the output $\hat{u}_L(m)$ is the predicted sales for the month to be forecasted. The number of neurons N_hidden and the p-value are obtained through experiments. According to the experimental results, the final selection of p is 12, and N_hidden is 17. Meanwhile, only one month's sales are forecast, so the number of neurons N_output is 1. The sliding window method is used to construct the sample:.

$$X = \begin{bmatrix} u_1(0), u_1(1), \cdots u_1(s - 1) \\ u_2(1), u_2(2), \cdots u_2(s) \\ \vdots \\ u_L(m - s), u_L(m - s + 1), \cdots u_L(m - 1) \end{bmatrix} \quad Y = \begin{bmatrix} u(s) \\ u(s + 1) \\ \vdots \\ u(m - 1) \end{bmatrix} \quad (2)$$

The Levenberg-Marquardt algorithm [29] is used to train the samples, and the hyperparameters are set as follows: $lr = 0.05$ (learning rate), momentum $= 0.9$ (momentum factor), $epoch = 10000$ (maximum number of iterations), and $error = 1e - 5$ (training target error).

(2) Support vector machine: SVM avoids the phenomenon of requiring a large number of samples in the traditional derivation process and training process and is an ideal small-sample regression method [30].

The input sample of SVM is $\{(x_1, y_1), ..., (x_m, y_m)\} \in R^n \times R$. The mathematical model can be described by a formula:

$$y = (\omega \cdot x_i) + b \quad (3)$$

where y_i is the output, vector x_i is the input of n dimensions. The weight vector ω and the threshold b are estimated by the following linear optimization problem:

$$\min \frac{1}{2} \|\omega\|^2 + c \sum_i^l (\xi_i + \xi_i^*) \quad (4)$$

$$s.t. \begin{cases} ((\omega \cdot x_i) + b) - y_i \leq \varepsilon + \xi_i, i = 1, ..., m \\ y_i - ((\omega \cdot x_i) + b) \leq \varepsilon + \xi_i^*, i = 1, ..., m \\ \xi_i^{(*)} \geq 0, i = 1, ..., m \end{cases} \tag{5}$$

where c is punishment coefficient, ξ^* is slack variable, ε is insensitive loss function.

For nonlinear regression, assume that there is such a transform: $R_n \to H, x \to \varphi(x)$, making $K(x, x') = \varphi(x) \cdot \varphi(x')$. The nonlinear model can be determined:

$$y = \sum_{i=1}^{l} (\overline{\alpha}_i^* - \overline{\alpha}_i) K(x_i, x) + \overline{b} \tag{6}$$

$$s.t. \begin{cases} \sum_{i=1}^{l} (\overline{\alpha}_i^{(*)} - \overline{\alpha}_i) = 0 \\ 0 \leq \alpha_i^{(*)} \leq C, i = 1, ..., m \end{cases} \tag{7}$$

where $\overline{\alpha}_i^{(*)} = (\overline{\alpha}_1, \overline{\alpha}_1^{(*)}, ..., \overline{\alpha}_i, \overline{\alpha}_i^{(*)})^T$ is the solution of Eq. (6).

Theoretically, when using support vector machines to predict cigarette sales [15], many kernel functions can be selected. However, the experimental results show that the performance of support vector machines based on radial basis functions is the best in this study. The samples are constructed by using the same formula (2) sliding window, and the Sequential Minimal Optimization algorithm (SMO) is used.

(3) Extreme learning machine: During the execution process, there is no need to adjust the input weight of the network and the bias of the hidden layer, and a unique optimal solution is generated. Therefore, ELM has fast learning speed and good generalization performance. At the same time, it can overcome the problems encountered by BP such as high learning rate, local minimum, over-tuning [31, 32]. The structure of ELM model is shown in Fig. 2.

First, (x_i, t_i) is the training sample of the model, where x_i is the n-dimensions input and t_i is the m-dimensions output. Then $\tilde{N}$ is the hidden nodes of ELM and the output is $g(w_i, b_i, x_j)$. The mathematical model can be described by a formula:

$$f_{\tilde{N}}(x_j) = \sum_{i=1}^{\tilde{N}} \beta_i G(a_i, b_i, x_j) = t_j, j = 1, ...N. \tag{8}$$

Input weight $[w_1, w_2, ..., w_N,]$ and bias $[b_1, b_2, ..., b_n,]$ are referred to as the parameters of hidden layer, and β_i is ith connection weight between hidden node and output node. Function $f_{\tilde{n}}$ can be equivalent to describe (8): $H\beta = T$,

$$H(w_1, ..., w_{\tilde{n}}, b_1, ..., b_{\tilde{n}}, x_1, ..., x_{\tilde{n}}) = \begin{bmatrix} g(a_1, b_1, x_1)...g(a_{\tilde{N}}, b_{\tilde{N}}, x_j) \\ g(a_1, b_1, x_N)...g(a_{\tilde{N}}, b_{\tilde{N}}, x_N) \end{bmatrix}_{N \times \tilde{N}} \tag{9}$$

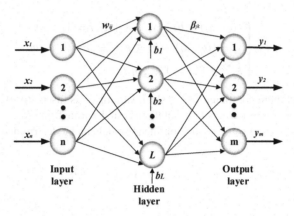

Fig. 2. Structure of the ELM model

$$\beta = \begin{bmatrix} \beta_1^T \\ \vdots \\ \beta_{\tilde{N}}^T \end{bmatrix}_{\tilde{N} \times m}, \quad T = \begin{bmatrix} t_1^T \\ \vdots \\ t_{\tilde{N}}^T \end{bmatrix}_{N \times m} \tag{10}$$

where H is the output matrix of hidden layer of ELM. H can be obtained according to the training set and the randomly assigned parameters (w_i, β_i) of the hidden layer. The sample is trained according to the zero-error approximation principle, and the least squares solution of the formula (11) is obtained.

$$\min_{\beta} \left\| H_{N \times l} \beta_{l \times m} - T'_{N \times m} \right\| \tag{11}$$

The weight β can be calculated by the following formula

$$\hat{\beta} = H^+ T \tag{12}$$

where H^+ is the Moore-Penrose generalized inverse of matrix H.

Fusion Method. The reliability of the prediction can be improved by fusing the results of different algorithms. BP algorithm can satisfy any precision requirements in the training process of the nonlinear model, but ELM has better generalization and computational performance than BP. SVM can deal with small sample problems, both ELM and SVM can obtain the global optimum solution. It is hard to determine which method can get better prediction results of cigarette sales when limiting a specific region and time. Therefore, we combine them to enhance the performance of the sales forecasting model.

The essence of the fusion process is the weighted sum of different forecasting results. However, many factors can affect the high-frequency components of cigarette sales, so it is unfair to determine the weight of each algorithm in advance. To avoid the arbitrariness of human decision, the weight is assigned by minimizing the L-P norm.

$$F = \min \left(\left\| u - w\hat{u} \right\|_p \right) \tag{13}$$

where w is 1×3 the size of the weight vector, $\hat{u}$ is $3 \times N$ the prediction result matrix, and $\hat{u}$ represents the results of the three different methods for the prediction of the moment N, the vector u is the actual high-frequency sales component of the N moment length, $\|\cdot\|_p$ is the p norm.

$$\|x\|_p = \left(\sum_K |x_k|^p \right)^{1/p} \tag{14}$$

Due to the strong similarity of cigarette sales each year, N is set to 12. The weight of p is chosen from the 1.2, 1.4, 1.5, 1.6, 1.8, 2, and it is found that the choice of norm has a slight effect on the result, and the comprehensive effect is better when the norm is $p = 1.8$.

3 Experimental Results

3.1 Data Collection

The data set for this study is provided by tobacco companies in Xianning, Hubei Province [26]. Sales data are collected from retailers according to geographical location, and ultimately the month as a unit to form a sales time series. There are 62 townships and 6 offices in Xianning, and the time series of sales between January 2010 and April 2016 are collected from the above 30 townships. According to the actual sales situation in Xianning, if the price of a carton of cigarettes is less than or equal to 40 (CNY, Chinese Yuan), it is considered to be grade 1 cigarettes, the price between 41–80 (CNY) are grade 2 cigarettes, the price between 81–150 (CNY) are grade 3 cigarettes, the price between 151–240 (CNY) are grade 4 cigarettes, the price between 241–400 (CNY) are grade 5 cigarettes, and the price higher than 400 (CNY) are grade 6 cigarettes. Considering that the data to be processed have the same physical meaning and unified dimension, so first identified the maximum and minimum values of sales data between January 2010 and October 2015, then all sales data are normalized to [0,1] according to these two values for normalization [33]. The training set is composed of sales data between January 2010 and December 2014, and the test set is sales data between January 2015 and October 2015.

4 High and Low Frequency Decomposition Results

DB6 is used as the mother wavelet, and Mallat algorithm is utilized to decompose the cigarette sales sequence into 4-layers and reconstruct the cigarette sales sequence. The contour of the fourth part is labeled as the low-frequency component, and the sum of the details of the first to fourth layer is labeled as the high-frequency component. The results of the decomposition of the first to the fourth cigarette in a township are given in Fig. 3.

Fig. 3 shows the decomposition of forecast sales into high and low frequency components. As can be seen from Fig. 3, the low-frequency component reflects the overall trend of sales, while the high-frequency component reflects seasonal fluctuation. The

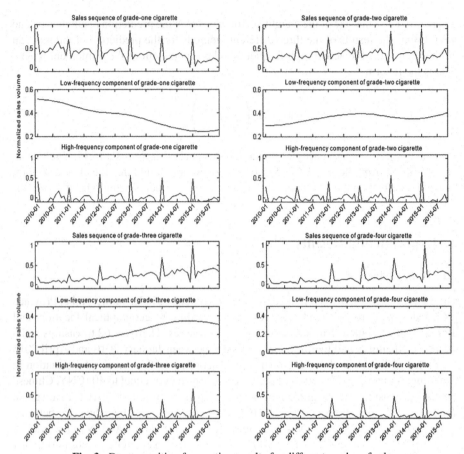

Fig. 3. Decomposition forecasting results for different grades of sales

method of unit root test is used to examine the stability of low-frequency components, the obtained unit root statistics ADF (Augmented Dickey-Fuller) is -13.2523 and -0.1572, and the critical value of ADF is significantly less than the level of 1% to 10%. Therefore, the original hypothesis is rejected and the sequence is considered to be stable. The LB (Ljung-Box) model is used to test white noise of high-frequency components. When the lag order is the first to the fourth, the Q-statistic (the Q-statistic is a test statistic output by the LB test) is smaller than the p-value, so the original hypothesis is rejected, and the high-frequency component is not a random sequence.

5   The Forecast Results of the Cigarette Sales

The final forecast result can be obtained by mixing the forecast results of high-frequency and low-frequency components. BP, SVM, and ELM algorithms are respectively used to predict the results of sales sequence, and the prediction results of the four methods are compared, as shown in Table 1. Two indicators, mean absolute percentage error (MAPE) and variance of mean absolute error (VMAE), are selected to quantify the performance of the four methods [34]. The forecast results for grade 1 to 6 cigarettes are shown in Table 1. The MAPE is defined as:

$$r = \frac{1}{N} \sum_{m=1}^{M} \sum_{n=1}^{N} \left| \frac{\hat{f}p(n) - fp(n)}{fp(n)} \right| \times 100\%$$ (15)

For town m in the nth month, $\hat{f}_p(n)$ is the forecast result of the model, and $f_p(n)$ is the true value.

The definition of the VMAE is shown in the formula (16).

$$s^2 = \frac{1}{n} \sum_{i=1}^{n} (r_i - \mu)^2$$ (16)

where r_i is the mean absolute error, and $\mu = \frac{1}{n} \sum_{i=1}^{n} r_i$ is the mean of r_i.

Table 1. The prediction results of multiple models.

Algorithms	Grade-one		Grade-two		Grade-three	
	MAPE	VMAE	MAPE	VMAE	MAPE	VMAE
ELM	7.05%	0.0035	11.57%	0.0056	6.57%	0.0065
BP	8.62%	0.0023	11.33%	0.0044	7.05%	0.0046
SVM	8.98%	0.0035	11.71%	0.0071	7.45%	0.0055
Fusion	6.93%	0.0030	8.65%	0.0026	6.45%	0.0029
Algorithms	Grade-four		Grade-five		Grade-six	
	MAPE	VMAE	MAPE	VMAE	MAPE	VMAE
ELM	5.08%	0.0264	13.09%	0.0258	4.17%	0.0057
BP	5.22%	0.0034	14.45%	0.0213	4.73%	0.0028
SVM	6.62%	0.0051	14.23%	0.0137	4.99%	0.0071
Fusion	4.20%	0.0028	12.15%	0.0192	3.58%	0.0017

6 Discussion

6.1 Evaluation of Low-Frequency Component

Fig. 3 shows the sales prediction results of the low-frequency components of the R town. An obviously decreasing tendency of the grade 1 cigarette can be seen as above, but the grade 2 cigarette, the grade 3 cigarette, and the grade 4 cigarette are present an increasing tendency of the low-frequency components. The reason for this change may be

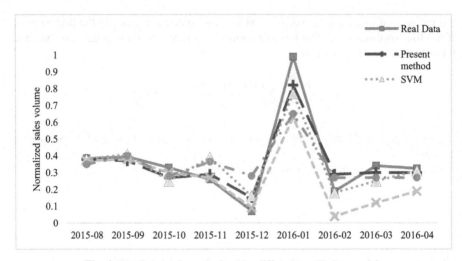

Fig. 4. Predicted values obtained by different prediction models

that consumers pay more attention to the quality of tobacco with economic development, and another reason may be the rise in cigarette prices or changes in the strategy of tobacco companies [35]. However, the forecast results of the high-frequency component of R town are shown in Fig. 3, it can be seen that the high-frequency component has only a fluctuating trend but no overall trend. Therefore, by the wavelet decomposition approach, it is proved that the sales sequence component has a clear physical meaning, which can be used to simulate the overall trend and seasonal fluctuation of cigarette sales.

7   Evaluation of Low-Frequency Component

Figure 4 shows that the fusion results of multiple models are more stable and precise than other single models. Considering that sales are influenced by the external factors, the fusion results of several models can be closer to the forecast results of the wavelet decomposition. Based on MAPE and VMAE indicators, Table 1 summarizes the forecast accuracy and model stability of the four models BP, SVM, ELM, and fusion. The MAPE of the fusion model of grade-one cigarettes is 1.28% less than the average MAPE of the above three methods. The forecast accuracy of grade 2 to 6 cigarette sales increase by 2.89%, 0.57%, 1.44%, 1.77% and 1.05%. Meanwhile, the forecast stability of the overall trend of cigarette sales increases by 37.21%.

Furthermore, according to the forecast results of cigarette sales from grade 1 to 6, it can be seen that the minimum MAPE is 3.58%, which accounts for about 1.07 days sales on average of the town, and the maximum MAPE is 12.15%, which is equal to the actual sales of the township for 3.65 days. The results show that the forecast accuracy is already within the acceptable range. Even for the state-of-the-art method ELM, it can be found from Table 2 that the stability of ELM is 0.0123, and it is not the best. In terms of parameter VMAE, compared with ELM, BP and SVM, the average stability of the fusion model is improved by 56.09%, 16.92% and 22.85%, respectively. Furthermore, the hybrid forecasting model presented in this paper decreases the predicted deviation rate by 0.93%, 1.58%, and 2.01%, respectively. Hence it is proved that the fusion method can be effectively used for cigarette sales prediction based on source data.

In this study, the cigarette sales sequence can be decomposed by the mother wavelet DB6 to obtain the corresponding high-frequency and low-frequency components. Different methods are used to predict sales based on the fluctuation trend of the high-frequency component, and different prediction results are weighted according to the norm $p = 1.8$. Figure. 5 shows the average absolute percentage error of different cigarette prediction models. It can be found that in terms of mean square error, the fusion method is smaller than the above single model, especially the grade 6 cigarette with the best prediction accuracy of 96.42%. The proposed method can perform well in predicting the sales volume of cigarettes of different grades, and the forecast results of fusion are better than any single algorithm.

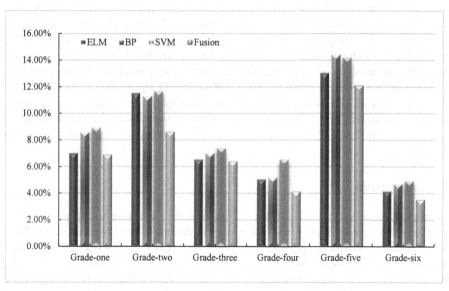

Fig. 5. Mean absolute percentage error with different prediction models for kinds of cigarettes

Table 2. The mean VMAE and MAPE of four methods

Algorithms	Fusion	ELM	BP	SVM
VMAE	0.0054	0.0123	0.0065	0.007
MAPE	6.99%	7.92%	8.57%	9.00%

Table 3. The MAPE of fusion method for six products

Product	1	2	3	4	5	6	mean
Fusion	6.93%	8.65%	6.45%	4.20%	12.15%	3.58%	6.99%

8   Further Discussions

The forecast results of six grades of cigarettes are listed in Table 3. The grade 6 cigarettes obtained the highest forecast accuracy, while the MAPE of the fusion approach is only 3.58%. Moreover, the average MAPE of the fusion method is 6.99%, which is equal to 2.01 days of actual sales in the town, and it is within the acceptable error range. Actually, the forecast accuracy of a method is usually affected by many factors such as the inherent attributes and the sales model of the product. Wavelet decomposition as a pre-processing approach is used to extract time-series features of any product sales, and to structure different frequency components into the input and output sequences. The change in low-frequency components is slow and stable, so the AR model is used to predict the overall trend of cigarette sales. The fluctuation of high-frequency components is drastic and unstable, and thus the detailed fluctuation in sales can be predicted by a fusion approach which is based on SVM, BP, and ELM algorithms. What's more, the unit root statistics ADF is used to estimate the stability of low-frequency components, and the Q-statistic of LB model is used to examine the white noise of the high-frequency components. Therefore, the proposed method is also applicable to the sales forecast for other products.

9   Conclusion

Improving the accuracy and stability of the cigarette sales forecasting model is the purpose of this paper. Firstly, wavelet is used to decompose the time series to obtain low-frequency and high-frequency components. Secondly, the AR model and the hybrid model are used to predict cigarette sales in these two components, respectively. In the end, the forecast result is obtained by mixing the results obtained from the two components: high-frequency and low-frequency. The main innovations are as follows:

Firstly, the sales sequence is essentially a time sequence, which can be decomposed by wavelet, and its high-frequency and low-frequency components can be obtained respectively. These two components are used to simulate the seasonal fluctuations and the overall trend of the sales sequence, respectively. The characteristics of the components are known, and thus we can choose the appropriate forecast methods for different components. Secondly, by fusing the results of different models to predict high-frequency components, the accuracy and stability of the forecast model can be greatly improved.

The proposed method not only increases the accuracy and stability of the model but also provides a theoretical basis for formulating an allocation strategy. Furthermore, this method is also applicable to other fields and can be used to deal with similar problems.

Acknowledgement. This work was supported by National Natural Science Foundation of China under Grant nos. of 61801454; Natural Science Foundation of Zhejiang Province under Grant nos.

of LQ18F010006; and Fundamental Research Business Expenses of Central Universities under Grant nos. of CZY19002.

References

1. Fildes, R., Goodwin, P., Lawrence, M., Nikolopoulos, K.: Effective forecasting and judgmental adjustments: an empirical evaluation and strategies for improvement in supply-chain planning. Int. J. Forecast **25**(1), 3–23 (2009)
2. Ton, Z., Raman, A.: The Effect of Product Variety and Inventory Levels on Retail Store Sales: A Longitudinal Study. Prod. Oper. Manag. **19**(5), 546–560 (2010)
3. Juan, R.T., Nikolaos, K., Robert, F.: On the identification of sales forecasting models in the presence of promotions. J. Oper. Res. Soc. **66**(2), 299–307 (2015)
4. Wan, X., Sanders, N.R.: The negative impact of product variety: Forecast bias, inventory levels, and the role of vertical integration. Int. J. Prod. Econ. **186**, 123–131 (2017)
5. Joossens, L., Merriman, D., Ross, H., Raw, M.: The impact of eliminating the global illicit cigarette trade on health and revenue. Addiction **105**(9), 1640–1649 (2010)
6. Teunter, R.H., Syntetos, A.A., Babai, M.Z.: Intermittent demand: linking forecasting to inventory obsolescence. Eur. J. Oper. Res **214**, 606–615 (2011)
7. Ye, M.L.: Cigarette sales forecasting model based on combined forecasting method of time series. J. Fujian Comput. **36**(02), 63–67 (2020)
8. Babu, C.N., Reddy, B.E.: A moving-average filter based hybrid ARIMAANN model for forecasting time series data. Appl. Soft Comput. **23**, 27–38 (2014)
9. Sun, X.S., Chen, F.F., Wu, M.L., Qi, Q., Jia, J.: Discussion on prediction method of photovoltaic power generation of AR model based on EMD decomposition. Electr. Eng. **11**, 8–14 (2019)
10. Shi, Y.Y., Li, M., Fu, Y., Wang, L.W., Sun, M.X., Hao, J.M.: Multi-scenario traffic land demand forecasting based on grey system-BP neural network model: a case study of urban agglomeration in the middle reaches of the Yangtze River. J. China Agric. Univ. **25**(06), 142–153 (2020)
11. Zhang, S., Chen, S.J., Ma, G.W., Huang, W.B., Tao, C.H.: Power spot market clearing price forecasting based on DPBIL-SVM hybrid model. Water Resourc. Power **38**(04), 197–200 (2020)
12. Guo, T.T., Chen, X.G.: The prediction study on China's grain price based on PCA-ELM. Prices Monthly **12**, 21–26 (2015)
13. Short-Term Traffic Flow Forecasting based on Hybrid FWADE-ELM: Chen, R.Q., Li, J.C., YU, J.S. Control Decis. **12**, 1–8 (2020)
14. Kim, T.Y., Oh, K.J., Kim, C., Do, J.D.: Artificial neural networks for non-stationary time series. Neurocomputing **61**, 439–447 (2004)
15. Björn, W., Jan, K., Elke, L., Oliver, K., Detlev, H.: Comparing support vector regression for PV power forecasting to a physical modeling approach using measurement, numerical weather prediction, and cloud motion data. Sol. Energy **135**, 197–208 (2016)
16. Zhao, C.L., Li, L.: Analysis of factors influencing cigarette sales based on SVAR model-taking yichang city, hubei province as an example. China Circul. Econ. **17**, 6–9 (2016)
17. Han, Y.L.: Discussion on cigarette market sales forecast-taking cigarette market sales in a city of shandong province as an example. Modern Bus. **33**, 26–27 (2014)
18. Hamzaçebi, C.: Improving artificial neural networks' performance in seasonal time series forecasting. Inf. Sci. **178**(23), 4550–4559 (2008)
19. Daubechies, I.: The wavelet transform, time-frequency localization and signal analysis. IEEE Trans. Inf. Theory **36**(5), 965–1005 (1990)

20. Zhang, L., Fang, Z., Ma, T.F.: AQI prediction of bp neural network based on wavelet decomposition of time series. J. Xuzhou Inst. Technol. (Nat. Sci. Edn.) **35**(01), 45–52 (2020)
21. Chen, Z.C., Wang, S.H., Zhao, S.C., Liu, Z.C., Wang, G.: Short-term power forecasting of wind farms based on GA-BP and wavelet -SVM algorithm. J. Shijiazhuang Tiedao Univ. (Nat. Sci. Edn.) **33**(01), 104–109 (2020)
22. Azimi, R., Ghofrani, M., Ghayekhloo, M.: A hybrid wind power forecasting model based on data mining and wavelets analysis. Energy Convers. Manage. **127**, 208–225 (2016)
23. Vaccaroa, A., Mercoglianob, P., Schianob, P., Villaccia, D.: An adaptive framework based on multi-model data fusion for one-day-ahead wind power forecasting. Electric Power Systems Research **81**(3), 775–782 (2011)
24. Li, X., Wang, J., Wang, Z., Zhou, H.: Construction study on application of decomposition and reconstruction for energy price analysis. J. China Univ. Petrol. (Edn. Soc. Sci.) **35**(04), 1–8 (2019)
25. Upadhyaya, B.R., Mehta, C., Bayram, D.: Integration of time series modeling and wavelet transform for monitoring nuclear plant sensors. IEEE Trans. Nucl. Sci. **61**(5), 2628–2635 (2014)
26. Zhu, J.J., He, X.Z., Wang, J.S.: Cigarette sales volumes forecast based on hybrid model. Acta Tabacaria Sinica **22**(5), 120–125 (2016)
27. Wang, L., Peng, L., Xia, D., Zeng, Y.: BP neural network incorporating self-adaptive differential evolution algorithm for time series forecasting. Comput. Eng. Sci. **37**(12), 2270–2275 (2015)
28. Cybenko, G.: Approximation by superpositions of a sigmoidal function. Mathematics of Control Signals, and Systems (MCSS) 2(4), 303–314 (1989)
29. Fun, M.H., Hagan, M.T.: Levenberg–Marquardt training for modular networks. IEEE Int. Conf. Neural Netw 1, 468–473 (1996)
30. Li, G.Q., Liu, Z., Jin, G.B., Quan, R.: Ultra short-term power load forecasting based on randomly distributive embedded framework and BP neural network. Power System Technol. **44**(02), 437–445 (2020)
31. Wang, X.S., Zhao, J.J., Chen, Y.H.: Unsupervised domain adaption classifier via ELM. Control Decis. **35**(04), 861–869 (2020)
32. Zhang, Y., Yan, Q.S.: Forecasting model for shanghai composite index closing price based on ABC-ELM. Comput. Simul. **37**(05), 154–160 (2020)
33. Morsi, W.G., El-Hawary, M.E.: Time-frequency non-sinusoidal currentdecomposition based on the wavelet packet transform. In: IEEE Power Engineering Society General Meeting, pp. 1–8 (2007).
34. Wallström, P., Segerstedt, A.: Evaluation of forecasting error measurements and techniques for intermittent demand. Int. J. Prod. Econ **128**, 625–636 (2010)
35. Gallet, C.A., List, J.A.: Cigarette demand: a meta-analysis of elasticities. Health Econ. **12**(10), 821–835 (2003)

Research on Adaptive Adjustment Algorithm of Isolated Word State Number Based on ANN-HMM Mixed Model

Qiuming Yin[⊠] and Tianfei Shen

College of Mechanical Engineering and Automation,
Shanghai University, Shanghai 201900, China
eric_ming96@163.com

Abstract. In order to overcome the contradiction between the recognition accuracy rate, training speed, and storage space caused by using the fixed state number as the mixed model state number, this paper conducts an experimental analysis on the traditional method of using the fixed state number as the model state number. A method for adaptively adjusting the number of model states is proposed above. Through theoretical analysis and simulation research, the method of selecting model state number is given, and a set of best state number suitable for the speech sample library used in this paper is determined. The simulation results show that the adaptive adjustment state number method proposed in this paper is in recognition accuracy and training speed. And the storage space is superior to its fixed state number method, which effectively improves the accuracy, stability and real-time performance of the speech recognition system.

Keywords: Mixed model · Hidden markov model · Speech recognition · Adaptive state number

1 Introduction

With the development of speech recognition technology and the increase in the volume of speech data, the accuracy of speech recognition models based on template matching has dropped [1], and the hidden Markov model based on statistical pattern recognition is widely used, which uses state transition probability and observation Probability models the speech rate and speech changes [2], which improves the recognition accuracy, but this acoustic model is obtained by preparing the training based on maximum likelihood, and needs to know the a priori probability, so the classification decision-making ability of this model is weak. The state observation probability distribution of the hidden Markov model is a mixture of multidimensional Gaussian distributions [3]. For the continuous hidden Markov model used in this paper, the matching relationship between the prior choice of this model and the true Gaussian mixture density is relatively It is weak, and due to various noise interferences in the actual use environment, it will seriously affect the recognition of voice commands. The model is shown in Fig. 1. Article [4] uses artificial neural network (ANN) modeling with anti-noise, strong learning ability, and

© Springer Nature Singapore Pte Ltd. 2020
M. Fei et al. (Eds.): LSMS 2020/ICSEE 2020 Workshops, CCIS 1303, pp. 483–497, 2020.
https://doi.org/10.1007/978-981-33-6378-6_36

strong robustness, but the neural network is not strong in modeling dynamic time series. The general framework of article [5] uses a hidden Markov model, and the probability estimation and contextual correlation are realized by ANN. The mixed model of ANN and HMM at the speech frame level discards a series of unreasonable assumptions of HMM and overcomes the defect of poor decision-making ability of maximum likelihood criterion. Therefore, the ANN/HMM hybrid model [6–8] is a recognition problem that is very suitable for processing the HMM model of time series such as speech, which makes the accuracy of speech recognition greatly improved, but ANN replaces the original HMM The Gaussian mixture distribution in the model makes the training speed slower, so the number of ANNs in the ANN/HMM hybrid model used in the article "Application of ANN/HMM hybrid model in speech recognition [9] The size of the network, the amount of resources occupied by the training process, and also affects the recognition effect, so the number of states is a very important parameter in the hybrid model [10]. The current research on the selection of the number of states in the hybrid model is There is a hot spot, and there is still little research on the strategy for determining the number of states in the hybrid model with ANN and HMM. Therefore, this paper studies the strategy for selecting the number of states to make the model achieve a better modeling effect.

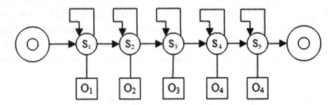

Fig. 1. Hidden Markov model diagram

2   Traditional Fixed State Number Method

The fixed state number method means that the number of hidden Markov model states selected is fixed before training the hybrid model. The number of ANN (Artificial Neural Network) in the hybrid model is fixed. In the previous literature [11], the number of states of the HMM is determined before training, and remains unchanged throughout the training process to achieve parameter sharing, so this method of fixed state number is widely used, but this The method also has disadvantages. When the differences in speech samples are relatively large, such as swallowing and continuous reading, the model with a fixed number of states at this time will have a large error in modeling these differences, and if the number of states is fixed, Then the strategy for determining the number of states, and the number of states shared by each mode to be recognized or the different number of states for each mode cannot be determined. The states in the model are given a clear meaning according to the voice sample, such as a basic unit. The basic unit included in each mode may be different, and the number of states of the mode is different [12–14]. If all modes use the same number of states, then for continuous reading, swallowing, and normal sounds, the fixed state number method cannot achieve a balance between accurate modeling and the smallest network size.

3   Improved Adaptive Adjustment State Number Method

The model network structure based on the above fixed number of states is relatively simple, but it is not able to adapt to more morphological speech samples, and has a relatively large dependence on the modeled speech samples themselves during the training process. The form automatically increases or decreases the number of states, and adjusts the structure of the mixed model accordingly. A method for adaptively adjusting the number of states in the mixed model is proposed.

First of all, the number of states of the model is not fixed, but automatically increases or decreases the number of states according to the specific situation of the speech sample. Among them, the increased state is based on the fact that the modeling accuracy of the selected basic unit is insufficient based on the state, that is, the effective number of frames of speech is long, and the state needs to be increased to build the model more carefully. The specific method is to divide the state with insufficient modeling accuracy into two states, and connect the divided two states. In the early stage of model training, because the results of the neural network training are very different from the real results, the model will continue to improve the modeling accuracy, resulting in too many split states and redundant states. This situation may be very late in the model training. Seldom appear. In the ANN/HMM hybrid model, each state of HMM is an ANN. The network size of the model will increase as the state increases, and the system storage overhead will also increase. When the storage space of the speech recognition application platform is small, the system may not have enough space to store each model. It is very necessary to delete the state at this time, so when the length of the corresponding voice primitive is short, the redundant state is merged into the adjacent state, so as to achieve the purpose of reducing the state and saving overhead.

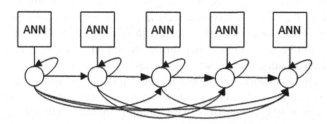

Fig. 2. Improved hidden Markov fully connected hybrid model

Before adaptively adjusting the number of states, the traditional hidden Markov transition from left to right without jump structure to full connection structure is shown in Fig. 2: Based on the order, each state can be transferred to itself or to all other states. There are three main reasons for improving the structure of the HMM to the state of Fig. 2. First, in the modeled speech sample, due to the irregularity of the speech sample, it may have a complete sample shape, or it may contain some missing basic units sample.

Therefore, in order to make the model adapt to these forms of speech samples at the same time, this paper improves the traditional model by adding state spanning. If the speech sample does not lack the basic unit, the state will be transferred from beginning to end in order. The basic unit is missing, then these missing states are spanned while ensuring that the left and right states can still be connected. Second, in order to increase the versatility of the model and algorithm, that is, regardless of the selection of different basic units such as syllables and phonemes, the model structure used is the same. Whether each speech sample to be recognized has a missing basic unit, and which specific basic unit is missing are uncertain. Therefore, each state may be crossed. Third, considering that the difference between the basic unit represented by the model obtained by adding or deleting the state and the model before adding or deleting the state will not become very large, so that each state and all subsequent states, there is a direct transfer relationship. That is, when a state is added or deleted, try to maintain the transfer relationship between other states, and it will not change because of the presence or absence of a state. For all the speech samples being modeled, although the number of states will be increased or decreased during the training process, at the moment when the model is initially trained, the number of states of the model is the same.

In the improved fully-connected model, adding the state essentially divides the original state into two parts, and divides the original state that the modeling accuracy of the speech sample is not enough. Evaluating a state is to evaluate the effectiveness of an ANN modeling effect. It is mainly judged according to the square error E of the output error of the ANN. If E is large, it indicates that the modeling accuracy of this state is not enough. At the same time, in order to avoid a large number of redundant states due to the large errors at the initial stage of training, this paper introduces another important evaluation standard is the speed of error decline. If the error E declines slowly during a certain period of time, it means that the neural network has reached the accuracy and will not continue to learn. Continued training will no longer produce the desired effect, and it is easy to cause overfitting. Therefore, the two parameters are considered together to decide whether to split a certain state. Since the weights of ANN are initialized randomly at the beginning, in the initial stage of training ANN with BP algorithm, the output error square sum E of the model will be relatively large in the initial period of time. If only the value of E is referenced, when E is greater than a certain threshold, it will be split, and the state will be split continuously during the initial period of training, resulting in many redundant states, thereby wasting storage resources. According to the learning theory of the neural network, it is normal for the output error to be too large in the early stage of the neural network training, and in the initial stage, the error decline rate is relatively fast. In the later stage of network training, the error will drop to a relatively small value, and continue to decline at a slower rate, and this phenomenon is normal in the later stage of training. Therefore, these two parameters should be comprehensively measured as a criterion for judging whether the state should be split. For all modeled objects, although the number of states will be increased or decreased during the training process, the number of states of the model is the same at the beginning of model training. Or for the sake of versatility, no matter what the modeled object is, the modeling method is unchanged.

3.1 Ways to Increase Status

Assuming that the nth state meets the following conditions, then split it.

$$E_n^i \wedge > \delta_1, \Delta E_n^i < \delta_2 \tag{1}$$

$$\Delta E_n^i = \left| E_n^{i-1} - E_n^i \right| \tag{2}$$

where i, n, and E represent the sum of squared output errors of the nth RBF after the iteration. As shown in Fig. 3, the speech sequence is x_1-x_T, assuming that the initial state number is equal to 4, when the state 3 needs to increase the state through algorithm calculation, the state number is split from itself into two, 3–1, 3–2, And connected to states 1, 2, 4 in the fully connected model to form a new connection model.

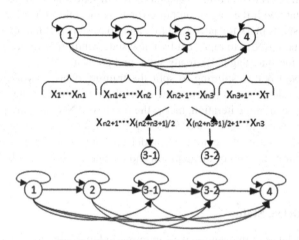

Fig. 3. Training method of splitted state

After the state split is completed, the split state is not used to replace the original state immediately, but the two newly split states are separately trained several times. When it is determined that state 3 needs to be split, the original model is still retained, and the data is split under this model using the Viterbi algorithm. First, perform random initialization of the weights of the two new ANNs. As shown in the figure, after the training data is divided by the Viterbi algorithm, the input data is divided into 4 parts, where $X_1...X_{n1}$ represents the training data corresponding to state 1, and similarly, it represents the training corresponding to state 3. We hope that after the state is split, the changes in the basic units represented by the states that have not been split are as small as possible. Therefore, for the training of the two new ANNs, we still use the training data corresponding to state 3 in the original model. And, when two new ANNs are trained separately, the rest of the state remains unchanged.

The training data of state 3 divided by the Viterbi algorithm is divided into two parts evenly. The first part is $X_{n2+1}...X_{(n2+n3+1)/2}$. The second part is $X_{(n2+n3+1)/2} + 1...X_{n3}$.

Use the first part to train state 3–1, and use the second part to train state 3–2. This process is iterated several times, so that the two new ANNs have a preliminary model of the primitive represented by the original state 3. Then proceed to the next steps. After completing this iterative process, state 3 is deleted from the original model and replaced with 3–1 and 3–2 to form a new model. Several iterations of training are also required for this new model. It should be noted that in this iterative process, adding or deleting states again is not allowed. After the iteration, the entire model has a preliminary modeling of the entire symbol before judging the adding or deleting states. Otherwise frequent addition or deletion of states will seriously affect the modeling effect.

The specific algorithm is as follows:

1. Initialize the current state marked as "undivided".
2. If the current state is "undivided", perform an iterative training and check the output error of each ANN to determine whether there is a state that needs to be split. If yes, select the state with the largest output error E as the split object for this time, and change the status mark to "need to split", and go to step 3. If not, exit.
3. Create two new ANNs to represent two new states, initialize the weights randomly, and change the state flag to "Already split" and go to step 4.
4. Use the model before splitting to split the training data of two new ANNs, and perform the calculation of the two new ANN. Line training. Iterate this process m1 times. At the end of the iteration, merge the two new ANNs into the original model, mark the status as "merged", and go to step 5.
5. Keep the structure of the combined model unchanged, and perform m2 iteration training. Change the end of the iteration and change the status mark to "undivided" and exit.

3.2 Ways to Delete Status

The method of deleting the state makes a simple judgment based on the shape of the voice sample. What is deleted is the state where the corresponding feature vector frame sequence is short. Assuming that the frame sequence of a sample contains T frames, set a threshold = 10, and the number of states of the type is N, then the average number of frames included in each state is The minimum number of frames in the i-state is compared with, as shown in formula 4, if the formula 4 is satisfied, the state is deleted.

$$\bar{n} = \frac{T}{\alpha N} \tag{3}$$

$$f_i < \bar{n} \tag{4}$$

As shown in Fig. 4, suppose a model contains N = 4 states. When a frame sequence X of T = 120 feature vectors represented by this model passes through the model, the optimal state sequence calculated by Viterbi algorithm is shown in the figure. Among them, state 1 contains two frames of and, state 2 contains-4 frames in total, state 3 contains-111 frames in total, and state 4 contains-3 frames in total. Then the average value of the number of frames included in each state is n = 30. Among these four states, the state with the least number of frames is state 1, the number of frames contained in it is less than 2, so state 2 is deleted. If the minimum number of frames calculated last time is equivalent to the average value, you do not need to delete and continue training, such as state 2 and 4. In this way, the state with a relatively small number of frames is included and merged into the previous or next state. Where exactly it is merged is automatically determined and completed with the continuous training of ANN.

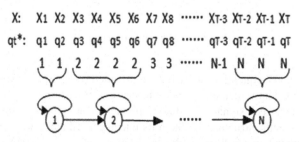

Fig. 4. Delete state principle

The specific algorithm is as follows:

1 Initialize the current state marked as "not deleted".
2 If the current state is "not deleted", perform an iterative training and check whether any state needs to be deleted. If there is, delete the status and change the status mark to "deleted", go to step 3, if not, exit the delete module.
3 Keep the deleted model structure unchanged and perform m iteration training. At the end of the iteration, change the status mark to "Undeleted" and exit.

4 Hybrid Model Training Method

To complete the training of the entire mixed model, we must first obtain the various parameters in the model. One of the more important parameters here is the output probability matrix B in the HMM model. Suppose there are N ANN models in the mixed model, numbered as ANN(1), ANN(2),......, ANN(N). Take a sample $X = \{x_1, x_2, ..., x_T\}$ from the training sample set, where x_i is the F dimensional feature vector corresponding to each sampling point of the sample, and T is the number of sampling points in the sample.

Let X go through ANN(1), ANN(2), ..., ANN(N) in turn, then for the elements in X, through each ANN model, a corresponding output error square sum E will be obtained. It is represented by dots in Fig. 5. After we arrange each ANN model and the eigenvector sequence X of the input sample in a graphical manner, it happens to have the same structure as the output probability matrix B. From the previous analysis, we can conclude that the corresponding elements in E and B are inversely proportional in value. In this way, we take the reciprocal of each E obtained and normalize it according to the rules of the formula, and place it in the corresponding position of matrix B to obtain matrix B.

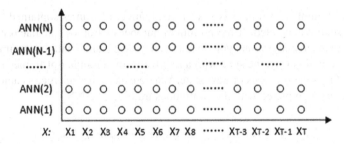

Fig. 5. Computation of matrix B

In some previous sequential pattern recognition methods, the pattern to be recognized is generally divided into small units, that is, basic units. For example, speech recognition is to segment the speech signal according to phonemes or syllables, and then calculate according to a specific structure to obtain its intrinsic attributes and interrelationships, and then recognize them as domain objects. In many previous documents, the method of dividing the basic unit is generally artificial forced division, that is, according to the temporal or spatial relationship characteristics of the object to be identified, it is divided into several basic units. The basic units segmented in this way are intuitive and easy to understand. However, this method also has its shortcomings, that is, the segmentation method is too blunt, and it is more difficult to use programs to judge the segmentation point. Moreover, this kind of artificial segmentation may not be the most suitable segmentation method for the recognizer, and may not achieve the best results. Therefore, this article does not use this artificial method of dividing the basic unit, but an automatic method of dividing. The implementation of this segmentation method mainly uses the Viterbi algorithm, one of the basic algorithms of the HMM model. The Viterbi algorithm can be used to obtain a best state sequence for a given observation sequence under the current model. This sequence is the basis for dividing the basic unit. As shown in Fig. 6, the curve represents a backtracking path calculated using the Viterbi algorithm, which corresponds to the final optimal state sequence q_t^*. Taking the situation in Fig. 6 as an example, this state sequence is:

$$1, 1, 1, 2, 2, 2, 2, 3, \ldots\ldots, N-1, N, N, N$$

These numbers correspond to the ANN number, the same numbers are grouped into the same group, and the input frame sequence corresponding to the q_i with the same value is the training data of the ANN with the number q_i value. So far we have completed the segmentation of the ANN training data. In this segmentation method, the system can automatically segment the training data of each basic unit according to the specific shape of the model each time.

In this way, the training data of each ANN is segmented, and each ANN is trained once. After the training, the weight of ANN has been adjusted. Therefore, before the next training starts, the value of the element in B calculated again will change. When the Viterbi algorithm is used again, a different path will be obtained. Repeat the above process, as the training progresses, the state sequence q_i^{t*} will get closer and closer to the shape of the training sample, and finally the training goal is achieved. At the beginning

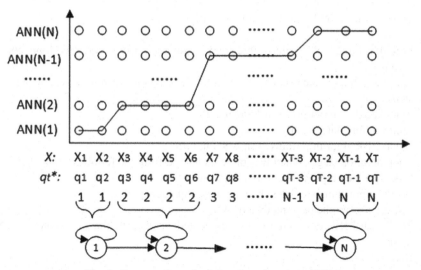

Fig. 6. Segmentaion of training datas for every ANN

of training, the weights of the neural network are initialized randomly, so there may be a big difference from the training sample. In this case, the initial state sequence q_i^{t*} is not very ideal, and it may even appear as follows (assuming that the model has 4 states):

$$q_t* = \{0, 0, 0, 0, \ldots\ldots, 0, 1, 2, 3\}$$

Such a state sequence is obviously unreasonable if it is used as the basis for segmenting ANN training data. If such data is used as the input of ANN, then it is always difficult for the model to approximate the true shape of the sample. In order to avoid this situation, during the first several iterations of model training, we adopt a method of forcing the training data to be equally divided. In other words, according to the number N of states contained in the model, $X = \{x_1, x_2, \ldots\ldots, x_T\}$ is divided into N parts, and each part is used to train an ANN. After the iteration, the model will have a preliminary shape, and then continue training on this shape, the model is more likely to approach the actual situation of the sample. Each frame of a training sample will get an output error E through the ANN of the current model.

Suppose the output error obtained at time t is E_t, then we use the average value of the output error $\bar{E}$ as the criterion for judging whether the iterative process of model training is over.

$$\bar{E} = \frac{1}{T} \sum_{i=1}^{T} E_i \tag{5}$$

When $\bar{E}$ is greater than the threshold δ, iterative training continues, and when $\bar{E}$ is less than the threshold δ, the model training ends.

The specific ANN/HMM hybrid model training algorithm is as follows:

1. Initialize model parameters, threshold δ, output error average E, loop control parameters loop.
2. Split the training data evenly. When loop is less than n, perform the following three steps:
   Calculate the output probability matrix B;
   Split training data equally;
   Batch-EBP algorithm trains each ANN in turn;
3. Loop iteration stage. When $\bar{E}$ is more than δ, perform the following four steps:
   Calculate the output probability matrix B;
   Viterbi algorithm calculates the best state sequence q_t^*;
   Use q_t^* to split the training data of ANN;
   The Batch-EBP algorithm uses the segmented data to train the corresponding ANN in turn;
4. End.

5   Experiment

The voice samples in the experiment are ten voice commands to be recognized: on, off, default, acceleration, deceleration, left turn, right turn, forward, backward, stop. It contains five female voices and five male voices. Each person pronounces each command 100 times, of which 80 are used as training samples and 20 are used as recognition samples. The sampling rate of the voice command used in the experiment is 16 kHz, the frame length is 25 ms, and the frame shift is 12.5 ms, where one frame refers to a piece of voice data. The MFCC (Mel Frequency Cepstrum Coefficient) [7] is used. The feature parameter extracted in this paper is a 39-dimensional vector. It contains 12 Mel cepstrum coefficients, 1 short-term energy parameter and 13 first-order and second-order difference coefficients totaling 39 dimensions. The 13-dimensional features obtained directly are static features, and the first-order and second-order differential features are dynamic features.

In this paper, we have conducted an experimental study on the fixed number of states. The overall algorithm flow of the ANN/HMM hybrid model with automatic increase and decrease state algorithm is shown in Fig. 7. Adding an adaptive increase and decrease state algorithm is different from the traditional algorithm in the process. In the early stage of training, it is necessary to perform several iterations on the model, so that the model has a preliminary modeling effect on the sample, to avoid the situation of undesirable increase or deletion. Several trainings have reduced the absolute output error of the model to a certain extent, and after the decrease, whether to increase the number of state judgments. When the model is in the middle of training, it enters the alternating phase of adding and deleting states. This alternation also needs to be repeated several times, so that the model can automatically retain the state that is most suitable for the sample according to the shape of the sample and the change of the model. When the process of adding and deleting states reaches a balance, the iteration ends and enters the later training process. Later training is similar to the training process with a fixed number of states. In this process, the number of states has reached the optimum. Similarly, a threshold ε is set. When the output error square sum E < ε, the training is ended, otherwise, the above iteration process is continued.

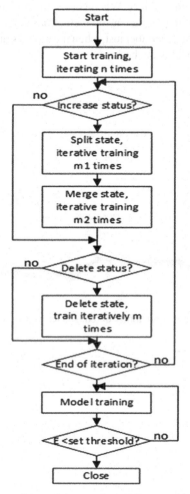

Fig. 7. Flow chart of the training method with auto-add-and-delete states

According to the voice commands studied, the words are isolated words, so the number of states is set to N = 2, N = 3, N = 4, N = 5, N = 6, N = 7, N = 8.are trained to obtain seven fixed state number models. Using the Viterbi algorithm for recognition, you will get the probability value of ten words respectively. According to the probability value, you can judge the recognition result of the test speech. The greater the probability, the better the recognition performance. In order to prevent the occurrence of data underflow, the probability value is logarithmic, amplified, etc. The new value describing the probability is negative because it takes a logarithmic operation. The experimental results of the fixed state number are shown in Fig. 8. From the experimental results, it can be seen that when N = 6, the highest two probability differences are the largest, and the highest burst style probability difference is larger, indicating that the system is judging the maximum probability. The more accurate the decision, the higher the recognition rate and the better the system stability. When N = 8, the probability value is the largest. This is because the modeling accuracy is too meticulous, which leads to the occurrence of overfitting. In the fixed state number test, when the model state number is 6, the overall modeling effect of the model is the best.

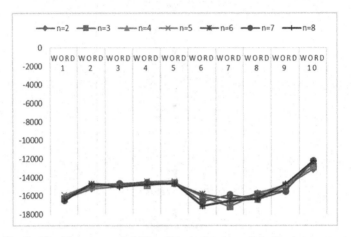

Fig. 8. Viterbi output probability of ten isolated words in fixed state count

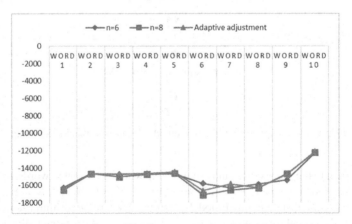

Fig. 9. Comparison of fixed state number and adaptive adjustment

The data in the test of the automatic increase and decrease state number method uses the same training data as the fixed state number method. We take the initial state number $N = 6$ and the coefficient $= 10$. The final total number of states of the automatic increase and decrease state number method is $7 + 5 + 6 + 4 + 3 + 2 + 5 + 7 + 6 + 8 = 53$. In the fixed state number method, the state with the most balanced modeling effect is 6 and the total number of states is $6 * 10 = 60$, the state with the finest modeling effect is 8 and the total number of states is $8 * 10 = 80$. The state number method achieves a good modeling effect while the main function is to save a lot of redundant storage space. We increase or decrease the state in this way, the purpose is to improve the modeling effect of the model, at the same time, in the case of ensuring that the modeling effect is basically unchanged, try to reduce the requirements of the system storage space and shorten the training time of the model. In Table 1, the output probability ranking of ten isolated words in seven fixed States and adaptive adjustment States is counted, and the total ranking of output probabilities of all words in each state is calculated. It can be seen that the method of adaptive adjustment of the number of States ranks the highest, that is, the highest output probability. The recognition probability of the automatic increase and decrease state number method is shown in Table 2. The average recognition rate is 91.5%. The Viterbi probability output of the automatic increase and decrease state number method is shown in Fig. 9, when compared with the fixed state number $N = 6, 8$. It can be seen that compared with $N = 8$, the automatic increase/decrease state number method saves a lot of space. Compared with $N = 6$, the difference between the two highest output probabilities is the largest. At the same time, the training time, storage space and the specific difference between the two highest probabilities are counted in the experiment. As shown in Table 3, it can be seen that the training time at $N = 8$ is almost twice that of the adaptive state method. The state method is close to the detailed modeling $N = 8$ in recognition accuracy and system stability, and is close to the fixed state number $N = 6$ in training speed, and the storage space is less than the fixed state

Table 1. Output probability ranking and sum of ten words in different states

	Word 1	Word 2	Word 3	Word 4	Word 5	Word 6	Word 7	Word 8	Word 9	Word 10	Sum of rankings
$N = 2$	4	8	7	6	6	2	7	1	5	8	54
$N = 3$	5	6	6	8	4	3	8	2	4	7	53
$N = 4$	2	7	3	7	3	6	5	8	6	6	53
$N = 5$	1	4	5	1	1	7	4	3	3	5	34
$N = 6$	3	2	4	4	8	1	3	4	7	4	40
$N = 7$	8	1	8	5	7	8	6	6	1	3	44
$N = 8$	8	1	8	5	7	8	6	6	1	3	53
Adaptively adjust the number of states	6	3	2	3	2	4	1	5	2	1	29

Table 2. Modeling effect of adaptively adjusting the number of states

	Open	Close	Accelerate	Decelerate	Forward	Backward	Turn Left	Turn Right	Default	Stop
Recognition rate	89%	94%	92%	90%	88%	91%	96%	90%	93%	92%

Table 3. Comparison of adaptive adjustment and fixed state number methods

	The difference between the top two in probability	Space occupied storage	Training time
N = 2	1509.1	2 * 10 = 20	609.8 s
N = 3	1725.8	3 * 10 = 20	914.7 s
N = 4	1738.8	4 * 10 = 20	2047.9 s
N = 5	1938.3	5 * 10 = 20	5928.8 s
N = 6	2371.8	6 * 10 = 20	6710.2 s
N = 7	2244.2	7 * 10 = 20	7629.3 s
N = 8	2419.7	8 * 10 = 20	8030.3 s
Adaptively adjust the number of states	2379.4	7 + 5 + 6 + 4 + 3 + 2 + 5 + 7 + 6 + 8 = 53	3028.5

number of better performance N = 6, 8 Occupied space resources, so it can be seen from the above that the automatic increase and decrease state number method greatly improves the performance of the speech recognition system.

6 Conclusion

The final total number of states of the automatic increase and decrease state number method: 7 + 5 + 6 + 4 + 3 + 2 + 5 + 7 + 6 + 8 = 53 and the fixed state number method has better modeling effect N = 6 and N = 8 The total number of states is 60 and 80, indicating that the automatic increase and decrease state number method can achieve a good modeling effect and save a lot of redundant storage space. The purpose of increasing or decreasing the state adaptation is to improve the modeling effect of the model, and at the same time, to ensure that the modeling effect is basically unchanged, try to reduce the requirements of the system storage space, and shorten the training time of the model, so that the voice The accuracy, stability and real-time performance of the identification system have been improved.

References

1. Burchard, B., Romer, R., Fox, O.: A single chip phoneme based HMM speech recognition system for consumer applications. IEEE Trans. Consum. Electron. 46(3), 914–919 (2000)

2. Fook, C.Y., Hariharan, M., Yaacob, S., Adom, A.: A review: Malay speech recognition and audio visual speech recognition, In: 2012 International Conference on Biomedical Engineering (ICoBE), Penang, pp. 479–484 (2012)
3. Kumari, D., Saini, I., Sood, N.: Detection of heart rate through speech using Mel frequency cepstrum coefficients. In: 2019 3rd International Conference on Trends in Electronics and Informatics (ICOEI), Tirunelveli, India, pp. 455–458 (2019)
4. Guerid, A., Houacine, A.: Recognition of isolated digits using DNN–HMM and harmonic noise model. IET Signal Process. **13**(2), 207–214 (2019)
5. Yamaguchi, K.: A neural network controlled adaptive search strategy for HMM-based speech recognition. In: 1993 IEEE International Conference on Acoustics, Speech, and Signal Processing, Minneapolis, MN, USA, vol. 2, pp. 582–585 (1993)
6. Lounnas, K., Satori, H., Hamidi, M., Teffahi, H., Abbas, M., Lichouri, M.: CLIASR: a combined automatic speech recognition and language identification system. In: 2020 1st International Conference on Innovative Research in Applied Science, Engineering and Technology (IRASET), Meknes, Morocco, pp. 1–5 (2020)
7. Zhao, L., Han, Z.: Speech recognition system based on integrating feature and HMM. In: 2010 International Conference on Measuring Technology and Mechatronics Automation, Changsha City, pp. 449–452 (2010)
8. Ogawa, T., Kobayashi, T.: Hybrid modeling of PHMM and HMM for speech recognition. In: 2003 IEEE International Conference on Acoustics, Speech, and Signal Processing, 2003 Proceedings (ICASSP 2003), Hong Kong, pp. I–I (2003)
9. Paramonov, P.: Fast algorithm for isolated words recognition based on Hidden Markov model stationary distribution. In: 2017 IEEE 4th International Conference on Soft Computing & Machine Intelligence (ISCMI), Port Louis, pp. 128–132 (2017)
10. Karpagavalli, S., Chandra, E.: Phoneme and word based model for tamil speech recognition using GMM-HMM. In: 2015 International Conference on Advanced Computing and Communication Systems, Coimbatore, pp. 1–5 (2015)
11. Paramonov, P., Sutula, N.: Simplified scoring methods in HMM based speech recognition. In: 2014 International Conference on Soft Computing and Machine Intelligence, New Delhi, pp. 154–156 (2014)
12. Ahcène, A., Aissa, A., Abdelkader, D., Khadidja, B., Ghania, D.: Automatic segmentation of Arabic speech signals by HMM and ANN. In: 2016 International Conference on Electrical Sciences and Technologies in Maghreb (CISTEM), Marrakech, pp. 1–4 (2016)
13. Abdelaziz, A.H.: Comparing fusion models for DNN-Based audiovisual continuous speech recognition. IEEE/ACM Trans. Audio Speech Lang. Process. **26**(3), 475–484 (2018)
14. Dubagunta, S.P., Magimai.-Doss, M.: Segment-level training of ANNs based on acoustic confidence measures for hybrid HMM/ANN speech recognition. In: ICASSP 2019–2019 IEEE International Conference on Acoustics, Speech and Signal Processing (ICASSP), Brighton, United Kingdom, pp. 6435–6439 (2019)

An SEIR Model for Assessment
of COVID-19 Pandemic Situation

Peiliang Sun[1], Kang Li[1(✉)], Zhile Yang[2], and Dajun Du[3]

[1] School of Electronic and Electrical Engineering, University of Leeds,
LS2 9JT Leeds, UK
k.li1@leeds.ac.uk
[2] Shenzhen Institute of Advanced Technology, Chinese Academy of Sciences,
Shenzhen 518055, Guangdong, China
zyang07@qub.ac.uk
[3] School of Mechatronic Engineering and Automation, Shanghai University,
Shanghai 200072, China
ddj@shu.edu.cn

Abstract. The ongoing COVID-19 pandemic spread to the UK in early
2020 with the first few cases being identified in late January. A rapid
increase in confirmed cases started in March, and the number of infected
people is however unknown, largely due to the rather limited testing
scale. A number of reports published so far reveal that the COVID-19
has long incubation period, high fatality ratio and non-specific symptoms, making this novel coronavirus far different from common seasonal
influenza. In this note, we present a modified SEIR model which takes
into account the latency effect and probability distribution of model
states. Based on the proposed model, it was estimated in April 2020
that the actual total number of infected people by 1 April in the UK
might have already exceeded 610,000. Average fatality rates under different assumptions at the beginning of April 2020 were also estimated.
Our model also revealed that the R_0 value was between 7.5–9 which is
much larger than most of the previously reported values. The proposed
model has a potential to be used for assessing future epidemic situations
under different intervention strategies.

Keywords: COVID-19 · SEIR model · Coronavirus pandemic
assessment

1 Introduction

Accurate estimation of the infection prevalence plays a vital role in the study
of an infectious disease in assessing its dynamics and potential risks to public
health. While it often unlikely to test a large proportion of population to estimate the infection prevalence, mathematical modelling becomes an irreplaceable
tool to gain insights into the disease transmission. The Susceptible-Infectious-
Removed (SIR) model, as one of the simplest compartmental models originated

© Springer Nature Singapore Pte Ltd. 2020
M. Fei et al. (Eds.): LSMS 2020/ICSEE 2020 Workshops, CCIS 1303, pp. 498–510, 2020.
https://doi.org/10.1007/978-981-33-6378-6_37

in the early 20th century [12], has been widely used to model infectious diseases. Several SIR derivatives which contain more states have been used to analyse the COVID-19 pandemic all over the world [9,11,16]. In these SIR derivatives, the dynamics of each state is governed by several parameters such as: contact rate, infection rate, and cure rate etc. These averaged parameters are estimated using statistic methods, but their value distributions are rarely taken into account in the SIR modelling, thus the complex dynamics of the COVID-19 may not be captured.

The COVID-19 has shown to have long incubation time, and the time from being infected to recovery/death varies in a large scale. Moreover large proportion of the infected people only show mild symptoms or even have no symptoms, implying that the spread of the virus could be far more broad. In [10], authors claimed that a majority of infections (>80%) are undocumented, due to most of the infections experienced mild, limited or no symptoms [15]. Therefore in our model only a certain proportion of people exposed to COVID-19 are classified as 'quarantined cases' and the infections with mild symptoms are assumed to have a longer infectious period.

Authors in [18] analysed the probability distribution of onset to death and onset to recover time, and the proportion of all infections that would lead to hospitalisation according the data in China. These statistic data are used in this paper to formulate the latency of COVID-19 status transition with the probability distribution over time. This paper was deposited in medRxiv in April 2020 [17].

2 Modelling

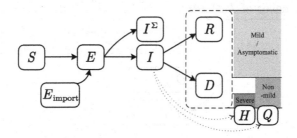

Fig. 1. A modified SEIR dynamic model

Figure 2 illustrates the basic structure of the SEIR model, a SIR derivative, where $S(t)$ represents the susceptible cases at day t, $E(t)$ is for the exposed cases at day t, $I(t)$ stands for the infectious cases at day t and $I^{\Sigma}(t)$ denotes the total infected population at day t. $R(t)$ and $D(t)$ represent the cumulative recovered cases and cumulative deaths till day t respectively. The total of cases that may need hospital treatment at day t is denoted as $H(t)$, while $Q(t)$ represents the quantined cases at day t. The time variable is $t = 1, 2, \cdots, K$ and K is the prediction horizon.

The model dynamics are formulated as follows:

$$\Delta S(t) = S(t+1) - S(t) = \frac{-\beta}{N(t)} S(t) (I(t) - Q(t)) \tag{1}$$

$$\begin{aligned}\Delta E(t) &= E(t+1) - E(t) \\ &= -\Delta S(t) + E_{\text{import}}(t) - \Delta I^{\Sigma}(t)\end{aligned} \tag{2}$$

$$\begin{aligned}N(t+1) &= N(t) + E_{\text{import}}(t)\Delta I^{\Sigma}(t) \\ &= I^{\Sigma}(t+1) - I^{\Sigma}(t)\end{aligned} \tag{3}$$

where $\frac{\beta}{N(t)}$ is the contact ratio and $N(t)$ is the current population in the whole country. In (1), the unquarantined people can contact and spread the virus to the some of the susceptible people $\Delta S(t)$. The imported case $E_{\text{import}}(t)$ and $\Delta S(t)$ add up to the increment of $E(t)$. At each time step a certain proportion of $E(t)$ will be converted to $I(t)$ and the increment of total infected people $I^{\Sigma}(t)$ is subtracted from $E(t)$ as shown in (2).

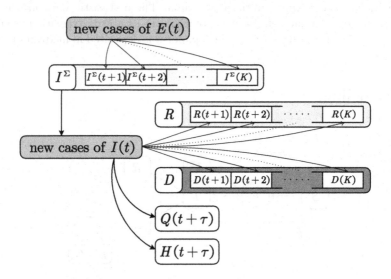

Fig. 2. Model state transitions

However once a person is exposed to the virus and get infected, the virus starts to incubate in his/her body and that person can then become infectious. Therefore a cumulative distribution function of the incubation time in [14] is introduced and is shifted two days[1] before the symptoms onset to the new cases of $E(t)$ to

---

[1] According to WHO report people can shed COVID-19 virus 24–48 h prior to symptom onset [2].

$I^{\Sigma}(t)$ as shown in Fig. 2,

$$
\begin{aligned}
\left[I^{\Sigma}\left(t+1\right) I^{\Sigma}\left(t+2\right) \cdots I^{\Sigma}\left(t+K\right)\right] \\
= \boldsymbol{P}_{E2I} \cdot \left(-\Delta S(t) + E_{\text{import}}(t)\right)
\end{aligned}
\tag{4}
$$

where $\boldsymbol{P}_{E2I}$ is a row vector of dimension K representing the cumulative probability.

Similarly the number of recovered and death cases can also be derived by introducing two separate cumulative probability vectors $\boldsymbol{P}_{I2R}$ and $\boldsymbol{P}_{I2D}$ as shown in (5) and (6),

$$
\begin{aligned}
\left[R\left(t+1\right) R\left(t+2\right) \cdots R\left(t+K\right)\right] \\
= (1-\kappa)\boldsymbol{P}_{I2R} \cdot \Delta I^{\Sigma}(t)
\end{aligned}
\tag{5}
$$

$$
\begin{aligned}
\left[D\left(t+1\right) D\left(t+2\right) \cdots D\left(t+K\right)\right] \\
= \kappa\boldsymbol{P}_{I2D} \cdot \Delta I^{\Sigma}(t)
\end{aligned}
\tag{6}
$$

where κ is the average infection fatality rate (IFR).

The state $Q(t)$ is estimated three days[2]

$$
Q(t+\tau) = \lambda_Q I(t) \tag{7}
$$
$$
H(t+\tau) = \lambda_H I^{\Sigma}(t) \tag{8}
$$

where τ is the time period from becoming infectious to having developed non-mild symptoms[3] that need either admission to hospital or self-isolation at home, λ_Q and λ_H are ratio of non-mild symptoms and ratio of people needing hospital admission.

Finally, the infectious people i.e. state I at day t can be updated by (9).

$$
I(t) = I^{\Sigma}(t) - R(t) - D(t) \tag{9}
$$

Apparently at day t, $N(t) = S(t) + E(t) + I(t) + R(t) + D(t)$ is automatically satisfied.

3 Modelling Assumptions and Parameter Setting

The real epidemiological dynamics is extremely complex and we made several assumptions to set parameters and to simplify the model:

---

[2] In average, patients may need hospital admission on the 3rd day after symptom onset [3] after onset of non-mild symptoms, i.e. after people develop fever/dry cough symptoms they need either be under quarantine in hospital or start self-isolation at home. State $H(t)$ is used to represent the accumulation of people that need hospital treatment.

[3] According to the WHO report, 80% of the patients experienced mild illness [1].

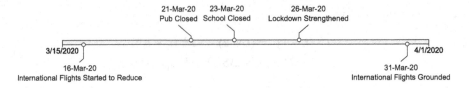

Fig. 3. Intervention timeline

1. The genetic difference in human race is not considered. The cumulative probability functions used in the model are acquired from the analysis of the COVID-19 data from China, and are applied to the UK situation under the assumption that this virus spreading mechanism remains unchanged.
2. About 20% of the infected people will develop symptoms and can be classified as quarantined cases.
3. Infected people can become infectious two days before symptom onset.
4. The virus tests can be conducted to each patient in severe situation on the third day after COVID-19 symptom onset.
5. The officially published death case number is accurate [5].
6. The total confirmed case number is larger than the number of cases that require hospital treatment (severe cases).
7. The earliest imported cases occurred on 15 Jan 2020 [7] and the imported cases kept growing at an exponential rate till international flights were reduced.

Based on published results [8,18], official statistics and announcements, the parameters used to build the UK pandemic model are illustrated in Fig. 4 and Fig. 5.

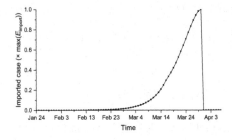

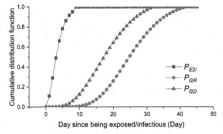

Fig. 4. Imported exposed cases

Fig. 5. Cumulative distribution functions of P_{E2I}, P_{I2R}, P_{I2D}

By the end of March 2020 international flights are greatly reduced from the normal capacity, we assume the number of imported cases grew exponentially before 16 March. The international flight capacity was decreased linearly afterwards and by the end of March most of the international flights are grounded.

The number of imported cases is used as an adjustable parameter and is also used as an initial condition for future projection.

As shown in Fig. 3, from 23 March, pubs and schools are closed and people dispersed to their home to work and study. So we simply assume the infection rate[4] as:

$$\beta(t) = \begin{cases} \beta_0 & \text{before 23 March} \\ \beta_0/50 & \text{after 23 March} \end{cases} \tag{10}$$

Table 1. Model prameters

Symbol	Description	Value
$N(1)$	Initial population	110 Million
$E(1)$	Initial exposed cases	5
$\Delta E_{\text{import}}^{\text{max}}$	Maximum imported cases	100–10k
β_0	Infection rate without intervention	0.15–0.5
λ_Q	Ratio of people who have non-mild symptoms	20%
λ_H	Ratio of people who need hospital treatment	1–6%
κ	Infection fatality ratio	0.5–3%
τ	Time period from being infectious to having non-mild symptom	5 days

The actual infection fatality ratio κ and the severe case ratio λ_H can not be accurately calculated, while the reported total case number might be way below the actual infected cases, therefore we have adopted the following procedures to fit the model to the UK death toll published so far (1 Feb to 10 Apr):

1. Select a set of λ_H and κ satisfying $\lambda_H > \kappa$.
2. Search feasible combination of $\Delta E_{\text{import}}^{\text{max}}$ and β_0 such that the following conditions (a)–(c) are met:
   (a) The average of daily reported cases must be greater than the average of estimated $H(t)$ over the whole data fitting period, and the number of reported cases should be greater than the model estimated $H(t)$ in the last day of the whole period, i.e. 10 April;
   (b) The increment of $H(t)$ on 8 April was lower than the increment of $H(t)$ on 7 April based on a news report [4] that the growth in the number of hospitalised coronavirus patients is definitely getting slower;
   (c) $D(t)$ on 10 April should be greater than reported death toll on the same day, i.e. 10 April.
3. The best combination of β_0 and $\Delta E_{\text{import}}^{\text{max}}$ are selected such that the estimated curve of $D(t)$ fits the daily reported total death number over the data fitting period with the least normalised mean square error.

---

[4] This model is used to fit the death case curve, the variation on $\beta(t)$ after lockdown does not affect death number from February to the early April. So for convenience, $\frac{1}{50}\beta$ is used.

4   Discussions on Nowcasting and Forecasting

4.1   Epidemic Prevalence Estimation on 1st April 2020

By performing the aforementioned modelling procedures, estimates of the number of actual infected people with different infection fatality ratio κ and different severe case ratio λ_H are plotted in Fig. 6. The figure indicates that although the actual IFR κ cannot be calculated accurately, based on our modelling results shown in Fig. 6, the total number of infected cases could have been greater than 0.61 million by 1 April 2020. If the IFR κ was lower than 2%, the total case number could be even greater than 1 million. Several reports indicate that IFR is below 1% [10], but based on our model, the lower IFR κ implies lower severe case ratio λ_H and larger infection number. Hence, the total number of infections would exceed 2 million if IFR κ is below 1%. However, according to the UK demographic data, about 5% infected individuals may need hospital treatment, therefore a higher IFR ($\kappa > 2\%$) is more likely to be true in the early stage of this virus outbreak.

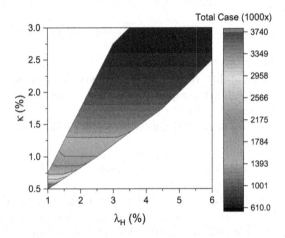

Fig. 6. Estimation of the number of actual infected people by 1 April 2020

By curve-fitting the results illustrated in Fig. 6, the estimate of the infection rate β is around 0.41. The basic reproduction ratio R_0 can be estimated by calculating $S(\infty) - S(1)$ while setting $E(1) = 1, \Delta E(t) = 0$. In our estimate $R_0 \approx 8.8$ which is much larger than the results obtained in other published papers. However if we assume anyone who is infected by the virus will be under quarantine 7 days after s/he becomes infectious, then our model estimate for $R_0 \approx 2.8$–3.6 which is comparable to the current popular estimates of R_0 reported in the literature [13,19]. The reason for our estimated R_0 being much larger than the R_0 estimates reported in the literature might be due to that there is a large proportion of people who only got mild symptoms or even were asymptomatic.

They can spread the virus before symptom onset till recovery over a long period of time before the lockdown measure was put in place.

4.2   Assessment of Future Epidemic Situation (from 10 April)

As presented in the previous section, although the severe case ratio λ_H and the infection fatality ratio κ cannot be accurately obtained, our model produces some consistent estimates on the total death toll if community spread is greatly suppressed after 23 March. If the UK government continued the lockdown strategy to the end of June, the most optimistically estimated total death number was about 21000 (if the reproduction ratio was suppressed to 0.2). Figure 7 and Fig. 8 illustrate the total number of deaths and infections to the end of April if the reproduction ratio is controlled to be less than 0.2 after the lockdown, and assume the severe case ratio $\lambda_H = 4\%$, infection fatality ratio (IFR) $\kappa = 2\%$, and the maximum imported cases $E_{\text{import}}^{\max} = 2100$. These figures suggest that about 1 million people would be infected, and 81.2% of infected people will be recovered by the end of April in this scenario.

However the death toll might be higher if the lockdown rules were not strictly obeyed by people unless improved medical treatment was available in the forthcoming weeks. On the other hand, a long period of lockdown may damage the economy to an unacceptable extent, and the wishes to lift the lockdown measure as early as possible may grow with time.

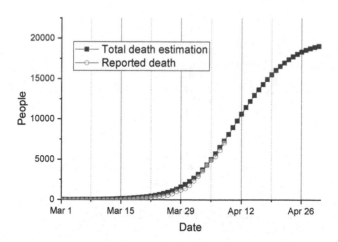

Fig. 7. Total death toll projection

Figure 9 and Fig. 10 show different curves representing possible daily increase in death toll and daily increase in hospital treatment requirements under different reproduction rates. Clearly if the reproduction rate is suppressed below 1, both numbers will keep decreasing. However even the viral transmission is managed

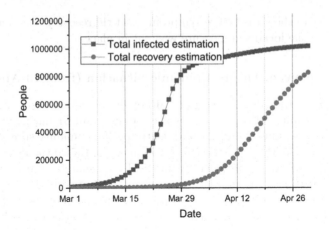

Fig. 8. Total infection number projection

to the minimum, the daily death toll will still likely keep at a high level (>500) for two weeks. Based on the current data[5] in the UK, our model also projects that the inflection point of current pandemic wave in the UK is likely to occur in the following week (between 12–15 April 2020) if the lockdown rules are strictly obeyed by people.

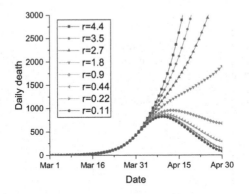

Fig. 9. Daily death toll projection

The model shows that in the ideal situation where everyone stays at home until the end of April, then the daily death number can be suppressed to less than 200 or even less afterwards. Similarly the curve of the number of infected people who need hospital treatment can also be flattened by mid of April.

---

[5] Up to 10 Apr 2020.

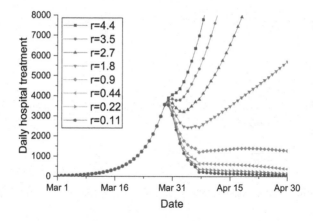

Fig. 10. Daily increase in hospital treatment

According to Fig. 11[6], if people start to return to work in May, then the reproduction ratio has to be suppressed below 2 in order that the NHS resources are sufficient to cope with the medical treatment requirements in the next two months (May and June).

However if the reproduction ratio can not be suppressed well enough while the lockdown measure is lifted in May, one feasible solution to slow down the transmission might be to allow people work every other day. This kind of intervention measures (e.g. only allow people to go outside every other day) has a similar pattern (r = 4, 0.2)[7] as the curve with a halved reproduction ratio (r = 2).

5   Conclusions and Discussions

In this paper, a modified SEIR model has been proposed to allow the state transitions to be governed by the cumulative probability vector rather than by averaged values, so the COVID-19 data can be utilised to reflect the distribution characteristics over a long period of time. Time variant parameters such as the number of imported cases and infection rate are also introduced and designed to reflect the changes introduced by adopting intervention measures such as reducing the number of international flights and the lockdown measures.

The model was fitted using the officially published data. Because the number of total reported cases may not accurately indicate the real infection cases, the death toll was used instead to assess the pandemic progression in the UK. The results revealed that more than 610,000 people might have been infected by

---

[6] Because the hospitalised number is based on Chinese data. It can be different from the UK scenario where the hospital admission procedures are different. In Fig. 11, λ_H is set to 4%.

[7] When people go out for work the reproduction ratio r may become 4. On the next day all people should stay at home and r is suppressed to 0.2.

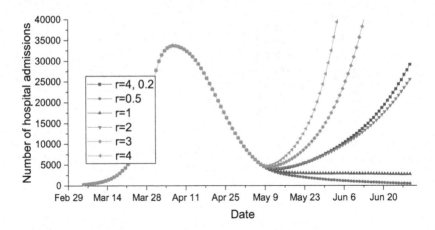

Fig. 11. Projection of hospital beds demand for COVID-19

1 April 2020 and the average death rate in last month was very likely to be greater than 1% and may be even higher than 2% in the early stage of this pandemic. Also according to the proposed model, the value of R_0 was much greater than the previously published values in the literature, if no intervention measure was implemented. The estimated range of R_0 is between 7.5–9 due to large proportion of patients with mild symptoms. However, if people could keep self-isolated after developing COVID-19 symptoms, then the reproduction ratio is about 2–3, which is comparable to the values published in the literature.

Our model revealed that if people fully comply the 'stay at home' order to the end of April, the total death toll was likely to be less than 21,000. To lift the lockdown measure from May, people will need to keep the strict social distance rule and apply other personal protection measures to ensure that the reproduction ratio was suppressed below 2 in order that the NHS has the sufficient capacity to cope with the medical treatment demand from people having severe COVID-19 symptoms. However, if this was not achievable then interventions measures such as people stay at home every other/third day may equally reduce the reproduction ratio in the near future.

The above projections and discussions were based on the assumption that there was no international imported cases after May. However it was likely that the current global pandemic may not reach its end in the next few months, therefore how to limit international imported cases would become a crucial issue to tackle in the future epidemic prevention.

This paper was written in early April 2020 and posted on medRxiv.org [17] on 17 April 2020. The data up to 10 April 2020 was used for modelling and discussion. At the time of submitting the final paper to this conference, the real death data in the prediction is available. The total death toll on 10 June reached 39,013 which is sadly greater than the most optimistic estimation of 21,000 reported in our paper posted in early April. However our earlier estimation was based on the ideal assumption that people fully comply with the lock down measures with the best means to stopping the spread. Because the actual daily death number at the end of April by 7 days average was 650 and base on the result in Fig. 9, it is clear that $r \approx 0.75$ in this case. This implies that the spread was not fully contained in the lock-down period time. And this reproduction number 0.75 is consistent with the official announcement which is 0.7–0.9 [6].

6 Disclaimer

All the data used in the paper are based on publicly available resources, including references to official and professional websites and peer-reviewed journals. The model, data and discussions presented in this paper is for research and education only. The model may not represent the real situation, and it may fail due to inadequate model elements and incorrect initial settings.

References

1. Coronavirus disease 2019 (covid-19) situation report 41. https://www.who.int/docs/default-source/coronaviruse/situation-reports/20200301-sitrep-41-covid-19.pdf?sfvrsn=6768306d_2
2. Coronavirus disease 2019 (covid-19) situation report 46. https://www.who.int/docs/default-source/coronaviruse/situation-reports/20200306-sitrep-46-covid-19.pdf?sfvrsn=96b04adf_4
3. Coronavirus disease 2019 (covid-19) situation report 46. https://patient.info/news-and-features/coronavirus-how-quickly-do-covid-19-symptoms-develop-and-how-long-do-they-last
4. Covid-19 hospital admissions 'flattening'. https://www.hsj.co.uk/coronavirus/covid-19-hospital-admissions-flattening/7027364.article
5. Number of coronavirus (COVID-19) cases and risk in the UK. https://www.gov.uk/guidance/coronavirus-covid-19-information-for-the-public, library Catalog: www.gov.uk
6. The r number and growth rate in the UK. https://www.gov.uk/guidance/the-r-number-in-the-uk#history
7. UK patient zero? East Sussex family may have been infected with coronavirus as early as mid-January. https://www.telegraph.co.uk/global-health/science-and-disease/uk-patient-zero-east-sussex-family-may-have-infected-coronavirus/
8. Backer, J.A., Klinkenberg, D., Wallinga, J.: Incubation period of 2019 novel coronavirus (2019-nCoV) infections among travellers from Wuhan, China, 20–28 January 2020. Eurosurveillance **25**(5), 2000062 (2020)
9. Calafiore, G.C., Novara, C., Possieri, C.: A modified sir model for the COVID-19 contagion in Italy. arXiv preprint arXiv:2003.14391 (2020)
10. Famulare, M.: 2019-nCoV: preliminary estimates of the confirmed-case-fatality-ratio and infection-fatality-ratio, and initial pandemic risk assessment (2020)
11. Ivorra, B., Ferrández, M., Vela-Pérez, M., Ramos, A.: Mathematical modeling of the spread of the coronavirus disease 2019 (COVID-19) considering its particular characteristics. The case of China. Technical report, MOMAT, 03 2020 (2020). https://doi-org.usm.idm.oclc.org

12. Kermack, W.O., McKendrick, A.G.: A contribution to the mathematical theory of epidemics. Proc. R. Soc. London Ser. A **115**(772), 700–721 (1927). Containing Papers of a Mathematical and Physical Character
13. Kucharski, A.J., et al.: Early dynamics of transmission and control of COVID-19: a mathematical modelling study. The Lancet Infectious Diseases (2020)
14. Lauer, S.A., et al.: The incubation period of coronavirus disease 2019 (COVID-19) from publicly reported confirmed cases: estimation and application. Annals Internal Med. **172**, 577–582 (2020)
15. Li, R., et al.: Substantial undocumented infection facilitates the rapid dissemination of novel coronavirus (SARS-CoV2). Science **368**, 489–493 (2020)
16. Lopez, L.R., Rodo, X.: A modified SEIR model to predict the COVID-19 outbreak in Spain: simulating control scenarios and multi-scale epidemics. medRxiv (2020)
17. Peiliang, S., Li, K.: An SEIR model for assessment of current COVID-19 pandemic situation in the UK. medRxiv (2020)
18. Verity, R., et al.: Estimates of the severity of coronavirus disease 2019: a model-based analysis. The Lancet Infectious Diseases (2020)
19. Wu, J.T., Leung, K., Leung, G.M.: Nowcasting and forecasting the potential domestic and international spread of the 2019-nCoV outbreak originating in wuhan, china: a modelling study. Lancet **395**(10225), 689–697 (2020)

A Regression Model for Short-Term COVID-19 Pandemic Assessment

Xuan Liu[1], Kang Li[1(✉)], Zhile Yang[2], and Dajun Du[3]

[1] School of Electronic and Electrical Engineering, University of Leeds,
Leeds LS2 9JT, UK
k.li1@leeds.ac.uk
[2] Shenzhen Institute of Advanced Technology, Chinese Academy of Sciences,
Shenzhen 518055, Guangdong, China
zyang07@qub.ac.uk
[3] School of Mechatronic Engineering and Automation, Shanghai University,
Shanghai 200072, China
ddj@shu.edu.cn

Abstract. COVID-19 has rapidly spread around the world in the past few months, researchers around the world are working around the clock to closely monitor and assess the development of this pandemic. In this paper, a time series regression model is built to assess the short-term progression of COVID-19 pandemic. The model structure and parameters are identified using COVID-19 pandemic data released by China within the time window from 22 January to 09 April 2020. The same model structure and parameters are applied to a few other countries for day ahead forecasting, showing a good fit of the model. This modeling exercise confirms that the underlying internal dynamics of this disease progression is quite similar. The differences in the impact of this pandemic on different countries are largely attributed to different eternal factors.

Keywords: COVID-19 · Regression model · FRA · Data driven

1 Introduction

COVID-19 is an infectious disease which was first reported in December 2019 in Wuhan, China. This novel coronavirus rapidly swept the world in just a few months. Unlike most coronaviruses, COVID-19 is a new strain which has not been previously identified in humans. The symptoms of COVID-19 are mainly manifested in the respiratory system, such as coughing, shortness of breath, difficulty breathing and fever, etc. [10]. COVID-19 is highly contagious, and it has been widely reported that the incubation period of the virus symptoms is long and asymptomatic infection is difficult to prevent [1]. For most common cold, the symptoms usually manifest within three days of infection. However, the effects of COVID-19 generally appear in 2 to 14 days. Up to April 20, 2020, there are more than 2.4 million people have been infected with COVID-19 and the total number of deaths has exceeded 165,000.

© Springer Nature Singapore Pte Ltd. 2020
M. Fei et al. (Eds.): LSMS 2020/ICSEE 2020 Workshops, CCIS 1303, pp. 511–518, 2020.
https://doi.org/10.1007/978-981-33-6378-6_38

In order to assess and forecast the development of COVID-19 pandemic, great efforts have been made by researchers around the world to closely monitor and assess the latest development of this pandemic. Among various approaches to model infectious diseases, compartmental models such as Susceptible-Infectious-Removed (SIR) model [4, 9] and its derived ones are most popular [2, 7]. However, the performance of these models are highly dependent on initial parameter settings and cannot address the causal factors in the development of epidemics [3]. Regression models [5, 8] are also widely used for epidemics assessment and forecasting because the causal relationships between the dependent and independent variables could be inferred. In this paper, a time series linear regression model is developed for inferring the correlations among the epidemic data. Once the correlations are inferred, this model can be used for short-term epidemic forecasting without model adjustment, under the assumption that the underlying dynamics of this infectious disease is unchanged, and the differences in the outcomes are largely attributed by external factors such as intervention measures.

2   Modelling

Although the first outbreak of COVID-19 occurred in China, the spread of this infectious disease was then effectively controlled in March due to a series of intervention measures being implemented consequently. Therefore, public available data for the whole period can be analyzed as a reference for countries which are still suffering from this pandemic. It was noted that the number of counted daily new infections is highly dependent on the testing scale in a particular country. Therefore, it is highly likely that the number of counted daily new infections is much smaller than the actual infection number. In contrast, the counted number of daily deaths is more likely closer to the actual number than the confirmed infection cases, therefore it is more meaningful to analyze internal dynamics of this infectious disease using the number of daily reported deaths. Given the aforementioned considerations, the daily death number reported from China in the time window from 22 January to 09 April 2020 is used in this paper for building a time-series model.

Although the incubation period of COVID-19 and the time from symptom onset to final exit (death or recovery) vary for each infected individual, their expectations may be correlated to the daily death (DD) data. In other words, the DD value of day n, denoted as *Day* (*n*) can be projected using the DD data of some specific days in the past. Thus, this problem can be converted to a regression modelling problem.

In order to determine the structure and parameters of the time-series model, a subset selection method namely Fast Recursive Algorithm (FRA) [6] is applied in this case to select and determine the model regression terms and parameters simultaneously. For a linear-in-the-parameter model to represent a time series, its discrete time form generally can be represented as

$$y = \Phi\,\Theta + \Xi \tag{1}$$

where $y = [y(1), ...y(m)]^T$ are the observations, $\Phi = [\phi_1, ...\phi_n]$ is the regression matrix and each $\phi_i = [\phi(1), ...\phi(m)]^T$ $(i = 1, ...n)$ contains all candidate regression terms. $\Theta = [\theta_1, ...\theta_n]^T$ is the set of unknown parameters to be identified. $\Xi = [\xi_1, ...\xi_m]^T$ is the residual matrix of the model. In FRA, there are two predefined recursive matrices, M_k and R_k to fulfil the forward model selection procedure as

$$M_k = \Phi_k^T \Phi_k \tag{2}$$

$$R_k = I - \Phi_k M_k^{-1} \Phi_k^T \tag{3}$$

where Φ_k contains the first columns of the full regression matrix. Then, it has

$$R_{k+1} = R_k - \frac{R_k \phi_{k+1} \phi_{k+1}^T R_k^T}{\phi_{k+1}^T R_k \phi_{k+1}}, \qquad k = 0, 1...(n-1) \tag{4}$$

Define E_k as the cost function, as the first k columns in Φ are selected, E_k can be expressed as

$$E_k = y^T R_k y \tag{5}$$

Then, using Eq. (4) and (5), it has

$$E_{k+1} = y^T R_{k+1} y = E_k - \frac{y^T R_k \phi_{k+1} \phi_{k+1}^T R_k^T y}{\phi_{k+1}^T R_k \phi_{k+1}} \tag{6}$$

Therefore, the net contribution of the selected model term ϕ_{k+1} to the cost function can be calculated as

$$\Delta E_{k+1} = -\frac{(y^T \phi_{k+1}^k)^2}{(\phi_{k+1}^k)^T \phi_{k+1}^k} = \frac{(a_{k+1,y}^T)^2}{a_{k+1,k+1}} \tag{7}$$

The net contribution of each term can be calculated and the terms with maximum contributions will be selected. Finally, the model parameters can be identified by the procedure as

$$\hat{\theta}_j = \frac{aj, y - \sum_{i=j+1}^{k} \hat{\theta}_i a_{j,i}}{a_{j,j}} \qquad j = k, k-1, ...1 \tag{8}$$

The last step is to determine how many of the most contributing terms should be selected. If too few terms are selected, some important candidates may be abandoned which may lead to underfitting. If too many terms are selected, the terms which have little contribution will be selected. This not only increases computing cost, but may also lead to overfitting the model. In order to address this problem, the DD data in China is divided into two parts. 80% of the data (22 Jan 2020 to 25 Mar 2020) is used for model training and the other 20% of data is used for model validation. Therefore, a set of models that have various number of most contributing terms can be generated. These models will be validated on the validation data and their root mean square errors (RMSE) will be compared as the an indicator. The model which produces the minimum RMSE in validation will be considered as the most suitable model in this paper.

3   Results and Discussion

3.1   Model Analysis Based on Pandemic Data from China

As mentioned earlier, the models generated in Sect. 2 are trained with 80% of the DD data and validated with the other 20% of the DD data from China. The validation results indicate that the model which selected 3 most important terms ($\phi(n-1)$, $\phi(n-2)$ and $\phi(n-6)$) has the best performance. While $\phi(n-1)$ and $\phi(n-2)$ imply the persistence of the pandemic progression, $\phi(n-6)$ implies the inherent latency effect of this disease. The model parameters are identified as θ_1 = 0.6728, θ_2 = 0.4076 and θ_3 = −0.1114.

In order to determine the system stability, Z-transform technique is applied to the developed model to convert the discrete-time expression into a frequency domain representation. The Z-transform of the developed model can be expressed as

$$z^6 = 0.6728 - z^5 + 0.4076 * z^4 - 0.1114 \tag{9}$$

Therefore, the poles of the system can be calculated as shown in Table 1, and their locations are presented in Fig. 1.

Table 1. System poles

Poles	P_1	P_2	P_3	P_4	P_5	P_6
	0.7523	0.96104	0.05873 + 0.59618i	0.05873−0.59618i	0.579 + 0.30677i	0.579−0.30677i

All of the system poles are located within the unit circle, therefore, the internal stability and convergence of the system can be guaranteed. It worth noting that there is a pole (P_2) very close to the unit circle, which implies the stability margin of the system is small. The system may approach to its critical point of stability at the peak region, which may lead to oscillatory fluctuations. The oscillatory fluctuations are particularly visible at the inflecting point and after the external interventions are introduced. Moreover, two complex-conjugate pole pairs (P_3 and P_4, P_5 and P_6) represent the system oscillating behaviours due to the external disturbances. The locations of P_3 and P_4 are almost on the imaginary axis, there will be little damping in their generated oscillations. P_5 and P_6 have larger phase angles with the real axis than the ones of P_3 and P_4, which implies they may cause oscillatory fluctuations with higher frequency.

The model predictions compared with the real data from China is shown in Fig. 2. The RMSEs of model training and validation are 15.8306 and 2.1756 respectively. It is shown that the output of the model is in good agreement with the actual data. The mean absolute error (MAE) is 0.962 which is less than 2.3% in relative error. However, it is worth noting that the maximum absolute error (MaxAE) is 54 which occurs on 24 February. The reason for the large error on this day is that the data on this day has a 52.67% sharp drop compared to the previous day. According to Fig. 2, it is worth noting that large oscillations occur

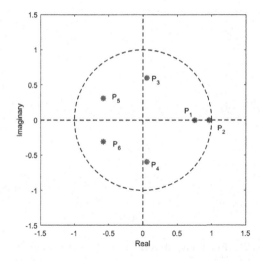

Fig. 1. The locations of the system poles in Z-plane

at the peak region, and oscillations are also observed at the downhill region. As discussed earlier, the oscillatory fluctuations are likely caused by the latency effect of the internal process.

From the system perspective, relatively strong oscillations occur in the region near the peak (12 to 27 February). The model parameters are more sensitive to measurement errors in this region than others. Thus, it may be inferred that when the system generates large oscillations, it is a sign of the apex.

3.2 Applications of the Trained Model

To validate our assumption that the pandemic progress in other countries has a similar internal dynamics as being observed in China, we applied the trained model presented in Sect. 3.1 directly to other six countries which include UK, Italy, Germany, Spain, France and USA. The same model structure and parameters in Sect. 3.1 are applied to the data (15 February to 08 April 2020) of these countries. The model outputs and the actual DD data of these countries are presented in Fig. 3. Four indicators namely root mean square error (RMSE), mean absolute error, maximum absolute error and coefficient of determination (R^2) are used to evaluate the model performance. The results of model performance are summarized in Table 2.

It is shown that the developed model performs well for these countries. The coefficient of determination for all of these countries are above 0.5 and most of them are greater than 0.87. This implies that the developed model has a good day-ahead forecasting performance, and internal dynamics of this pandemic progression is quite similar. Thus, the model developed using the data reported in China can be directly transferable to other countries. Through the internal dynamics is similar, but the outcomes of this pandemics vary

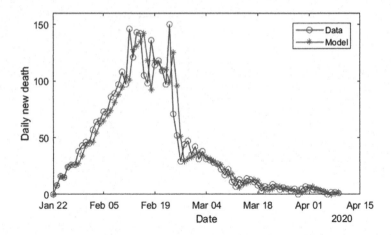

Fig. 2. Model outputs based on the data from China

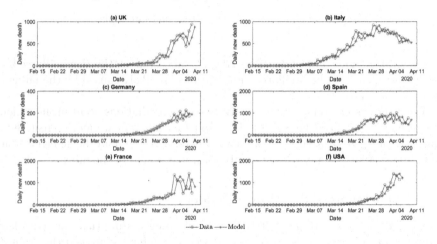

Fig. 3. One-day-ahead daily death toll forecasting of COVID-19. (a) UK, (b) Italy, (c) Germany, (d) Spain, (e) France, (f) USA.

significant in different countries. Some countries have high fatality rate, while others have less. This is largely due to external factors, such as the differences in the culture, social background, health care systems, and infections disease control and intervention measures introduced by different governments. Different levels of system oscillations are also observed in these six countries at the peak and down-hill regions.

Table 2. Results of model performance

Country	RMSE	Mean absolute error	Maximum absolute error	R^2
UK	83.4715	37.8551	300.2394	0.879
Italy	78.3751	48.5672	249.9158	0.941
Germany	18.6790	9.4191	65.8328	0.9095
Spain	119.8564	66.9247	319.1592	0.872
France	194.9673	73.6698	842.4610	0.6874
USA	112.3940	45.3986	507.3930	0.9016

4 Conclusion

In this paper, a time series regression model is developed for assessing the internal dynamics of the COVID-19 pandemic. The model structure and parameters are determined using the FRA technique and are based on the data from China. The regression terms $\phi(n-1)$ and $\phi(n-2)$ are favored to reflect the persistence of the pandemic progression, and the other regression term $\phi(n-6)$ is selected, implying the inherent latency effect of the pandemic. Since all the system poles locate inside the unit circle in the Z-plane, the internal stability and convergence of the system can be ensured. The position characteristics of the system poles can reflect the negative latency effect which may causes oscillatory fluctuations around the inflecting point or due to measurement errors and system disturbances. Since the oscillatory fluctuations are particularly evident at the inflection point, the developed model may be used as an indicator to analyze if the pandemic progression is approaching the inflection point. This model is directly applied to 6 other countries. The application results reveal that the model fits fairly well with the data from other countries, and can be used for short-term forecasting of the pandemic progression. The model coefficients of determination (R^2) of all the simulated countries are greater 0.5 and most of them are greater than 0.87. Moreover, as the pandemic reaches the peak, large oscillations are observed.

5 Disclaimer

All the data used in the paper are based on publicly available resources, including references to official and professional websites and peer-reviewed journals. The model, data and discussions presented in this paper are for research only. The model may not represent the real situation, and it may fail due to inadequate model elements.

References

1. Bai, Y., et al.: Presumed asymptomatic carrier transmission of COVID-19. JAMA **323**, 1406–1407 (2020)
2. Caccavo, D.: Chinese and Italian COVID-19 outbreaks can be correctly described by a modified SIRD model. medRxiv (2020)
3. Chen, D.: Modeling the spread of infectious diseases: a review. In: Analyzing and Modeling Spatial and Temporal Dynamics of Infectious Diseases, pp. 19–42 (2014)
4. Fanelli, D., Piazza, F.: Analysis and forecast of COVID-19 spreading in China, Italy and France. Chaos, Solitons Fract. **134**, 109761 (2020)
5. Harrell Jr., F.E., Lee, K.L., Matchar, D.B., Reichert, T.A.: Regression models for prognostic prediction: advantages, problems, and suggested solutions. Cancer Treat. Rep. **69**(10), 1071–1077 (1985)
6. Li, K., Peng, J.X., Irwin, G.W.: A fast nonlinear model identification method. IEEE Trans. Autom. Control **50**(8), 1211–1216 (2005)
7. Peng, L., Yang, W., Zhang, D., Zhuge, C., Hong, L.: Epidemic analysis of COVID-19 in China by dynamical modeling. arXiv preprint arXiv:2002.06563 (2020)
8. Wang, Y., Beydoun, M.A.: The obesity epidemic in the united states-gender, age, socioeconomic, racial/ethnic, and geographic characteristics: a systematic review and meta-regression analysis. Epidemiol. Rev. **29**(1), 6–28 (2007)
9. Yang, Z., et al.: Modified SEIR and AI prediction of the epidemics trend of COVID-19 in china under public health interventions. J. Thorac. Dis. **12**(3), 165 (2020)
10. Zheng, Y.Y., Ma, Y.T., Zhang, J.Y., Xie, X.: COVID-19 and the cardiovascular system. Nat. Rev. Cardiol. **17**, 259–260 (2020)

Author Index